Water Bankruptcy in the Land of Plenty

Water Bankruptcy in the Land of Plenty

Editors

Franck Poupeau
UMI iGLOBES, CNRS/University of Arizona, USA

Hoshin Gupta
Department of Hydrology and Atmospheric Sciences, University of Arizona, USA

Aleix Serrat-Capdevila
UMI iGLOBES CNRS/Department of Hydrology and Atmospheric Sciences, University of Arizona, USA

Maria A. Sans-Fuentes
Biosphere 2, University of Arizona, USA

Susan Harris
Department of Hydrology and Atmospheric Sciences, University of Arizona, USA

László G. Hayde
UNESCO-IHE, Institute for Water Education, Delft, The Netherlands

CRC Press is an imprint of the
Taylor & Francis Group, an **informa** business

A BALKEMA BOOK

Cover illustration: László G. Hayde, *Landscape with Saguaros*, Tucson region, Southern Arizona, USA, April 2012
Cover design: Peter Stroo, UNESCO-IHE, Institute for Water Education, Delft, The Netherlands

© 2016 UNESCO-IHE Institute for Water Education, Delft, The Netherlands

Print edition published by: CRC Press/Balkema
P.O. Box 11320, 2301 EH Leiden, The Netherlands
e-mail: Pub.NL@taylorandfrancis.com
www.crcpress.com – www.taylorandfrancis.com

CRC Press/Balkema is an imprint of the Taylor & Francis Group, an informa business

Typeset by V Publishing Solutions Pvt Ltd., Chennai, India
Printed and Bound by CPI Group (UK) Ltd, Croydon, CR0 4YY

Although all care is taken to ensure integrity and the quality of this publication and the information herein, no responsibility is assumed by the publishers nor the author for any damage to the property or persons as a result of operation or use of this publication and/or the information contained herein.

Library of Congress Cataloging-in-Publication Data

Applied for

ISBN: 978-1-138-02969-9 (Pbk), Taylor & Francis Group
ISBN: 978-1-4987-7699-8 (eBook PDF), UNESCO-IHE, Delft, The Netherlands

All rights reserved.
A pdf version of this work will be made available in open access via http://repository.tudelft.nl/ihe/. This version is licensed under the Creative Commons Attribution-NonCommercial 4.0 International License, http://creativecommons.org/licenses/by-nc/4.0/

Table of contents

List of Acronyms vii
Preface ix

Introduction 1

1 The idea of a transatlantic dialogue 3

2 Organization of the book and mind map 7

Maps 17

Socio-historic perspectives on water in the American southwest 25

3 The Tucson basin 27

4 Laws of the river 45

5 Water for a new America 65

6 Sharing the Colorado River 79

7 The making of water policy 101

Narratives of urban growth 119

8 The social logic of urban sprawl 121

9 Water and urban development challenges of urban growth 141

10 Comprehensive urban planning	159
11 Potential impacts of the continuing urbanization on regional climate	179

Ecosystem services and biodiversity   195

12 Quantification of water-related ecosystem services	197
13 Qualitative assessment of supply and demand of ecosystem services	223
14 The role of biodiversity in the hydrological cycle	249

Water use and groundwater management   289

15 Implications of spatially neutral groundwater management	291
16 Groundwater dynamics	321
17 Alternative water sources towards increased resilience	337
18 Differentiated approaches of groundwater management	363

Stakeholders' perspectives   379

19 Presentation	381
20 Texts	383

Conclusion   399

21 Bringing all the stories together: Beyond the Tucson case study	401
22 Next steps: Collaborative research and training towards transdisciplinarity	417
Contents (full titles and authorship)	423
Acknowledgments	427
Subject Index	429

List of Acronyms

ADEQ   Arizona Department of Environmental Quality
ADWR   Arizona Department of Water Resources
AMA   Active Management Areas
ARS   Arizona Revised Statutes
AWBA   Arizona Water Banking Authority
AWS   Assured Water Supply (AWS) certificate

BCPA   Boulder Canyon Project Act
BOR   Bureau of Reclamation
BSC   Biological Soil Crusts

CAGRD   Central Arizona Groundwater Replenishment Districts
CALS   College of Agriculture and Life Science
CAP   Central Arizona Project
CAPA   CAP Association
CAPLA   College of Architecture, Planning and Landscape Architecture
CAWCD   Central Arizona Water Conservation District
CGMI   Citizen's Growth Management Initiative
CICES   Common International Classification of Ecosystem Services
CLS   Conservation Lands System
CNRS   Centre National de la Recherche Scientifique

DEM   Digital Elevation Model
DOI   Department of the Interior

EIS   Environmental Impact Statement

GCASE   Groundwater, Climate and Stakeholder Engagement
GIS   Geographical Information System
GMA   Groundwater Management Act of 1980
GSFs   Groundwater Saving Facilities
GUAC   Groundwater Users Advisory Councils

HOAs   Home Owners' Associations
HRUs   Hydrological Response Units
HWB   Human Well-Being Submodel

IID   Imperial Irrigation District
INAs   Irrigation Non-expansion Areas
IPAG   Institutional and Policy Advisory Group

IUCN   International Union for Conservation of Nature

LSM   Land Surface Model
LTSC   Long-Term Storage Credits
LULC   Land Use and Land Cover

MA   Millennium Ecosystem Assessment
MAF   Million Acre Feet
MLP   Market Land-Price Submodel
MuSIASEM   Multiscale Integrated Analysis of Societal and Ecosystems Metabolism

NARR   North American Regional Reanalysis
NEPA   National Environmental Policy Act
NIMBY   Not In My Back Yard

PAMA   Phoenix Active Management Area
PDI   Precipitation Drought Index
PSWP   Pacific Southwest Water Plan
PVA   Public Values Assessment

RAMS   Regional Atmospheric Modeling System
ROD   Record of Decision
RWH   Rainwater Harvesting

SALC   Southern Arizona Leadership Council
SBS   College of Social Behavioral Sciences
SCWEPM   Santa Cruz Watershed Ecosystem Portfolio Model
SDCP   Sonoran Desert Conservation
SDWA   Safe Drinking Water Act
SPRC   Southern Pacific Railway Company
SRP   Salt River Project
SWAN   Sustainable Water Action Network Project
SWAT   Soil and Water Assessment Tool

TDS   Total Dissolved Solids
TDW   Transatlantic Dialogue on Water
TEEB   The Economics of Ecosystems and Biodiversity
TEP   Tucson Electric Power

UCM   Urban Canopy Model
UK NEA   UK National Ecosystem Assessment
UMI   International Centre for "Water, Environment and Public Policy" CNRS-University of Arizona
UMI-iGLOBES   Interdisciplinary and Global Environmental Studies, CNRS-University of Arizona
USFs   Underground Storage Facilities

WAAs   Water Accounting Areas
WCPA   Water Consumer Protection Act
WFD   Water Framework Directive
WRDC   Water Resource Development Commission
WRES   Water-Related Ecosystem Services
WRF   Weather Research and Forecasting model

Preface

Editors

By all standards, water is today's most coveted resource, and it will continue to be so in the future. Most observers generally agree that, with continued population growth, conflicts around water are likely to harden, and will involve severe risks of social and political unrest, both in the South and in the North. Worrying trends include recurring flooding, increasing volatility of resource availability, the melting of glaciers (and consequent sea level rise), resource contamination due to industrial pollution, degradation of soils due to intensive farming, and insufficient access to adequate sanitation, but also, and most of all, drought. In this context, the semi-arid Southwestern United States, which is currently enduring its most severe "drought" to date, is of considerable scientific and political interest.

Droughts are not uncommon in the Southwest. Advances in paleo-climate reconstruction and instrumental records have revealed that several major droughts have occurred in the region during the past 200 years. However, projected changes in climate and an over-exploitation of resources are generally considered as primary causes of ecological disasters that may be expected to follow. Of course, to reduce the complexity of this phenomenon to simply a matter of *"scarcity of natural resources"* would ignore the fact that the character of a drought has many dimensions, including *meteorological* (prolonged below-average precipitation), *hydrologic* (the manifestation of meteorological drought as reduced streamflow and depleted aquifers), *agricultural* (driven by, and impacts to, agriculture demand) and *socioeconomic* (driven by, and impacts to, other socio-economic sectors). While drought can be viewed as a perturbation imposed upon a coupled natural and human system, the resulting scarcity of water is clearly the product of a complex interplay between physical availability, the operation of the environment, and the behaviors of human and the demands they impose.

In other words, the public narratives of "drought" and "water scarcity" are, in today's world, largely a social construct associated with progressive economic growth and a widespread adoption of consumptive lifestyles. Regardless of whether the scarcity of water is actually due to natural climatic variability, global warming, hydrologic change, land cover change, or the ever growing urban and agro-industrial pressures placed on a finite resource, the public focus is most often on the insufficiency of physical supply and the perceived "scarcity" of natural resources, rather than on the analysis of human processes that mediate the governance and management of that water.

This book proposes and explores the purposely provocative notion of "water bankruptcy" so as to emphasize the socio-economic dimension of water issues in the

Southwestern US (and primarily Arizona), between the narratives of growth and the strategies or policies adopted to pursue competing agendas and circumvent the inevitable. Given the long-term trend of development in this region, the current drought might indeed present a window of opportunity in which to induce change, and to challenge the hegemonic discourse that governs the management of water resources in the American Southwest. Importantly, the situation may present an opportunity to deal with threats that derive from imbalances between growth patterns and available resources, the primary cause of scarcity.

A first of its kind, developed through close collaboration among a broad range of natural scientists, social scientists, and resource managers from Europe and the United States, this book is a committed step towards the collective implementation of a transdisciplinary approach to unveiling the inner workings of how water is fought for, allocated and used in the Southwestern US. It offers an innovative scientific perspective that dissects the conflicted relationship that societies engage in with the environment. It produces a critical diagnostic evaluation of water problems in the West, with a particular view to identifying risks for the Tucson area in Arizona (which is facing continuous urban sprawl and economic growth). The book presents a diversity of complementary perspectives, including a discussion of natural resources, biodiversity & their management in Arizona, an analysis of the stalemates in drought management and their roots in the history of water policy, and an assessment of ecosystem services in the context of both local biodiversity and the economic activities (such as mines and agriculture) that sustain economic growth. Finally, this book is a concerted effort to explore the interplay between a variety of related scientific disciplines including climatology, hydrology, water management, ecosystem services, societal metabolism, water governance, political economy and social science.

Franck Poupeau
UMI iGLOBES, CNRS/University of Arizona, USA

Hoshin Gupta
*Department of Hydrology and Atmospheric Sciences,
University of Arizona, USA*

Aleix Serrat-Capdevila
*UMI iGLOBES CNRS/Department of Hydrology and
Atmospheric Sciences, University of Arizona, USA*

Maria A. Sans-Fuentes
Biosphere 2, University of Arizona, USA

Susan Harris
*Department of Hydrology and Atmospheric Sciences,
University of Arizona, USA*

László G. Hayde
*UNESCO-IHE, Institute for Water Education,
Delft, The Netherlands*

Introduction

Chapter 1

The idea of a transatlantic dialogue

The SWAN Consortium

This book sits at the nexus of a broad range of disciplines, perspectives and geographic locations. It was developed in the course of a four-year international cooperation project entitled SWAN (*Sustainable Water ActioN: Building Research Links between European Union and United States*) that was funded by the European Union under its 7th Framework Program (FP7-INCOLAB-2011) to incentivize international collaboration on water related issues.

This introduction describes, briefly, how this collaborative cross-disciplinary exploration between multiple research areas, users, management agencies and institutions came about, and discusses how the participants collaborated to integrate methods, to identify overlaps and connections between research areas, and to arrive at the realization that a holistic approach can be much more than the sum of its parts. The setting for this collaboration was the Tucson Basin, which provides a natural basis for anchoring methods and approaches to a contextual reality with transdisciplinary needs. The various chapters in this book tell the stories of humans and their environment and how their interactions have unfolded, until being threatened nowadays by the risks of "water bankruptcy" in the American Southwest.

The general objective of the SWAN project was to strengthen European research capacity in the USA, to promote competitiveness of European research and industry, and to inform and involve policy-makers and the general public. It included participants from five member states of the European Union (Bulgaria, France, Netherlands, Spain and the United Kingdom) and from the University of Arizona (USA). The project was coordinated by the French CNRS (Centre National de la Recherche Scientifique), represented by the UMI iGLOBES.[1]

The scientific goal of this collaboration was to develop a Transatlantic Dialogue on Water (TDW), with a view to building a major international network that can facilitate the collaboration of scientists and students with stakeholders and communities. The idea of the TDW is to bridge across multiple scientific disciplines, institutional participations, and international perspectives, with the working hypothesis that *it is necessary to apply multifaceted approaches that combine natural and social sciences*

---

1 iGLOBES (Interdisciplinary and Global Environmental Research) is an international joint unit created by the French CNRS (Centre National de la Recherche Scientifique) and the College of Science of the University of Arizona. iGLOBES and the former Department of Hydrology (now Department of Hydrology and Atmospheric Sciences) collaborated in the frame of the SWAN project.

into a new paradigm that explores new governance perspectives and is capable of dealing with both the uncertainty and complexity inherent in water related issues. With this perspective, the TDW constitutes a platform for bringing together research, education and knowledge exchange at both national and international levels. The research component constitutes a major pillar for knowledge exchange via training of students and interaction with stakeholders. The knowledge exchange is being accomplished through periodic extended research stays of European students at the international joint unit iGLOBES at the University of Arizona, and by bi-annual meetings of the SWAN teams, thereby making possible the collaborative research endeavor that has given rise to this book.

During the project, the scientific perspective shifted progressively towards the use of "*big data*" in support of the management of water, by examining the (open) socio-technical conditions required to access such data, and by examining the (knowledge) capacities necessary to ensure their utilization (*Pedregal et al., 2015*). This idea of "*open knowledge*" now appears as a key concept underpinning the production of new forms of scientific work and public participation with stakeholders, thereby constituting the core of the project and supporting the objective of transdisciplinarity. In contrast with multi-disciplinary approaches, transdisciplinarity engenders a framework in which researchers can both work in parallel in a traditional disciplinary fashion and also in interactive and interdisciplinary fashion to address a common problem, while taking account of the multiple perspectives of stakeholders and the general public (*Rosenfeld, 1992*). While this book, the short-term product of a research grant, does not fully realize the ideal of transdisciplinarity, the desire to achieve such an approach has served as a guiding principle for the international research teams involved in the project. Certainly we share a common conviction that dealing with water related issues requires new approaches to knowledge production that incorporate multiple scientific, professional and public perspectives.

What has become clear to us is that the complexity inherent in the management of water increasingly necessitates a combination of approaches that draw from the physical, environmental and social sciences, and that are open to and validated by civil society. This awareness results from a need to acknowledge "*the unavoidable existence of non-equivalent perceptions and representations of reality, contrasting but legitimate perspectives found among social actors, and heavy levels of uncertainty*" (see *Funtowicz and Ravetz 1991, 1993*, and *Giampietro et al. 2012*, among others). The natural result is a paradigm shift in the management of natural resources (see *Pahl-Wostl et al., 2011, Del Moral et al., 2014*, among others) that is characterized by a reorientation in objectives, methodologies and evaluation criteria, by the involvement of a broad variety of agents, and by a significant restructuring of institutional frameworks.

So it is useful, while reading this book, to remember that human problems that resist easy solution are typically characterized by (*Hernández-Mora and Del Moral, 2015*):

- Complexity: The human dimension introduces reflexivity into the managed system, while ecological systems respond to pressures and interventions in non-linear and unpredictable ways, so that socio-ecological systems are often characterized by non-predictable and unexpected responses.
- Uncertainty: The technical solutions and tools provided by science cannot hope to accurately represent the total system and all of its interactions in all their

complexity, even with sophisticated models, modelers and computers (*Giampetro et al., 2006*).

- Incommensurability: It is, in practice, impossible to construct a single computational model that comprehensively represents the heterogeneity of information, different kinds of disciplinary knowledge and descriptions of reality, and different but legitimate values, perceptions and interests that can be ascribed to non-equivalent descriptive domains (*Funtowizc and Ravetz 1994*).

The complexity makes it necessary to develop dynamic and adaptive approaches to resource management (*Brookshire et al., 2012*). The uncertainty (arising from lack of data and/or background information regarding the system under study) unavoidably requires us to simplify our scientific models (*Gupta et al., 2012*). And the incommensurability makes it necessary to investigate the range of alternative potential solutions, without explicit or implicit a priori weighting of priorities and relevance (for instance by monetizing all aspects of existing alternatives).

If we add to these the facts that a) cultural, political and ideological frameworks implicitly condition the context within which such model development occurs (i.e., the roles of "meaning" and "value" cannot be ignored), and b) it is not uncommon (due to outright *ignorance*) for us to '*not know what we ignore*' (*Wynne, 1993*), it becomes inescapable that knowledge must necessarily be co-produced. So, it is not possible to produce satisfactory answers to water management challenges via the "old" approach of simply bringing "technical" expertise to bear. Further, in that approach, claims for the legitimacy of specific interventions tend to reside exclusively in the realms of authority and privileged knowledge – the prevailing "state-engineering paradigm" that has over-determined water management for more than a century (*Staddon 2010*). We must instead adopt a participatory approach to governance that implies collaborative research at each step of the management process – in the definition of the problem, in establishing the range of options, in selecting the range of acceptable solutions, and in designing the indicators used to monitor and guide the process (*Lorrain & Poupeau, 2016*) – so that true legitimacy can be achieved via a shared vision of both the problem and the equitable solution set.

As expressed, in part, by this book, the TDW has sought to take this evolving water management paradigm into account while developing and supporting new forms of collaborative research that bridge across disciplines and incorporate the views of non-academic stakeholders. This book can be read as a first step in our journey towards a realization of the ideal of transdisciplinarity.

REFERENCES

Brookshire, D., Gupta, H.V. and Matthews, O.P. (Editors) (2012). *Water Policy in New Mexico: Addressing the Challenge of an Uncertain Future*, RFF Press, Resources for the Future Book Series: Issues in Water Resources Policy Series. ISBN 978-1-933115-99-3.

Del Moral, L., Pita, M.F., Pedregal, B., Hernández-Mora, N., Limones, N. (2014) Current paradigms in the management of water: Resulting information needs. In: Antti Roose (ed.) *Progress in water geography- Pan-European discourses, methods and practices of spatial water research*, Publicationes Instituti Geographici Universitatis Tartuensis 110, Institute of Ecology and Earth Sciences, Department of Geography. University of Tartu, pp: 21–31. ISBN 978-9985-4-0825-4.

Funtowicz, S.O. and Ravetz, J.R. (1991). A New Scientific Methodology for Global Environmental Issues, in Robert Costanza (ed.) *Ecological Economics: The Science and Management of Sustainability*, New York: Columbia University Press: 137–152.

Funtowicz, S.O. and Ravetz, J.R. (1993). *Uncertainty and quality in science for policy*, Dordrecht: Kluwer.

Funtowicz, S. and Ravetz, J.R. (1994). The worth of a songbird: ecological economicsas a post-normal science, *Ecological Economics*, 10: 197–207.

Giampietro, M., Allen, T.F.H. and Mayumi, K. (2006). The Epistemological predicament associated with purposive quantitative analysis, *Ecological Complexity*, 3(4): 307–327.

Giampietro, M., Mayumi, K. and Sorman, A.H. (2012). *The Metabolic Pattern of Societies. Where Economists Fall Short*. London and New York: Routledge.

Gupta, H.V., Brookshire, D.S., Tidwell, V. and Boyle, D. (2012). Modeling: A Basis for Linking Policy to Adaptive Water Management, Chapter 2 in Brookshire D., Gupta H.V. and P. Matthews (Editors), *Water Policy in New Mexico: Addressing the Challenge of an Uncertain Future*, RFF Press, Resources for the Future Book Series: Issues in Water Resources Policy Series.

Hernández-Mora, N. and Del Moral, L. (2015) Evaluation of the Water Framework Directive Implementation Process in Europe, SWAN Project, Deliverable 3.2, online: https://swanproject.arizona.edu/sites/default/files/Deliverable_3_2.pdf.

Lorrain, D. and Poupeau, F. (2016). The protagonists of the Water Sector and their Practices. Socio-technical Systems in a Combinatory Perspective, Introduction to: Lorrain, D. and Poupeau, F. (Editors), *Water Regimes: Beyond the Public and Private Sector Debate*, London, Routledge, Earthscan Series.

Pahl-Wostl, C., Jeffrey, P., Isendahl, N. and Brugnach M. (2011). Maturing the New Water Management Paradigm: Progressing from Aspiration to practice, *Water Resources Management*, 25: 837–856.

Pedregal, B., Del Moral, L., Cabello, V., Hernández-Mora, N. and Limones, N. (2015). Information and knowledge for water governance in the networked society, *Water Alternatives* 8(2): 1–19.

Rosenfield, P.L. (1992). The potential of transdisciplinary research for sustaining and extending linkages between the health and social sciences, *Social Science and Medicine*, 35: 1343–57.

Staddon, C. (2010). *Managing Europe's Water: 21st century challenges*, Farnham, Ashgate Press.

Wynne, B. (1993). Public uptake of science: a case for institutional reflexivity, *Public Understanding of Science*, 2(4): 321–337.

Chapter 2

Organization of the book and mind map

Editors

This book is about the physical and socio-economic roles played by water in the Southwestern US, with a primary focus on Tucson, Arizona. In the context of continued population growth, together with the fact that periods of drought are common in the Southwest and that the climate can be expected to change due to global warming, it is not unreasonable to expect that water will become increasingly scarce (leading to a "water bankruptcy") and that conflicts around water may increase. Such scarcity is, however, not a purely physical phenomenon, but results from a complex interplay between physical availability, the dynamics of the environment, and the behaviors of human and the demands they impose.

The chapters in this book explore both the physical and the socio-economic dimensions of water issues. Accordingly, the material is organized into *four main sections*, dealing progressively with the *"Socio-Historic Perspective"* regarding the evolution of laws and water policy, a discussion of the implications of *"Urban Growth"* driven by expansion of the population, a discussion of *"Ecosystem Services"* and how water and the biodiversity it supports together serve the needs of both humans and the natural environment, and finally a discussion of how strategies for *"Water Use and Groundwater Management"* have evolved to deal with water scarcity, and of how successful such strategies have been.

The four sections are followed by a collection of perspectives on water issues offered by professionals from different sectors and stakeholder representatives. Finally, the concluding section describes how this collective investigation was built (*"Bringing The Stories Together"*), synthesizes the material provided herein, and reflects (*"Next Steps: Collaborative Research and Training for Transdisciplinarity"*) on what has been learned about the water problems of the Southwestern US and about the nature of transdisciplinary investigation and education.

Also provided is a '*Mind Map*' that helps to visually link all of the various research perspectives presented in this book.

I SECTION ONE: SOCIO-HISTORIC PERSPECTIVES ON WATER IN THE AMERICAN SOUTHWEST

The first section (Chapters 3–7) explores the human factors related to water supply and demand in the Southwestern US. Setting the stage where the research in this book unfolds, **Chapter 3 *(The Tucson Basin)*** provides an overview of the physical context of the Tucson region and southern Arizona, as well as its human history until the present. The basin and range landscape and climate of the region endow it with unique hydrologic and ecological characteristics that have conditioned the lifestyles and struggles of the local human populations and have influenced their evolving relationship with the land, water and the environment. The chapter ends by touching on some of the current management challenges faced by the Tucson region.

Chapter 4 (Laws of the River) then provides a historical account of laws and agreements framing water management in the West; it examines the primary legal doctrines and rulings that have affected water allocation, thereby constituting the so-called '*Law of the River*'. Important historical highlights include: (i) the 1908 Supreme Court ruling that established the concept of federal reserved water rights that provided water to Native American reservations and reserved senior water rights for beneficial uses such as agriculture, (ii) the doctrine of '*prior appropriation*', that asserts that water rights arise from beneficial use and established a priority system among water users, (iii) the Colorado River Compact of 1922 that governs the allocation of water rights among the US states of Colorado, New Mexico, Utah, Wyoming, Nevada, Arizona and California, and the country Mexico, (iv) the Colorado River Basin Project Act of 1968 that allowed Arizona to proceed with construction of the Central Arizona Project Canal, and (v) the Arizona Groundwater Management Act of 1980 that established the first meaningful groundwater management law in the state's history. Importantly, the account illustrates how Western water management has moved away from traditional forms of conflict (litigation and court action), and resulted in the development of novel institutional tools that stress cooperation and consensus. It shows clearly that water policy involves a great deal more than managing flows, it also involves managing trust, people, and political power and even the forces of domination, which may often be implicit and charged with a particularized and historical energy. The modern form of struggle in policymaking, therefore, is less about taming the waters of the Colorado River and more about the struggle to reach consensus.

Chapters 5 and 6 together discuss the historical and social forces that led to the construction of the Central Arizona Project (CAP), a canal that carries Arizona's share of Colorado River water to its major urban centers. They examine the social history of water policy in the western US, and pay particular attention to the consequent social conflicts that arose among the various economic, political and administrative coalitions that formed to advance their respective interests and visions of the world. These chapters make a distinction between two phases of Western water policy, a first phase (late 19th century to 1920s) corresponding to the genesis of federal action, and a second phase (1920 to 1970) characterized by a shift from federal to regional decision making, in which the battle between Arizona and California for Colorado River water occurred, culminating in development of the CAP.

Chapter 5 (Water for a New America) addresses the first phase, discussing: (i) the transition from subsistence to commercial agriculture, growing "market" orientation, and the dominance of banks, railway companies, large-scale manufacturers, and farm product suppliers at the end of the 19th century; (ii) the Reclamation Act of 1902 that was intended to usher in a "New America" of small agricultural landowners but which instead helped to shore up the political and economic power brokers of the region; and (iii) the Roosevelt administrations' focus on infrastructure projects as the way to clamber out of the Great Depression.

Chapter 6 (Sharing the Colorado River) continues the story by investigating the historical and social forces that contributed to the construction of the CAP. It discusses how the sociological struggle between the various coalitions gradually shifted from the level of disagreements between the states (and the federal government) to local tensions over the viability of the CAP. Major aspects include: (i) Arizona's use of the US Supreme Court as an arbiter in regards to its sovereignty and legitimacy over the Colorado River; (ii) the attitudes of Arizona's elites, in the face of unprecedented demographic growth and the significant seasonal migration in a region where groundwater aquifers were gradually drying up due to the needs of agriculture; (iii) the 1948 creation of the Arizona Interstate Stream Commission to fight for the state's share of Colorado River water; (iv) the economic shift in the 1950's away from agriculture as the main source of wealth; (v) the Colorado River Basin Act of 1968 that was produced by a compromise between the various forces at play; (vi) the 1970s rise of the environmentalist movement; (vii) the 1980 Groundwater Management Act that instituted an innovative approach to managing groundwater and introduced limits to the expansion of irrigation; and (viii) the eventual delivery of CAP water to Tucson in 1992 and the subsequent tensions that arose among economic leaders, citizen organizations, local politicians and the utility companies. As shown by this discussion, water policies in the West are the product of temporary alliances between various economic, political, and administrative coalitions who regard water as an engine for economic development and political power.

Finally *Chapter 7 (The Making of Water Policy)* of this section provides a sociological analysis of how water conflicts are inscribed within spaces of power. In contrast to narratives such as *Cadillac Desert (Reisner, 1986)*, which illustrate the brute (and indeed brutal) force of economics, this chapter points out that water development in the West has involved a struggle over what constitutes the legitimate principles of *vision and division* of the world and its development. Beginning with the John Wesley Powell vision of settlers organizing themselves into *'cooperative commonwealths'*, this chapter discusses: (i) the political maneuvering that resulted in the 1902 Federal Reclamation Act whereby the federal government and an array of powerful economic forces took over the development of water infrastructure, gradually transforming the West into a breadbasket and economic powerhouse; (ii) the resulting concentration of power in a politico-bureaucratic elite, with the attendant shift from water viewed as a biological necessity (as in subsistence economies characteristic of traditional societies) to water viewed instead as a commodity valued for its role in economic production; (iii) the growing concerns about the legitimacy of this system, its role in promoting increasing levels of inequality, and the need for a focus on the conservation of nature, leading to the National Environmental Policy Act of 1970; and (iv) the replacement of the old system by one which is more receptive to citizens, and enables the public to lay claim to cultural attachments that cannot be reduced to monetary evaluation. In summary, this chapter poses the issue of water management as occurring within a field of struggle wherein the dominant groups must constantly refine and demonstrate the legitimacy of their perspective(s) in the face of the continuing involvement of other stakeholders.

2 SECTION TWO: NARRATIVES OF URBAN GROWTH

The next section (Chapters 8–11) explores the role played by population growth, and in particular urbanization, in regards to the demand and supply of water in the region. It begins, *Chapter 8 (The Social Logic of Urban Sprawl)*, with a discussion of the how and why urban centers in Arizona tend to sprawl out over the local countryside due to social and environmental pressures, even though one might expect that the poor availability of water would tend to restrict growth. Certainly, sprawl is driven to a significant degree by the actions of a "pro-growth" coalition composed of public and private actors, including the real estate industry, and is enabled by new flows of water brought to the region via the CAP canal. Interviews conducted with both developers and city managers provide insights into their perspectives regarding "*sustainability*".

Chapter 9 (Water and Urban Development Challenges of Urban Growth) continues this discussion by examining whether sustainable urban growth is possible in the context of available supplies of water and wastewater, for different urban settings and environmental conditions. It reviews the economic and demographic changes that occurred in the Tucson Metropolitan Region after the end of World War II, and the implementation of multiple strategies to establish diversified water supply sources, including: a) use of reclaimed water on parks and golf courses; b) recycled wastewater for indirect potable use; and c) recharge of effluent into aquifers. It discusses three main patterns of urban growth that result from the combination of land development and water/wastewater access – "*urban expansion*", "*leap-frog development*" and "*wildcat development*".

Next, *Chapter 10 (Comprehensive Urban Planning)* examines the implementation of environmental policies in Pima County and the city of Tucson, by reviewing the role and evolution of *urban planning*. It supplements the discussion with practical insights provided via interviews conducted with the Pima Services Department, the Pima County Planning Division-Comprehensive Plan, and the City of Tucson Housing & Community Development Department. A core concept that emerges is that of the integration of scientific disciplines, approaches and experiences within a coordinated dialogue between the social, natural and engineering sciences. Further, the chapter points to the spatial mismatch that can occur between different planning scales, and the difficulties that can arise in relation to the adjustment of the different hydrographic, socio-economic and jurisdictional aspects involved. The chapter concludes that there is room for greater efforts to be made to effectively engage society in comprehensive planning decision-making, especially in relation to water in this arid region.

Finally, *Chapter 11 (Potential Impacts of Continuing Urbanization on Regional Climate)* discusses how the growth of the "*Sun Corridor*", which is rapidly filling in the space between Phoenix and Tucson, is likely to result in climatic changes that urban and regional managers will have to deal with. Urban expansion changes the physical environment by altering the albedo, heat capacity, and thermal conductivity of the land surface, thereby changing the energy balance of the region. Detailed simulations, conducted using a coupled model of the land surface and the atmosphere, show that while projected changes in urban land cover between 2005 and 2050 are unlikely to alter precipitation patterns, they will strengthen the "urban heat island" effect and increase the demand for water and energy supply to levels that are not sustainable.

3 SECTION THREE: ECOSYSTEM SERVICES AND BIODIVERSITY

The third section (Chapters 12–14) explores the interplay between humans, water, and the environment. It begins, *Chapter 12 (Quantification of Water-Related Ecosystem Services)*, with a discussion of Water-Related Ecosystem Services (WRES) provided to society by the Upper Santa Cruz watershed, and how these services are affected by changing land use. In particular, the study shows that forested lands provide the highest levels of supply of WRES in the region, and that a variety of urban growth scenarios all can be expected to result in a decreasing trend in the supply of almost all services provided by the current ecosystem.

Chapter 13 (Qualitative Assessment of Supply and Demand of Ecosystem Services) continues with a survey and interview-based assessment of the perceived current levels of supply and demand for ecosystem services in the Pantano Wash watershed, in both time and space. The resulting maps display spatial and temporal mismatches in supply and demand, that can inform water planning efforts, and facilitate the optimization of strategies for sustainable management in which a balance is sought between the provision of natural resources and the demands imposed by a myriad of interests. Moreover, they provide support for cooperative decision-making and resource planning by illuminating perceptions that exist regarding the importance of various ecosystem goods and services.

Finally, *Chapter 14 (The Role of Biodiversity in the Hydrological Cycle)* discusses the need for water management strategies in the Southwestern US to take into consideration the negative effects that increasing aridity (due to changing climate) is likely to have on biodiversity in the region. Loss of biodiversity can be expected to alter the balance of Ecosystem Services. However, surprisingly little is known about how soil-dwelling and burrowing species change the permeability of the soil and thereby affect the hydrological cycle, and this chapter points out the need for more research in this area so that such information can be incorporated into water management and biodiversity conservation programs.

4 SECTION FOUR: WATER USE AND GROUNDWATER MANAGEMENT

The fourth section (Chapters 15–18) investigates the attempts to achieve sustainability that have been implemented in the Tucson Basin. It begins, *Chapter 15 (Implications of Spatially Neutral Groundwater Management)*, with a historical perspective on water use in the area, and on the changes induced by the arrival of CAP water from the Colorado River, with attention to the impacts that conservation programs have had on municipal and agricultural water demand, and on the spatial distribution of groundwater dynamics (recharge, pumping and water levels). The study uses the Multi-Scale Integrated Analysis of Societal and Ecosystem Metabolism (MuSIASEM) framework to analyze available data on water use, a variety of socioeconomic variables, and groundwater management, showing that the CAP served as a tipping point in the water metabolism, by multiplying the sources available while increasing infrastructural and institutional complexity, thereby fueling economic development.

It reviews the impacts that strategies of *"conservation"*, *"growth control"* and *"replacement of groundwater with CAP supply"* have had on various sectors, and highlights the facts that a) vulnerability to potential Colorado water shortages and b) uncertainties regarding the ability to achieve and maintain distributed safe yield will continue to be core management issues over the next decade.

Chapter 16 (Groundwater Dynamics) investigates the problem of how the dynamics of groundwater aquifers that serve the Tucson Basin are affected by natural cycles of drought at irregular inter-annual and seasonal time scales, by analyzing water tables and stream flows datasets. While drought cannot be avoided, proper planning can help to mitigate its environmental and social effects. The study shows that in recent years, when CAP deliveries were used to substitute for pumping, the onset of a *'groundwater drought'* following a *'precipitation drought'* was delayed by about 3.5 years, which means that the time when a hydrogeological drought can be expected to occur in the Upper Santa Cruz can be anticipated. However, when groundwater was pumped instead of using CAP deliveries (1980–2000): a) the pattern is much less obvious and is masked by human pumping; and b) the onset of *'groundwater drought'* in response to *'precipitation drought'* tends to be much more immediate, with a more rapid decline in groundwater levels.

Chapter 17 (Alternative Water Sources towards Increased Resilience) assesses the sustainability of water use in the Tucson Basin, and discusses feasible alternative options that might be pursued to increase resilience and help to fill future gaps between demand and supply. The investigation, based on comments and observations solicited from a diverse group of local water managers and stakeholders, discusses problem solving approaches and management strategies that have been proposed to help balance the water budget. Further, it provides a critical analysis of current and future water resource uses, projects and policies, and examines use of innovative approaches such as rainwater harvesting, storm water capture, grey-water systems, and use of reclaimed water for indirect and direct potable re-use. The chapter concludes that these alternative water sources are underexploited and hold significant potential to offset groundwater pumping, and that sustainability can best be accomplished through water management approaches that combine gray- and green-infrastructure that recognizes and nurtures ecosystem services within urban landscapes and the broader basin.

Finally *Chapter 18 (Differentiated Approaches of Groundwater Management)* compares the changes in water use and current water practices that have occurred in the Tucson Active Management Area (TAMA), with those that have occurred in the neighboring Upper San Pedro (USP) basin. Whereas the TAMA operates under the state regulatory structure, the USP basin (which was not designated as an Active Management Area) benefits from a partnership established between governmental and non-governmental entities. In both cases, municipal demand has declined and, by that assessment, the management measures can be deemed successful. However, while agricultural demand has been reduced significantly in the USP Basin, there has been little change in the TAMA. Similarly, there have been differences in the growth of new development, and the effects of needing to certify an *'assured water supply'* must be more fully considered. In neither basin have the problems of groundwater depletion been solved, nor has either safe or sustainable yield been achieved. The chapter concludes that a) the Groundwater Management Act should be revisited to determine if it is achieving its policy goals, and b) consideration needs to be given to the use of water by natural ecosystems in the TAMA.

5 STAKEHOLDERS' PERSPECTIVES

The fifth section of the book (Chapters 19–20) is a collection of written perspectives that was solicited from various stakeholders who have been involved, in one way or another, in helping to guide the investigations reported in this book. Rather than a unified vision, this chapter represents the diverse and sometimes opposing views that reflect the opinions and interests of various stakeholder communities. It is clear from these perspectives that the task of finding a middle ground for the benefit of the community as a whole (in the form of tradeoff solutions that balance the wide spectrum of preferences and values) remains a major challenge for planning, policy and management.

6 CONCLUSION

Finally, the last section of the book (Chapters 21–22) integrates the main findings, insights and recommendations from the various book chapters, and reflects on what has been learned through the investigations reported herein.

Chapter 21 (Bringing all the stories together) summarizes important conclusions and recommendations from the book chapters, and discusses how the participants in the Sustainable Action Water Network (SWAN) project, drawn from a variety of social and natural science disciplines, collaborated in an effort to bring a *transdisciplinary* perspective to the study of water in the Tucson Basin. Major insights from the book's chapters are woven together here, providing recommendations that may be useful to planners and decision-makers. While it is arguable whether true transdisciplinarity was actually achieved, the collaboration provided a very valuable learning experience and also resulted in the materials that form the basis for this book.

Finally, *Chapter 22 (Next Steps)* concludes this book with a broad overview of what has been learned though this collaborative research endeavor, and provides some recommendations for others interested in pursuing such an endeavor.

MIND MAP

To aid in synthesis the information included in this book, *Figure 1* presents a '*Mind Map*' that provides a visual perspective on how the various issues discussed in this book are connected. While there are many ways in which these concepts can be arranged (in keeping with the reality of multiple perspectives), here we have conceived of the main areas of investigation being the Natural and Social Sciences, and the Natural and Human Systems. The severe risks associated with poor solutions to water management problems lead naturally to the need for an encompassing transdisciplinary perspective (as discussed extensively in the latter part of this book), and ultimately

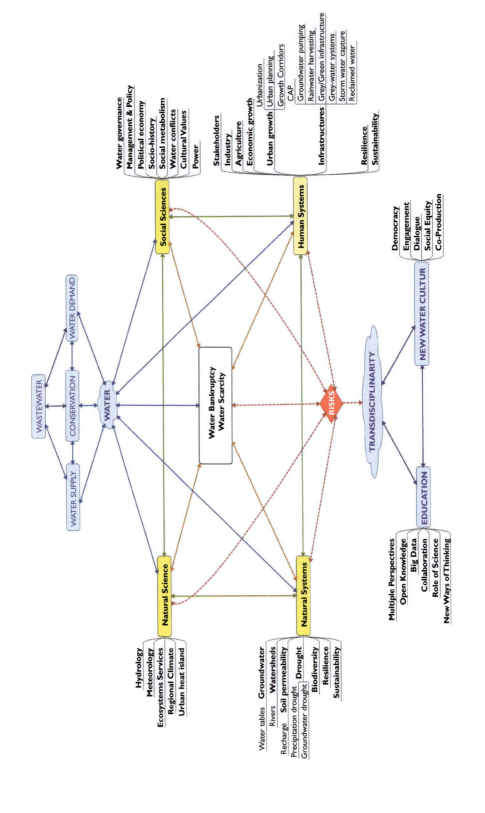

for the need to an enhanced approach to education and training (see Concluding Chapter). Through these, it may be possible to envision the emergence of a "new water culture", based in democratic principles, and with social equity at its heart.

Hoshin Gupta
Department of Hydrology and Atmospheric Sciences,
University of Arizona, USA

Maria A. Sans-Fuentes
Biosphere 2, University of Arizona, USA

Franck Poupeau
UMI iGLOBES, CNRS/University of Arizona, USA

Aleix Serrat-Capdevila
UMI iGLOBES CNRS/Department of Hydrology and
Atmospheric Sciences, University of Arizona, USA

Susan Harris
Department of Hydrology and Atmospheric Sciences,
University of Arizona, USA

László G. Hayde
UNESCO-IHE, Institute for Water Education,
Delft, The Netherlands

Maps

All the maps were realized by Rositsa Yaneva (*National Institute of Geophysics, Geodesy and Geography, Bulgarian Academy of Science*).

The SWAN project has implemented a tool for open access socio-environmental data at the website www.gis-swan.org.

GIS SWAN is a web-based viewer containing accessible water resources information. It aims at disseminating some of the research results obtained during the SWAN project that studied the Tucson Basin area (TAMA). It has been implemented by the University of Seville (Spain) and the National Institute of Geophysics, Geodesy and Geography (Bulgarian Academy of Science).

Several Geo-layers representing the water system, the land cover system and the territory system of Tucson Basin, as well as others produced by the SWAN teams, have been integrated into the SWAN GIS geo-viewer.

This tool represents an excellent example of how open knowledge can be disseminated and provides a way to connect citizens with ongoing scientific activities and results.

Map 1 United States of America.

Map 2 Colorado basin.

Map 3 State of Arizona.

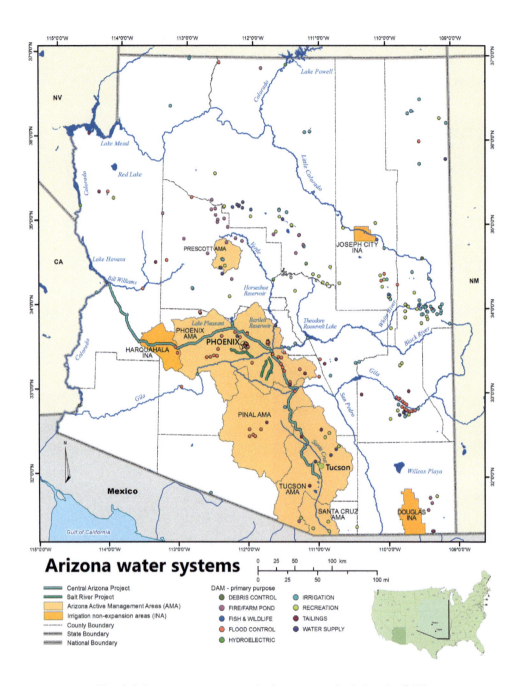

Map 4 Arizona water systems and infrastructures (including the CAP).

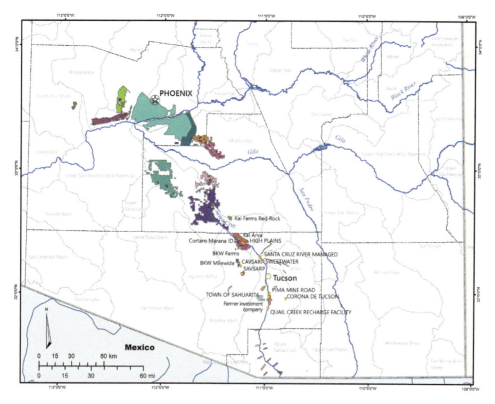

Ground water storage facilities
Southern Arizona

Map 5 Southern Arizona groundwater storage facilities.

Socio-historic perspectives on water in the American southwest

Chapter 3

The Tucson basin: Natural and human history

Aleix Serrat-Capdevila
UMI iGLOBES CNRS/Department of Hydrology and Atmospheric Sciences,
University of Arizona, USA

INTRODUCTION

The current state of a human-natural system and its management challenges cannot be understood without its historical evolution. This chapter describes the landscapes, geology, climate, hydrology and environment that have hosted the evolution of local human societies since the first arrival of people in the region. A discussion follows on the history of the interactions between these communities and the environment, with an emphasis on human events that introduced new technologies practices. These shaped both the environment and the societies in ways and scales not previously seen, and caused significant feedbacks to occur within the human-natural system. These new practices can be understood as metabolic transitions, as they represent changes in the way society uses resources to produce goods and well-being. The chapter ends with a brief overview of the current management challenges.

1 THE PHYSICAL SETTING

1.1 A semi-arid basin and range landscape

Most of the Southeastern Arizona landscape is dominated by *"basin and range"* systems caused by faulting and uplift 12 to 6 million years ago during the Miocene, resulting in a sequence of ranges (Horsts) and tectonic depressions (Graven). These depressions have – over millions of years from the Miocene to the Quaternary – been progressively filled with sediments eroded from the mountain ranges along the basin boundaries (*Figure 1*). The geomorphology that gives the Tucson basin its current shape is mostly dominated by the large alluvial fan of Cienaga Creek, surrounded by smaller alluvial fans over the rock pediment at the base of the Catalina Mountains to the north, the Rincon Mountains to the east and Santa Rita Mountains to the south.

The bodies of water that flow through and occupy the pore spaces of these basin sediments are called aquifers. Under pristine conditions (i.e., before the advent of extensive groundwater pumping using high-lift turbine pumps) water levels tended to be close to the surface, especially along the river channels. Due to slow and continuous

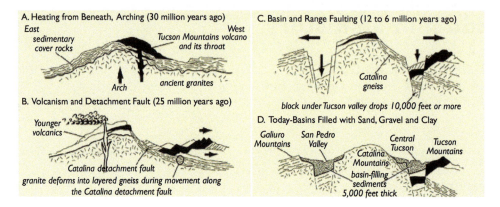

Figure 1 Recent geologic history of the Tucson region explained by four East-West cross-sections illustrating the formation of the basin and range landscape (adapted from Scarborough 2015).

replenishment through rainwater recharge, these aquifers directly intersected the river channels so that groundwater could drain out of the aquifer to flow down the slope of the river channel (which is the line of lowest elevation in the landscape) and support, during the long dry season, a lush and bio-diverse riparian corridor with cottonwoods, willows and mesquite forests.

1.2 Climate, hydrology and vegetation

The Southern Arizona region has a semi-arid climate due to its latitudinal position between the Hadley and Ferrel cells that contribute to global patterns of atmospheric circulation (*Figure 2*). In sub-tropical latitudes, cold and dry air masses from high atmospheric altitudes sink towards the land surface between the two cells and thereby limit the possibility of convection, i.e., the rising of moist air needed for cloud formation.

Daily normal temperatures range from 39°F low/65°F high (3.9°C/18.3°C) in the month of January to 70°F low/100°F high (21°C/38°C) in the month of June. While progressively hot and very dry throughout the spring and into the summer, relative humidity rises again with the arrival of the monsoon season (NWS-NOAA, 2015).

Rainfall in southeastern Arizona is characterized by a bimodal precipitation regime consisting of rainfall in both the winter and the summer, and high spatial and temporal variability (*Figure 4*). Being near the northern boundaries of the North American Monsoon system, summer monsoons in Tucson begin in early to mid-July and last until September, bringing convective thunderstorms of high intensity and short duration. In the months of September and October, moisture from dissipating tropical cyclones may also contribute some rainfall (Webb and Betancourt, 1992). In winter, during January and February, regional frontal storms originating in the Pacific Ocean provide rainfall of lower intensity but longer duration compared to

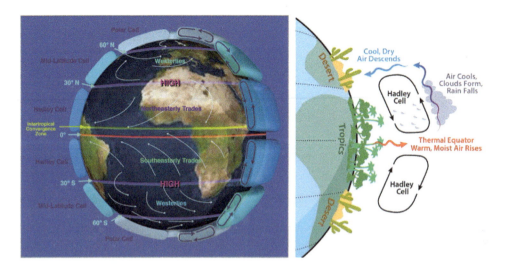

Figure 2 Global atmospheric circulation patterns (left) and the effect of the Intertropical Convergence Zone and the Hadley Cells on the aridity of sub-tropical latitudes (right). (Credit: left: NASA/JPL-Caltech; right: Moeller, 2013).

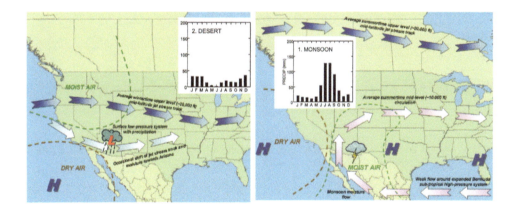

Figure 3 Average flow patterns and moisture air mass boundaries for winter (left) and summer (right) (adapted from Crimmins, 2006); and average monthly rainfall of the desert and monsoon regimes affecting the Tucson region (Comrie and Glenn, 1998).

the summer monsoons. Comrie and Glenn (1998) discussed the influence of various precipitation regimes in the US Southwest and Northern Mexico, and showed that the Tucson basin is influenced by two regional components. The "monsoon" component is characterized by important summer precipitation from June to October in the form of convective storms of short duration and high intensity, while the "desert" component is characterized by very low precipitation all year but with the lowest values in early summer and a slight increase in winter (*Figure 3*).

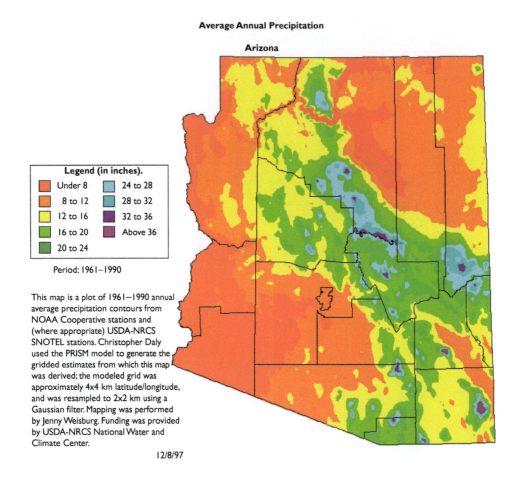

Figure 4 Average annual precipitation in Arizona for the period 1961–1990 (from PRISM Precipitation Maps, Courtesy of Oregon Climate Service).

Annual precipitation in the Tucson basin can range from over 5 inches (127 mm) to more than 20 inches (508 mm) with an average of approximately 12 inches (305 mm). Half of that precipitation typically occurs due to monsoon rainfall in the months of July, August and September, while the other half is spread over the rest of the year with higher values in the winter months (NWS-NOAA, 2015).

As the surrounding mountains reach elevations over 9,000 feet (2,743 m), a temperature and precipitation gradient exists with elevation, with yearly precipitation averages of 30 inches (762 mm) at high elevations, which include 65 inches (1.65 m) of snow in winter. This translates into a vegetation gradient that goes from the saguaros and cholla cacti and cottonwoods near springs and riparian areas down in the Sonoran Desert, up to mountain ecosystems with fir and spruce trees (*Figure 5*).

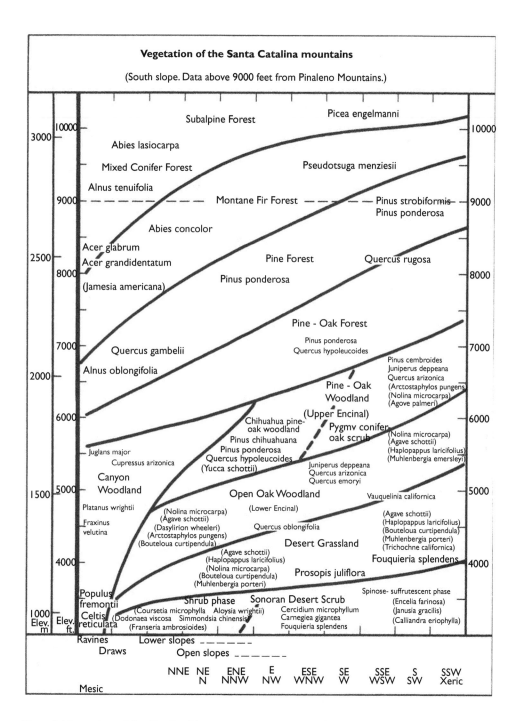

Figure 5 Vegetation of the Catalina Mountains across aspect and elevation gradients (from Whittaker et al., 1965).

1.3 Streamflow and recharge

As in many other parts of the world, in addition to seasonal and interannual variability, rainfall in Southeastern Arizona is also affected by natural decadal to multidecadal variability. This long term variability controls magnitudes and frequencies of droughts and floods. The extreme and damaging floods during the 1983 El Niño in Tucson motivated a study to better understand changing flood seasonality and frequencies at longer time scales. *Webb and Betancourt* (1992) looked at floods associated with different storm systems affecting the Tucson region. They found that for the period 1915–1986, monsoon storms dominate

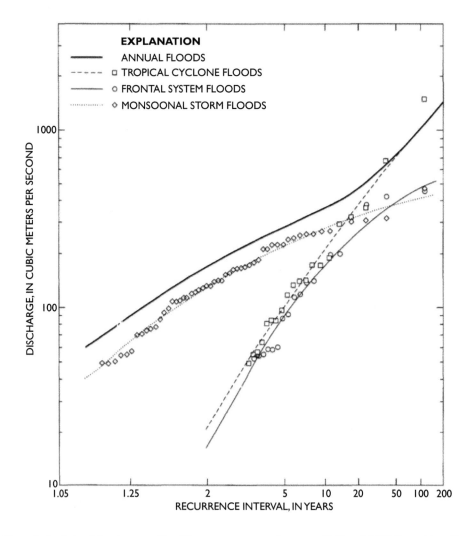

Figure 6 Analysis of floods caused by different storm types between 1915 and 1986, Santa Cruz River at Tucson, Arizona. Curves are presented for illustrative purposes only (Source: *Webb & Betancourt* 1992).

flood frequency for return periods of up to 10 years. Late summer and early fall storms related to residual moisture from tropical cyclones dominate flood frequencies for return periods above 20 years. The frequency of frontal storm floods follows that of the tropical cyclone floods through the 20 year recurrence interval, after which they diverge. However, the frequency of floods from each storm type is dependent on the period of record analyzed, and separate analysis of the first and second half of the period reveals significant changes in flood discharges for given return periods and increased flood variance, highly affecting the estimation of long recurrence events, such as the 100 year flood, and their uncertainty. The change in flood frequency is attributed to climatic variability, even if land use changes may have had some effect on the magnitudes of the flood events. Partly as a result of the study, Pima County adopted the estimate of 58,622 cubic-feet/second (1,660 m^3/s) as the design 100-year flood for the Santa Cruz River at its location near downtown Tucson.

During the rainy seasons, the ephemeral washes in the basin can flow during and shortly after storms, but are mostly dry between rain events. However, in the surrounding mountains, due to topography, geology, and the rainfall elevation gradient, streams can carry water almost year round, from raging floods to almost dry in the summer season. Due to their rocky and impervious nature, mountain canyons such as the Sabino, Bear, Pima, and Ventana in the Catalina mountains, and many others in the Rincon and Santa Rita mountains, collect sub-surface water draining from the hillslope soil layers, and continue to carry dwindling amounts of flow through part of spring and early summer, well after the winter rains, until they finally become trickles, but sufficient enough to support riparian canyon ecosystems until the arrival of the summer monsoons.

Aquifers in the Southeastern Arizona basin and range landscape are fed by recharge at the mountain-front and along the streambeds of ephemeral channels (*Figure 7*). Runoff from rainfall in the upper elevations flows down through creeks and hill-slopes and eventually infiltrates into the ground where the mountains meet the basin sediments. The water from storms in the lower elevations of the basins partially infiltrates along the stream channels. Because the Tucson basin is influenced by both summer monsoon and desert winter rains; mountain-front recharge and ephemeral channel recharge are of different annual and seasonal importance. Although winter rains constitute less than half the annual precipitation, they are responsible for a major portion of the annual recharge (*Eastoe et al.,* 2004), due to their lower intensity and longer duration, and the low rates of potential evapotranspiration (evaporative demand by the atmosphere) in winter. Basin wide recharge is impossible to measure directly and estimating it is a difficult task. Recharge estimates are obtained as averages calculated for periods of many years. This is almost always done through modeling by adjusting recharge rates until groundwater models replicate observed water table levels as closely as possible. In this manner, the Arizona Department of Water Resources has estimated recharge in the Tucson Active Management Area in the order of 95 Mm^3/year (76,600 acre-feet/year), of which 48 Mm^3/year (38,900 acre-feet/year) are from mountain front recharge and 46.5 Mm^3/year (37,700 acre-feet/year) are from streambed recharge. These numbers can vary significantly from year to year.

34 Water bankruptcy in the land of plenty

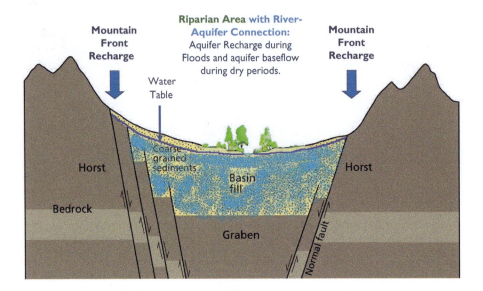

Figure 7 Cross-section illustrating the natural conditions of a typical basin and range aquifer system showing the areas of mountain front recharge (with runoff from the mountains) and the riparian area where the river is connected to the water table (created with materials from Thornberry-Ehrlich, T. (2015) and Eubanks 2004).

2 HUMAN HISTORY IN THE TUCSON BASIN

2.1 Prehistoric period

With the end of the Last Glacial Maximum and the subsequent melting of glaciers, humans followed herds of big game from Beringia into Alaska sometime after 14,500 BCE, giving rise to the population of the Americas. These *"Paleo-Indians"* arrived in the Tucson region around or before 11,000 BCE (*Haynes*, 2007). They subsisted by hunting large animals such as mammoth and bison in what were then desert grasslands. In the Murray Springs site in the Upper San Pedro, archeological research showed that a group of *Clovis hunters* killed a young female mammoth about 11,000 BCE while drinking from a water hole which the herd had dug out. They also killed 11 bison in the vicinity (*Haynes and Huckell*, 2007). However, much of the larger game became extinct during the Late Quaternary Pleistocene/Holocene Extinction. For example, the *Mamut americanum* (American Mammoth) became extinct in 10–8,000 BCE, due to a potential range of causes including hunting, climate change and environmental change (*Koch and Barnosky*, 2006).

The extinction of mega fauna progressively forced humans to reduce their dependence on big-game hunting and to develop a hunter-gatherer lifestyle. The appearance of grinding slabs marks the beginning of the *Archaic Period* (~7000 BCE to 300 CE). In the Tucson basin these hunter-gatherers are thought to have lived in the foothills and around streams, in seasonal camps between which they moved periodically, hunting and foraging wild foods, and drinking water from springs and streams.

During the last third of the Archaic Period, corn was brought in from what is now Mexico, likely through networks of trade and cultural exchange and it was planted along river banks and near bodies of water, most probably after floods, left to grow and later gathered, but not really cultivated. This type of flood farming was practiced along the banks of the Santa Cruz River by at least 800 BCE (*Mabry and Thiel*, 1994) in combination with the gathering of wild foods and hunting.

Around 200 CE, a few families likely arrived from the south and built homes in the Santa Cruz basin and along the Rillito Rivers. These newcomers brought with them the influence of Mesoamerican cultures, most notably a new technology that would change the way humans relate to the environment: irrigation canals and agriculture. This marked the end of the *Archaic Period* and the beginning of the Hohokam culture and "*rancheria*" lifeway. Inhabitants of the Tucson basin started building canals to irrigate fields of newly arrived crops: new varieties of corn, beans, squash, cotton and other domesticated crops such as amaranth. While the adoption of irrigation farming was a radical change, the Tucson basin inhabitants did not abandon their previous practices, and continued to combine agriculture with hunting and gathering. However, with the appearance of agriculture, populations began to grow and new cultural manifestations appeared, such as pottery and stone carved bowls, a sign of a less nomadic lifestyle, as well as human figurines.

Over the next millennium and a half, the Hohokam culture in Southeastern Arizona (covering the Santa Cruz, Gila and Salt river basins) developed as a part of the interwoven mosaic of native cultures in the Southwest. It evolved through the *Pioneer*, *Colonial* and *Classic* Periods (Casa Grande complex, 14th century) and was connected with neighboring cultures through networks of trade and exchange. In the Colorado Plateau region, the *Basketmaker* and then the *Pueblo* Cultures (aka "*Anasazi*") evolved to build the Chaco Canyon (10, 11th centuries), Betatakin and Ket Seel (10th, 11th centuries), and Hovenweep and Mesa Verde (12th, 13th centuries) settlements. Further to the south, the Mogollon Highlands saw the succession of *Early Pithouse*, *Late Pithouse* and *Mimbres Mogollon* Cultures, with the influence of the latter extending to southern New Mexico and Chihuahua. Wedged between all of the previous, and playing a key role in facilitating and perhaps controlling trade and cultural exchanges among cultures, the *Sinagua* also flourished between what is now Flagstaff, Prescott and Phoenix, founding the Wupatki and Tuzigoot (12th century), and Montezuma Castle (14th and 15th centuries) settlements (McNamee, 2014). One cannot imagine the Hohokam in the Tucson basin without thinking of the broader world they were connected to, including their southern Hohokam relatives and the Mesoamerican cultures further south. One can see interwoven cultural influences in the evolving pottery styles, agriculture and artifacts, such as seashells, and macaw feathers in Hohokam burials (Ferg, 2007). The Hohokam likely held a strategic position between Mesoamerica and southwestern cultures to the north, mediating trade and influence (Sheridan 1995; McNamee 2014).

The Hohokam culture in the Santa Cruz reached its peak during 950–1100 CE in terms of canal-building and living structures. This was followed by a period of decline between 1100 and 1450 CE, as happened with the large Hohokam centers of the Gila and Salt rivers where Phoenix is now. This period of crisis and change may have been due to a number of events such as a shift in climate, floods in the winter of 1358–59, which destroyed canals and fields, followed by a period of drought and yet another wet spell with more flooding. In the Santa Cruz, this led to the downcutting of some segments of the river channel, likely being the cause of a return to more dry farming and changes in settlements. It cannot be known to what extent this caused social unrest that contributed to a collapse of existing institutional structures. While most of the impressive mud-house centers and villages were abandoned and the cultural artifacts found for the previous period of Hohokam climax become scarce or non-existent during this period of crisis and malnutrition, irrigated agriculture may have continued near and downstream of marshy areas as was found by the first old world visitors arriving in the area. The centuries before the arrival of the Spaniards were a period of significant change for the cultures across the Southwest, and shifts in power and influence, trade networks, abandonment of entire communities and migrations (i.e., Puebloans) can be observed in the archaeological records. While the Hohokam society was forced to adapt to these environmental and social changes by moving their settlements within the region, they never left and there is consensus between experts and tribal members that the Hohokam are likely the ancestors of the O'odham people.

2.2 The arrival of the Spanish

When the Jesuit priest Father Kino first visited the Tucson Basin in 1692, he found three villages along the Santa Cruz River: Bac, to the South of Martinez Hill (where in 1701 he established the Mission of San Xavier del Bac); *shoockson* or *shoock-shon* (in Pima language meaning "at the foot of the black [...]" which the Spaniards then transcribed as Tucson) at the base of what is now *Sentinel Peak* or "*A Mountain*"; and Oiaur, to the North before the confluence of the Santa Cruz and the Rillito River. The Spaniards found many irrigation canals and farmland, with the villages of Bac and Oiaur having up to 800 inhabitants each. The villages were located in the vicinity of two areas where impervious underground volcanic formations forced groundwater to the surface, and where water would usually be available even in times of drought. For instance, the area along the Santa Cruz to the South of San Xavier del Bac used to be a wetland where it was easy to divert water from the river and from the many springs into irrigation canals.

While the cattle from a few Spanish settlers was already running in the grasslands of the Upper Santa Cruz more than ten years before the arrival of the Jesuits, the influx of colonists increased with the discovery of silver in a mining camp called *Arizonac*, in Sonora. Their livelihood was subsistence farming and ranching, with occasional prospecting. The name of the site, which later gave name to the state, could come from two Pima words *ali shonak* ("small springs") or from Basque pioneers: *arritza onac* ("valuable rocky places"), or *aritz onac* ("good oaks") (Sheridan, 1995, p. 31).

In 1695, Kino introduced wheat and cattle, and after the establishment of the San Xavier Mission, the Jesuits brought livestock and new crops (wheat, barley, chickpeas, lentils, onions, garlic). By 1772, after coping with the Piman revolts of 1751 and 1756 a new "*presidio*" (army garrison) was established by the Spanish Crown in Tubac in 1752, and the San Agustin mission "*visita*", a chapel and a residence were built; the first European buildings in what would later become Tucson. In 1775, the Spanish garrison in Tubac was transferred to what is now Tucson as part of the strategies and chain of events from the Bourbon Reforms to strengthen the Spanish presence to fight the Apaches and protect the Crown's mining interests in this frontier region and to the south. Soon after, Spanish settlers began irrigating lands on the east side of the Santa Cruz near Tucson, across from native fields on the west bank of the river, increasingly competing with the local Pima for the river water and thus creating conflict. This led to an agreement in 1776, which allocated three-quarters of the flow to the Indian villages and one quarter to the Tucson *Presidio*, which was modified 20 years later to "*half for the Indians and half for the Spanish*". After Mexico became independent from Spain in 1821, new settlers continued to arrive from the south and a Sonoran irrigation system was established in which shortages were shared equally and the scheduling was flexible depending on crop needs. By then, there were three *acequias madres* (main canals) maintained as common property and an elected overseer to supervise water distribution (*Mabry and Thiel*, 1994).

In 1825–26, the first contacts with Anglos occurred with fur trappers hunting beavers in the southern Arizona Rivers (such as the San Pedro, San Francisco, Gila and Salt Rivers), as well as with miners heading west to California during the 1849 gold rush. Later on, Anglo settlers would also arrive. The freighting industry started to develop, supplying the army garrisons, the prospectors, trappers and settlers, while the Apaches continued to raid local tribes, the Spanish and later the Anglos. Pimas and Europeans were often allied against the Apaches.

2.3 The incorporation of the Tucson area into the US

In 1854, with the Gadsden Purchase, the United States acquired (from Mexico) all of the land between the Gila River and the current US-Mexico border, and the influx of Anglos and their political and economic influence increased significantly in the region, as did the area of irrigated land and the number of irrigation canals and diversions from the river. In 1884–1885, the first law suit over competing interests for the Santa Cruz River water took place because new developers started irrigating old fields upstream (to the south) thereby creating shortages for the established farmers downstream. The new Anglo developers won the lawsuit, and this marked the decline of the traditional irrigation system. Over time, 33 new ditches were built by corporations and entrepreneurs, adding to the three main old canals, and as the competition for river flows intensified, the ditches were dug deeper to be able to divert diminishing water shares (*Mabry and Thiel*, 1994). In 1887, an event that had been waiting to happen was triggered. After Sam Hughes dug a new deep ditch to try to capture sub-surface flow to irrigate his fields, a large flood started downcutting the riverbed where the ditch was. The riverbed eroded down to the water table level, with a headcut that progressed quickly upstream, leaving all the diversion canals hanging above, disconnected from the river. The channel downcutting can be attributed to a combination of natural and human causes (*Betancourt*, 2015; *Webb et al.* 2014), including land use changes upstream and the lowering of the water table due to a continued increase of water diversions along the Santa Cruz River.

With the lowering of the water table and the riverbed, the next 15 years saw a number of developments based on the construction of wells near the river-banks and the floodplain (to tap artesian sources or the now 20 feet deep sub-surface flow of the river) and the construction of new canals and infrastructure improvements. Among others, a company of Chicago and British investors developed the "*Crosscut*" (a 19-well field) and installed electric pumps to bring the water up from the wells to the canal system. However, when a large flood destroyed most of the systems in 1940, irrigated agriculture in Tucson came to an end.

One of the oldest continuously cultivated sites in the US

The Middle Santa Cruz had been one of the oldest continuously cultivated sites in North America, from the late Archaic Period, through the Hohokam Era, to the Spanish, Mexico, and the Anglos. Efforts are currently taking place to preserve this heritage through the Mission Gardens Project of the Friends of Tucson's Birthplace, re-creating the garden that was part of Tucson's historic San Agustin Mission and "interpreting 4,000 years of Tucson agriculture" with the help of the Tohono O'odham Tribe.

Photos 1 & 2 View of Tucson from Sentinel Peak ("A" Mountain) during the late nineteenth century (above), and in November of 2015 (below). Left of the center are the adobe ruins of the San Agustin mission and its gardens, and behind it the Santa Cruz River flowed year-round between irrigated fields on both sides of the floodplain until its downcutting in 1887. Above is photo no. 12649 courtesy of the Arizona Historical Society, taken from Mabry and Thiel (1994). The photograph below was taken by the author on 11/11/2015 at 11am.

The arrival of investors from distant US cities was greatly facilitated by the railroad and the expansion of its network across the southwest. The Southern Pacific Railway Company (SPRC) and the Texas Pacific Company had been competing (politically, and for resources such as track steel) for railway construction to connect California with Texas and the East Coast, across Arizona. Imported Chinese workers were the main labor force that graded and laid track for the railroad; in 1878 they were paid $1/day.

In 1880, the railroad arrived to Tucson from the west by the hand of the SPRC and construction soon continued towards Texas. Railroad access brought new business ventures from outside investors and represents an important social and economic transition. As the Southern Pacific Railway Company controlled transportation, most freighter-merchant firms went bankrupt in Tucson, causing a depression and reorientation of the Arizona economy. In addition, the railroad also caused a cultural shift. Until then, Tucson-Mexico connections (North-South) had been the driver of culture and the economy: anglo and native-hispanic intermarriages and business partnerships were the social fabric of the Tucson region. With the arrival of the railroad, the connections were more strongly East-West, with a clear anglo-dominance of the economy and the end of intermarriages. Racial divides appeared. While the previous freighting industry was mainly bringing goods to supply the military forts, soldiers, prospectors, farmers and the service industry; the railroads marked the beginning of Arizona as an extractive colony inside the US, exporting goods out (*Sheridan*, 1995).

This shift can be seen in many sectors, such as in the ranching history. In the 1850's and 1860's (enabled by the Gadsden Purchase in 1853), cattle and sheep passed through Southern Arizona towards the California markets to supply beef, mutton and wool. During the American Civil War (1861–1865) Texas supplied most of the beef to the Confederacy and significantly expanded its market exports, which continued after the war. While a head of cattle was worth $5 to $15 in Texas, its price would rise up to $60 to $150 in California. Despite the fact that Apaches raided about one third of the 9,000 heads of cattle passing through, those years saw the rise of local ranching and homesteading in Arizona at the scale of hundreds of heads, small compared to future growth.

From the 1880's to 1891, cattle and sheep ranching skyrocketed in southern Arizona. The railroads were used to import cattle. In this ranching boom, the grasslands and ranges were over-stocked beyond capacity. With the increasing cattle numbers, it became evident that "*who controls the water controls the range*". The ranching industry peaked at 1.5 million head of cattle in 1891.

The following years (1892–93) witnessed the complete collapse of the ranching industry. The overstocking of ranges with the consequent de-vegetation and erosion combined with a series of drought years and El Niño years with intense storms that washed away the denuded and eroded topsoil in the ranges. This collapse of rangelands and pastures caused the death of 50% to 70% of the cattle, and the railroads then frantically exported cattle.

The mining sector is another of the "*bust and boom*" stories that characterize Southern Arizona. In 1877, prospector Ed Schieffelin arrived to the San Pedro Valley in the midst of Apache raids and was told that the only thing he would find was his tombstone. He found ore, and the tombstone mine was born. The mine and its

> **The Ranching Industry's Human-Environment Impacts**
>
> The overstocking of cattle in the rangelands led to severe rangeland degradation due to overgrazing and erosion. The combination of de-vegetation, soil degradation and a period of drought years punctuated by some El Niño years with heavy rains caused the progressive washout of topsoil, floods and severe erosion. The ranching sector's mismanagement or lack thereof of the ranges led to the collapse of ecosystem functions that severely affected its services, ranging from providing healthy grasslands to regulating functions regarding hydrologic partitioning: infiltration and aquifer recharge, water quality, erosion control, and flood regulation. These effects of the ranching boom are also thought to have played a role in the down-cutting of the Santa Cruz River during the 1887 flooding.

town saw their golden years between 1879 and 1886, with a population reaching 10–12,000 people. During this time, many mine discoveries in the region caused silver prices to plummet. The first major conflicts between unions and mine management erupted in 1884 with a four-month strike in Tombstone. The miners' unions lost the battle and saw a $4 to $3/day wage reduction. This was the beginning of big mines in the southwest. Strikes followed in the Clifton-Morenci mines in 1903; and in the mine of Cananea (across the border, in the headwaters of the San Pedro Basin) in 1906, which illustrated the labor struggles of the time and was one of the sparks leading to the Mexican Revolution of 1910 to 1920. Racial divides in the labor struggles in the workforce of the mining and agricultural sectors were a recurrent theme. Companies and government often exploited these divides to pit workers of different origins against each other to break strikes and other movements, and often led to mass deportations of Mexican American and Mexican citizens.

> **The Mine's Human-Environment Impacts**
>
> The need for wood to fuel the mills of Tombstone, Cananea, Bisbee, Clifton-Morenci and other mines leads to extensive deforestation in the San Pedro Basin and Southern Arizona. Such deforestation to fuel the mine industry occurred during the same decades as the overgrazing and cattle overstocking, likely adding significantly to environmental degradation.

The US National Depression hit in 1893 as the overbuilding and shaky financing of railroads took its toll. Many banks collapsed and went bankrupt. The price of silver dropped due to too many western mines and high supply. Agricultural prices also declined. Arizona became a State in 1912 with a progressive constitution. However, judges rulings tended to side with the mining companies and other large economic interests.

The dust bowl hit the Midwest in the 1930's caused by a combination of climate and agricultural management, and Arizona received thousands of desperate migrants from Oklahoma and Arkansas that came to work on the cotton fields. In a wave of racism due to too many workers, thousands of Mexican workers and Mexican-American citizens were deported to Mexico. Fearing that Hoover Dam would silt-up and have its storage capacity for irrigation reduced, the Soil Conservation Service convinced the Bureau of Indian Affairs to slaughter one million sheep and goats from Navajo and Hopi reservations to avoid overgrazing, devastating thousands of families (*Kleespie*, 2015).

Native American people endured enormous conflict and contradictions as their fate progressively transitioned (by means of war, massacres, negotiations and treaties, treason, incarceration, forced migrations, family separations, and cultural re-education) from free masters of these vast lands, to reservation life and an imposed shift towards strange and misplaced customs from western lifestyle and policy. Twenty Indian reservations exist today in Arizona. They have been and still are fighting in the courts for their share of Arizona water through various *water rights settlement acts* and in the surface water adjudications.

The Colorado River Compact was signed in 1922, governing the sharing of its waters among the seven basin states, and allocating 2.8 million acre-feet/year (3,454 mm^3/year) to Arizona. The discovery of the centrifugal pump and rural electrification lead to widespread groundwater pumping for agricultural irrigation, the progressive depletion of aquifers, and the subsequent disappearance of most of the riparian corridors that depended on shallow groundwater levels during the long dry seasons. Tucson grew significantly during the Second World War due to its good weather for training pilots year-round, and continued to grow in the next decades (The population of Arizona rose from 0.7 million in 1950 to 6.5 million in 2012) thanks to the invention of air-conditioning, its good weather, a growing economy, land availability and open spaces. The Groundwater Management Act was passed in 1980 and the Central Arizona Project was completed in 1994, bringing Colorado River water to Tucson. The details on the history of the conflicts around water of the last century in Southern Arizona are explained in detail in the following Chapters 4 to 7.

2.4 The Tucson region today

Water management in the Tucson basin occurs within a context of depleted aquifers due to past actions, a Colorado River water allocation that will depend on future flows, a CAP water recharge system, a complex market of groundwater credits, continuing projections of growth, and the last few surviving riparian areas – scattered in the fringes of the basin – whose roots still manage to reach groundwater. These new realities present a number of challenges, old and new, for the sustainability of water and environmental management in the Tucson region.

In addition to the disappearance of the riparian corridors in the Santa Cruz and Rillito Rivers, the effects of aquifer pumping of the last half century are still being felt today. Land subsidence is a process that continues to occur after pumping has stopped. Average land subsidence in the Tucson area is estimated at 8.1 cm (3.2 in) for the period of 1987–98, and 9.4 cm (3.7 in) for 1998–05. In Avra Valley, where most of the CAP water is being recharged, subsidence estimates are 10.6 (4 in) and 8.4 cm (3.3 in) for the same periods (*Carruth et al.* 2007).

The arrival of CAP water to the Tucson region was a needed and welcome resource, but its recharge into the aquifers is not devoid of issues. The basin sediments and the aquifer they host are providing the ecosystem service of filtering and blending Colorado River water of very different quality than the native groundwater. A recent report looking at water quality in basin-fill aquifers of the US Southwest presents evidence of increasing concentrations of total dissolved solids in groundwater. It has also been observed that artificial recharge and groundwater withdrawals are moving contaminants to deeper parts of basin-fill aquifers. The study reports that in the Upper Santa Cruz Basin and the Sierra Vista sub-basin, 13% and 5% of the drinking-water-well samples contained respectively a geologic contaminant (Radon, Arsenic, Uranium, etc.) and nitrates at a concentration exceeding human-health benchmarks (*Thiros et al.* 2014).

Regarding water quantity, the goal of safe yield (defined as pumping being equal or less than recharge) in the Tucson Active Management Area has been attained, as the natural and artificial CAP recharge exceeds pumping rates – at least for now. However, the safe yield goal does not consider the spatial distribution of pumping and recharge within the TAMA and thus many of the last riparian ecosystems supported by shallow groundwater areas continue to be vulnerable to pumping. As the Tucson AMA Safe Yield Task Force is aware of this issue, the concept of Water Accounting Areas was developed to help understand and mitigate negative impacts of spatial imbalances between recharge and abstractions (see chapter 15).

A recent study shows that pumped groundwater comes in part from aquifer depletion and from capture. In arid environments (i.e., recharge-limited) capture is mostly the interception of groundwater flows that would become streamflow or riparian evapotranspiration. Thus the fraction of pumping that comes from storage depletion or capture has major implications for environmental sustainability. In the basin-fill aquifers of Southern Arizona, the estimated long-term cumulative capture fraction of pumping is 0.4 to 0.6. In other words, about half of the groundwater pumped in this area would have become streamflow and riparian evapotranspiration (*Konikow and Leake*, 2014).

With projections of increasing growth and possibly dwindling water supplies (potential reductions in CAP deliveries due to over-allocation of Colorado River flows), water resources in the Tucson basin are becoming more and more valuable. During the period of 2008–2014, investment firms acquired more than twelve times the amount of long term storage credits acquired by municipalities. Effluent reuse is becoming a coveted resource. Many projects and initiatives, both at the institutional and community level, are promoting water conservation efforts by pushing the use of alternative water resources such as stormwater and rainwater, as well as through education and incentives for behavior change. In addition, urban growth and changes in land use cover add to changes in flood return frequencies to greater degrees than climate change. Encroaching developments on the Santa Cruz floodplain and sediment infill in the entrenched riverbed between banks lined with soil-cement, all increase the exposure to flood risk.

While water management institutions have little or no control over growth, they are tasked with maintaining safe yield and pursuing water supply sustainability. With relatively cheap municipal water rates, and being part of a developed economy, Tucson and other cities in the arid West have a high capacity to adapt to change through increased investment, technology and infrastructure if needed. While water will always flow out of people's faucets and it is extremely unlikely that anyone would go thirsty, it is not clear what will happen to the last remaining riparian ecosystems that depend on shallow groundwater. Will the adaptation strategies have enough foresight to ensure environmental sustainability? And what will be the technological advances and institutional arrangements that will continue to enable growth while ensuring a secure supply? The chapters of this book explore many of these issues and try to make sense of the past, present and future of the Tucson region as it relates to water and the environment.

REFERENCES

Betancourt, J. (2015): Webinar: Requiem for the Santa Cruz: An Environmental History of an Arizona River. At: https://www.doi.gov/ppa/seminar_series/video/environmental-history-of-an-arizona-river. (last visited Oct 1st, 2015).

Carruth, R.L., Pool, D.R. and Anderson, C.E. (2007). Land subsidence and aquifer-system compaction in the Tucson Active Management Area, south-central Arizona, 1987–2005: U.S. Geological Survey Scientific Investigations Report 2007–5190, 27 p.

Comrie, A.C. and Glenn, E.C. (1998). Principal components-based regionalization of precipitation regimes across the Southwest United States and Northern Mexico, with an application to monsoon precipitation variability. *Climate Research,* 10: 201–215.

Crimmins, M. (2006). Arizona and the North American Monsoon System. University of Arizona College of Agriculture and Life Science. http://extension.arizona.edu/sites/extension.arizona.edu/files/pubs/az1417.pdf.

Eubanks, E. (2004). Riparian Restoration, T&D Publications, National Forest Service, 0423-1201-SDTDC, at http://www.fs.fed.us/t-d/pubs/html/04231201/page04.htm. (last visited Oct 31st, 2015).

Ferg, A. (2007). Birds in the Southwest, Archaeology Southwest, Center for Desert Archaeology, Volume 21, Number 1.

Haynes, C.V. Jr. and Huckell, B.B. eds. (2007). Murray Springs: A Clovis site with multiple activity areas in the San Pedro Valley, Arizona (University of Arizona Press, Tucson).

Kleespie, T. (2015). *Arizona's Dust Bowl, Lesons Lost,* Arizona Public Media, PBS, NPR. Available at: https://originals.azpm.org/dustbowl/ (last visited Oct 31st, 2015).

Konikow, L.F. and Leake, S.A. (2014). Depletion and Capture: Revisiting "The Source of Water Derived from Wells", *Groundwater,* 52: 100–111.

McNamee, G. (2014). The Ancient Southwest: A Guide to Archaeological Sites. Rio Nuevo Publishers, pp. 88, ISBN 978-1-933855-88-2.

Moeller, K. (2013, July 24). Delving into Deserts, ASU – Ask A Biologist. Retrieved October 26, 2015 from http://askabiologist.asu.edu/explore/desert.

NWS-NOAA, National Weather Service Forecast Office, Tucson, Az. http://www.wrh.noaa.gov/twc/climate/tus.php (last visited Feb 4th, 2015).

Scarborough, R. (2015). The Geologic Origin of the Sonoran Desert; 2015 Arizona-Sonora Desert Museum, in http://www.desertmuseum.org/books/nhsd_geologic_origin.php.

Sheridan, T.E. (1995). Arizona, A History. University of Arizona Press; First Edition (February 1, 1995), 434 pages, ISBN-13: 978-0816515158.

Thiros, S.A., Paul, A.P., Bexfield, L.M. and Anning, D.W. (2014). The quality of our Nation's waters—Water quality in basin-fill aquifers of the southwestern United States: Arizona, California, Colorado, Nevada, New Mexico, and Utah, 1993–2009: U.S. Geological Survey Circular 1358, 113 p., http://dx.doi.org/10.3133/cir1358. (last visited Oct 1st, 2015).

Thornberry-Ehrlich, T. (2015). credit for illustration used in *Figure 6*, taken from the National Park Service, Geologic Illustrations, Plate Tectonics, Great Basin Horst and Graven, at https://www.nature.nps.gov/geology/education/education_graphics.cfm (last visited Oct 31st, 2015).

Webb, R. and Betancourt, J. (1992). Climatic Variability and Flood Frequency of the Santa Cruz River, Pima County, Arizona; U.S. Geological Survey Water Supply Paper 2379.

Webb, R.H., Betancourt, J.L. and Turner, R.M. (2014). Requiem for the Santa Cruz: An Environmental History of an Arizona River, University of Arizona Press, 296 pp., ISBN 978-0816530724.

Whittaker, R.H. and Niering, W.A. (1965). "Vegetation of the Santa Catalina Mountains, Arizona: a Gradient Analysis of the South Slope. *Ecology* Early Summer, 6(4): 429–452.

Chapter 4

Laws of the river: Conflict and cooperation on the Colorado River

Brian O'Neill
Water Resources Research Center, UMI-iGLOBES, University of Arizona, USA

Franck Poupeau
UMI iGLOBES, CNRS/University of Arizona, USA

Murielle Coeurdray
UMI iGLOBES, CNRS/University of Arizona, USA

Joan Cortinas
UMI iGLOBES, CNRS/University of Arizona, USA

INTRODUCTION

"The courtroom was overflowing. Dozens of lawyers were there, representing not only Arizona and California but also three other states having Lower Basin interests – New Mexico, Nevada, and Utah – all of which had joined the litigation, though not all voluntarily. Present also, of course, was the United States, a brooding omnipresence, which could have been going along for the ride but which, as it turned out, laid a claim to a large part of the vehicle and to the right to drive besides.... These and many, many more were present on that opening day, and an air of Armageddon pervaded the room-though of course there was sharp disagreement over the identity of the forces of Good and Evil. Men were present on both sides who had literally spent their lives on the battle over the Colorado. And for Arizona it was the long-awaited moment; it was now or never. For California, who had long played the waiting game, the crisis was now at hand -having contracts and works for 5.362 million acre-feet, she had nothing to gain, only water to lose."

The scene above was described in 1966 by Professor Charles J. Meyers (42–3), just a few years after the final decision in *Arizona v. California (1963)*, as the parties gathered to make the most important determinations regarding the flow of the Colorado River since the Colorado River Compact of 1922 (Compact). The description is vivid. It paints a stark reality, invoking a contentious atmosphere that may be unfamiliar to contemporary observers of the Colorado River Basin trials and tribulations, but it is certainly not forgotten. *Arizona v. California*, and the whole body of law, contracts, and agreements encompassing roughly a century of water disputes makes up the Law of the River (*Glennon & Culp, 2001: 912*). In essence, it governs and manages the river.

In this chapter we describe how the courts have often been utilized as a method of conflict resolution since the late 19th century. Our analysis begins with the Western US, and works toward specific elements related to Arizona. Our research hypothesis is as follows: *Whereas historically, the resolution of conflicts in the courts has created norms, rules, and instruments that take the form of legal and institutional tools, in more contemporary times, water management has political strategies at its core, often with the aim of avoiding court action, and producing social consensus between professionals of water management and the public.*

To say that there has been a consistent historical movement from conflict to cooperation would be far too simplistic: the water in the Colorado River Basin has not been managed in a smooth, linear fashion. Instead, conflict and cooperation are in constant flux (*Zeitoun & Murumachi, 2008*), and both of these components cohabitate in the complex world of water management. Conflict and cooperation necessitate a social, and therefore relational, way of thinking about water. Consequently, water management cannot be evaluated by a singular focus on the hydrologic cycle (*Linton & Budds, 2014*), groundwater-surface water interactions, legal structures, climate change, or the need for policy solutions to manage drought. The social agents involved, taken alongside the institutions and norms they created and continue to create, are of primary interest for a social analysis of the legal structures underlying water management in West United States.

First, we will set the historical backdrop of water management in the West in order to understand the roots of the present model of water management. Additionally, we will examine the primary legal doctrines affecting water allocation in the West. A discussion of the *Arizona v. California* decisions and related legislation will follow. Finally, some more recent examples will be presented – the Groundwater Management Act of 1980 (GMA) in Arizona, and the more recent Shortage Sharing Agreement of 2007 involving the Colorado River Basin states. By considering the aforementioned factors, it will be possible to identify how conflict and cooperation coexist, and how that cooperation toward consensus building can be seen as a logic of strategic political management, all the while illustrating the struggles of water management in the past and present.

I THE FRONTIER

Ditat Deus. "God enriches". Arizona's state motto holds much in its grasp. It is concise, but invokes a vast and involved history. Arizona's is a history that is similar to much of the West, and also much different. In the West we see a history full of struggle – early settlers struggled to acquire land, and then to carve out lives for themselves in previously inhospitable territories. Many of the earliest settlers and pioneers were disenfranchised individuals from the eastern US, who believed that the open spaces of the western frontier would provide them with a fresh start toward a better future (*Worster, 1992*). However, this future could only be made possible with sufficient water supplies: "*Everything depends on the manipulation of water -on capturing it behind dams, storing it, and rerouting it in concrete rivers over distances of hundreds of miles. Were it not for a century and a half of messianic effort toward that end, the West as we know it would not exist*" (*Reisner, 1986: 3*).

As the 20th century began, America saw conflict and controversy erupt frequently. Early evidence of such conflict is provided by *Winters v. United States (1908)*, in which the Supreme Court determined who had the right to the water from the Milk River in Montana: Native Americans or the settlers? In 1888, the US Government established the Fort Belknap Reservation in Montana. As settlers arrived, they diverted water from the nearby Milk River and after several years, the Native Americans were unable to irrigate their lands sufficiently. The settlers claimed "appropriative rights" to the water, stating that they had used the water for their own benefit prior to the Native American's beneficial uses. The water was just waiting to be used by the settlers, or so they thought. The Supreme Court ruled that when the federal government established the Native American reservation, it did so with the intention of allowing the tribes to become self-sufficient. Therefore, the right to water was "reserved" for the tribes' beneficial use, such as agriculture. This case established the Federal Reserved Rights Doctrine, commonly known as the Winters Doctrine. The doctrine would eventually come into play almost as an afterthought in *Arizona v. California (1963)*; however, the tribes were given senior priority rights to Colorado River water for their purposes (*Glennon & Kavkewitz, 2013: 26*). Although the main point of this paper is not to discuss the intricacies of Native American water rights, this early example illustrates the fact that land in the West is important, but it can be of little use without water.

Water disputes extend well beyond some of these initial determinations. Many issues stemming from the Compact and related to the allocation of water to each state are still being debated. These issues of allocation have recently come to the forefront, as Arizona and the other states of the Colorado River Basin may have to contend with an official declaration of a shortage in the Colorado River supply beginning in 2016 or 2017. Although the 2007 Shortage Sharing Agreement was a groundbreaking effort, exemplifying the ways in which parties with long-standing differences could come together for effective policy reform, the challenges into the future are great (*Grant, 2008; Glennon & Kavkewitz, 2013*), not the least of which will be how to reach consensus in the future, while working within the framework of the Law of the River.

2   FIRST LAW OF THE RIVER: SETTING THE FIELD OF PLAY IN WATER MANAGEMENT

If "*the story of the Colorado River is the great epic of water law and politics in America*" (*Thompson et al., 2013: 975*), one can say that everything starts with the Law of the River, which is the aggregate collection of various contracts, court decisions, and state or federal laws, constituting a complex, and at times seemingly incoherent, mire of history (*Meyers, 1966; Megdal et al., 2011*). Any major political or legal movement within the Colorado River Basin must deal with the Law of the River. As a system, it reveals a certain amount of coherence, involving all states within the Colorado River Basin. In a broader context, examining the Law of the River exemplifies how water management "*is not merely a technical field that can be addressed through infrastructure provisions and scientific expertise, but a political one that involves human values, behavior and organization*" (*Linton & Budds 2014: 170*). Therefore, "water is not an inert backdrop for social relations, but plays a positive role in social formations" (*Linton & Budds, 2014: 174*). The doctrines that underlie water law have influenced man's interactions with water. As people from the East populated the West, they gave water meaning and virtue that cannot be rivaled by any other natural resource in western American culture. All are aware, on a level foreign to the East, of the value of water. It is a cultural artifact of sorts and contemporary fascination. As one begins to see meaning in water, one becomes aware of the legal structures that give water practical definition. In the legal history of the West, one sees not so much the amalgam of the laws, or the many layers of a single corpus of law, but rather the creative imbrication that history has allowed. When faced with a shortage of water, the laws and structures of this imbrication are revealed as foundational and part of the doctrinal fabric of the Western spirit.

2.1 Doctrines: a not-so-equal appropriation

Many doctrines play a role in understanding the laws affecting the Colorado River, which collectively come to determine water rights. Over the years, the states have proven to be quite strong-willed in efforts to implement their own laws to govern water (*Getches, 2009; Thompson et al., 2013*). The doctrine common to most Western states is that of prior appropriations, which states that the first person to use the water has the right to it. The oldest user is legally known as a "senior" rights holder. When Arizona was allowed to proceed with the Central Arizona Project Canal, the catch was that it would hold "junior" priority through the Colorado River Basin Project Act of 1968. Therefore, if and when the Secretary of the Interior (SOI) declares a shortage on the Colorado River, Arizona will see reductions, but California will remain unaffected (*U.S. Department of the Interior, 2007; CAPa, 2014*). It has been argued many times that the prior appropriations doctrine needs to be reformed because it was originally conceived to encourage settlement of the West, and was highly conducive to creative water management, by utilizing diversions and dams for beneficial uses. In the end, the prior appropriation doctrine is not only indicative of the early settlers' views of nature, but also of their views regarding agricultural and industrial progress (*Glennon, 2012*).

Prior Appropriation

In the field of law, Arizona's Spanish heritage was reflected in the Latin principle *qui prior est in tempore, potior est in jure*, which means "he who is earlier in time, is stronger in law". When the Territory of Arizona was set up in 1863, the Howell Code passed by the Legislative Assembly incorporated Spanish and Mexican customs of prior appropriation. As more and more colonists gradually arrived in the Territory, and periods of flooding were repeatedly followed by severe drought, competition for land and water intensified, giving rise to numerous legal battles.

Regarding the Salt River, in 1882, Judge H. Kibbey of the Territory's Supreme Court was invited to resolve disputes between water consumers and the canal company of the Salt River Valley Land. The court's ruling reasserted the doctrine of prior appropriation. It said that water belonged to the land rather than the canal companies, and could not be sold as a separate commodity. The Kibbey decision, linking land and water, was used as the basis of water law in Arizona.

Approximately thirty years later, on March 1, 1910, Judge Edward Kent of Maricopa County, who worked in the Territory of Arizona, followed the Kibbey Decision with what has become known as the Kent Decree (1910). The Kent Decree opposed the owners of land irrigated by the Salt River Users Association who, under contract with the government, owned canals to the north of the river and received surplus water from those canals, as well as the individual owners to the south of the river, who had no contractual relationship with the federal government, and who claimed the right to access water. The Kent Decree reaffirmed the doctrine of prior appropriation concerning cultivated plots of land, including all of the Salt River Valley connected to the Salt River. The Kent Decree was the key element in terms of water law and the administration of Arizona at the moment that it ceased being a territory and became a state in 1912. The Kent Decree came into play again in the case of *Hurley v. Abbott (1966)* when "over a half-century later, in 1966, in the face of expanding Verde Valley water uses, SRP attempted to enlarge the Kent Decree to adjudicate Verde water rights, but the federal district court (the successor to the territorial court) held that the decree could not be reopened without joining all landowners and water users in the Verde watershed (*Feller, 2007: 406*)".

Furthermore, the prior appropriations system came into being as a result of miners, farmers and grazers moving west without owning land. Therefore, one was able to divert water for a beneficial use without needing to necessarily own property (*Getches, 2009: 6; Tarlock, 2001: 770–771*). Under this system, a certain inequality around water use developed, even though water is a public resource. As stated in A.R.S. 45–141,[1] "the waters of all sources, flowing in streams, canyons, ravines or other natural channels, or in definite underground channels, whether perennial or intermittent, flood, waste or surplus water, and of lakes, ponds and springs on the surface, *belong to the public* and are subject to appropriation and beneficial use..." In this way, 'prior appropriations' is similar to the riparian doctrine that governed water management in the Eastern US. According to the riparian doctrine, simply owning property that adjoins a watercourse endows one with a right to use the water; however, the prior appropriation doctrine operates differently (*Thompson et al., 2013: 167–181*). The riparian doctrine also provided that a "riparian", one who owned property along a river, must make some "reasonable use" of the water, relative to other users. This meant that there was some degree of equitability built into this method. However, not all Western states apply the same doctrines, for as Article 17 of the Arizona Constitution sets out most of these rules for that state, it clearly establishes that the riparian doctrine does not apply. In its modern form, prior appropriation water rights require permission from a state agency, which then monitors and considers the other users of the watercourse (*Getches, 2009*). The imbrication of riparian rights, prior appropriation and their foundations is illustrated nicely by *Wallace Stegner (1953: 226)* when he explains that:

"In general, American Law was based on English Common Law. But the Common Law, accumulated out of the experience of a rainy country where water was no problem, affirmed only what where known as riparian rights to the water of streams. The man who owned the bank could make any use that he pleased of the water, but he had to return it to the stream when he was through with it. That worked for running grist mills, but it did not work at all for irrigation, which used the water up instead of taking advantage of its passage. In the West, before and

1 Arizona Revised Statutes, see http://www.azleg.gov/ArizonaRevisedStatutes.asp.

since Powell's time, there have been heads broken with irrigation shovels because of someone's attempt to apply riparian law upstream, and take uncontrolled advantage of the water. In an irrigating country, appropriation becomes an essential criterion, and delicate refinements about more or less beneficial uses, and priority, and dipping rights, and a great many other complications still unheard when Powell wrote. There was nothing wrong with the riparian law for the West except when downstream bank-owners sooner or later found themselves with riparian rights to a dry creek bed. Water is the true wealth in a dry land; without it, land is worthless or nearly so. And if you control the water, you control the land that depends on it. In that fact alone was the ominous threat of land and water monopolies."

The battle to make the land worthy of settlement and development fell chiefly to the federal government. With respect to the major allocations that have been made for each state in the Colorado River Basin, large federal projects were built so that the states could make use of the water. This meant that over time, states could pay back the government, usually the Bureau of Reclamation (BOR); however, the states owned the water, rather than the government agency that had paid for the project. This process allowed the government agency to operate the projects under state laws rather than federal laws (*Trelease, 1960: 403; Thompson et al., 2013: 843–65*). Currently, there is a great deal of apprehension when states discuss the future of the Colorado River. Each state owns the rights to certain amounts of water from a specific source; however, the lines become blurred because the federal government represents Native Americans, who are legally seen as sovereign nations, and many of their rights are relatively senior on the Colorado River. Therefore, the states fear what may happen to their rights if the federal government intervenes, because it operates many of the diversion projects in crucial states such as Arizona and California. The federal government would prefer not to take sides in any water shortage negotiations, thereby allowing the states to continue to work within the parameters initially defined by the Law of the River, which began with the Compact in the early 20th century.

2.2 The compact and the players

The commonly accepted origin of the Law of the River is the Colorado River Compact of 1922, which was initially proposed by the states that have come to constitute what is known as the "Upper Basin" of the Colorado. These "States of the Upper Division", as described in Article II of the Compact, are Colorado, New Mexico, Utah and Wyoming, while the "States of the Lower Division" are Arizona, California and Nevada (*Meyers, 1966*). During these early years of Western settlement, a great deal of primitive canal building and irrigation was being performed. The settlers of California's Imperial Valley made their opinion known early when they asked the federal government to create the "All American Canal" (*Reisner, 1986:125*).

However, the turning point was in 1905 when a massive flood created the Salton Sea in the Imperial Valley, which eventually necessitated the construction of the Hoover Dam (*Reisner, 1986: 122–125*). Los Angeles saw great electricity potential in a dam project, and began organizing the Metropolitan Water District to lobby in Washington D.C., with help from Imperial Valley irrigators. Additionally, there was a drought in California at this time, which charged the whole situation with an extra impetus for reform (*Thompson et al., 2013: 978*). California approached its water problems aggressively by sending lobbyists to Washington D.C., and strategically placed itself in the best possible position for future management of its water supply into the 20th century. This initial political foresight has been a hallmark of California water politics.

As many water managers will say, even today the future remains uncertain in the Colorado River Basin. Arizona has historically been proactive with its conservation and management practices because it was politically late to the game, and has been unable to effectively push its agenda against that of the more powerful California. The Western states are concerned about California because its population continues to increase, as does its relative political power. Therefore, history may be repeating itself, as the water worries of the early 1900's are reflected in the concerns of the 21st century. "*What the representatives truly concluded to be sufficient water in the Colorado River for all needs for all time, based on the extensive studies of Colorado's R.I. Meeker and the Bureau of Reclamation's Arthur P. Davis, soon became a shortage. The seeds of controversy for the Colorado River Compact were sown*" (Tyler, 1998: 26). However, Delphus Emory Carpenter was one figure, during the negotiations of the Compact in 1922, who valued cooperation over conflict. This is not to say that he was not going to fight for his water, but he is one example of an individual who acknowledged the social and relational nature of water, and recognized that conflict and cooperation are not so easily separated.

Delphus Emory Carpenter: Origins of "win-win" Water Management

Trained as a lawyer at the University of Denver and graduating in 1899, Delphus Emory Carpenter had a constant penchant for finding himself in water law disputes as he practiced through 1908. In 1909 he ran for the Colorado Senate as a Republican, and won despite his opposition to the agrarian populist movement. Known as "Give-a-Dam Carpenter", he was truly a gentleman by all accounts, and always insisted that he get his way, but that the other parties received "a square deal" (*Tyler, 1998: 28*). He might even be considered as the father of modern interstate negotiations, and of the strategies of working towards "win-win" situations that many current politicians and negotiators utilize (*Mostert, 1998*).

Around 1910, he was made chairman of the Senate Committee on Agriculture and Irrigation, and he prepared a report on the Colorado River, its tributaries, and the legal criteria that should be applied to their use. He came to the conclusion that the doctrines of prior appropriation and beneficial use were key components of Western water law, and should remain so. Interestingly, he also felt that the states should take responsibility for the Colorado River, and prevent the federal government from intervening in the matters of the Basin. "*To Carpenter, the intervention of the Reclamation Service in Kansas v. Colorado was like a fire bell in the night. Even though the Supreme Court ultimately decided in 1907 that each state had full jurisdiction over the waters of its streams, the federal government appeared increasingly disposed to build its projects with scant attention to the statutes and judicial decisions of sovereign states*" (Tyler, 1998: 29–30).

> He was additionally appointed to work on *Wyoming v. Colorado (1922)*, which further provided a basis upon which interstate rivers might be divided. Despite initial infighting between Carpenter and Herbert Hoover (Warren G. Harding's Secretary of Commerce), they eventually came to work together quite closely. Once the final decision in *Wyoming v. Colorado (1922)* was made, Carpenter drew up the 50–50 (7.5 MAF-7.5 MAF) allocation scheme between basins upon Hoover's request for the Colorado River Compact. Million acre feet (MAF) were used as the measurement to accommodate the vast sums of water being used in the West.
>
> The "equitable" division of the Colorado has proven problematic into the future despite Carpenter's belief that it would decrease future litigation (*Tyler, 1997, 1998*). Mainly, Carpenter believed that the "equitable apportionment doctrine" that had been applied in *Kansas v. Colorado (1907)* was the best way to proceed. This doctrine has been criticized because despite the fact that prior appropriation may have many flaws, making attempts to divide waters in equal parts provides yet another level of uncertainty, however it can become a necessity for the courts when dealing with large disputes between states (*Patashnik, 2014*). However, one must consider that Carpenter, in one stroke, was attempting to make a sweeping reform in the West. By doing so, he attempted to be sensitive to Arizona's situation and understand that it would take many years to ratify the Compact, but he also wished to protect the Upper Basin's supply of water. At the time, everyone was concerned about the rapidity with which major Lower Basin cities were growing.
>
> In the decades following the Compact, almost all in attendance recognized the fundamental role that Carpenter played. However, it was Sims Ely of Arizona who spoke of the situation with great perspective and candor in 1920, and again in 1944 when he wrote that "*I shall never forget the prophetic look that came over your face, nor the clarity of your reasoning as you pointed out to me (in 1920) why that allocation [referring to one-half of the total flow of the Colorado River to the Upper Basin States] would be demanded by you when the time should come to frame the treaty.... You and I will not live to see it,' you said, 'but within the next one hundred years, perhaps' within fifty years, water for irrigation will have become so valuable that the easterly side of the Rockies will be pierced by a tunnel or tunnels, and water will thus be conveyed to the Plains below.' It was then that you became the prophet of great things to come*" (Tyler, 1998: 27).

Arizona eventually ratified the Compact in 1944, when Mexico was additionally guaranteed 1.5 MAF of Colorado River flow (*Glennon & Kavkewitz, 2013*). Carpenter's savvy legal ability and knowledge of the inner workings western water law, enabled him to successfully protect the Upper Basin from many future issues. However, one tricky element of the Compact was that it simply divided the Basin, and did not provide for intra-basin allotments. This issue eventually brought Arizona and California to the courtroom. The states would have to work out the inter-basin apportionments themselves.

3 INTERSTATE CONFLICT

California continued lobbying for the building of the All-American Canal and Boulder Canyon Dam. In 1928, the Boulder Canyon Project Act (BCPA) was finalized, clearing the way for construction of a new dam along with the All-American Canal. Even naming the dam became controversial, at first called Black Canyon Dam, later Boulder Dam, and finally Hoover Dam, prompting one observer to suggest the name "Hoogivza Dam". (*Thompson et al., 2013: 980*). In 1928, it was agreed that California would be allowed 4.4 MAF and 1 MAF of unused surplus on the river. The BCPA created a general provision for the entire Lower Basin, whereby Arizona would receive 2.8 MAF and Nevada would be allotted 0.3 MAF. At this point, the famous *Arizona v. California* lawsuits began. These decisions established various social norms that have taken the form of legal and institutional tools, even though Carpenter, and other proponents of the Compact, had hoped to avoid future legal action.

3.1 Arizona v. California (1963)

The set of decisions issued by the U.S. Supreme Court in the litigation between Arizona and California, from 1931 to 2000, if taken as a whole, is one of the longest lasting and most influential Supreme Court matters regarding the Colorado River. The issue stemmed from California's need to further divert water beyond that which was outlined by the BCPA in 1928. The 1963 decision is the best known, and the most well documented, because the court held that the SOI is responsible for contracting with states to allocate water according to various formulas outlined in the court's decisions. Arizona finally received its 2.8 MAF allocation and was able to begin the construction of the Central Arizona Project (see *infra* chapter 6). However, in the event of a water shortage, the SOI determines apportionment after the 1928 rights can be met.

Arizona v. California is also significant because the Federal Government has the final say in apportionment when water shortages occur. For many years, California took over 5 MAF because the SOI had declared a surplus on the River. This is not the case today as the river faces a potential declaration of shortage and prolonged drought conditions (*Glennon & Kavkewitz, 2013*). Additionally, in the subsequent 1964 and 1979 decrees, present perfected rights became further defined; therefore, apportionment to Native American nations and Lower Basin States became clearer. Native American allocations have increasingly been considered, and will continue to be an interesting twist to the Law of the River as the drought becomes severe. *MacDonnell et al. (1995)* argued early on that if droughts increase in the future, flaws in this aspect of the Law of the River would become glaringly apparent. Major cities such as Los Angeles, Phoenix, and Las Vegas may have to look elsewhere for water because they are not high enough on the totem pole of appropriation rights.

The federal government has had the largest impact on the policy process as it relates to the Law of the River for more than 100 years, and it will always have the final say in major water appropriation disputes. However, it is not an interest group in the traditional sense. The government's interest lies in the fact that more disputes over the Law of the River, and more drought in the Southwest, necessitate more government intervention. Historically, the SOI has been heavily involved and still holds a great deal of legally endowed power, especially in the way he/she can control water policy and frame debates. Arizona will continue to struggle with the allocation scheme that resulted from Arizona v. California and the agreements it made to secure construction of the Central Arizona Project (CAP), because the CAP has such a low priority on the Colorado River.

3.2 Arizona's groundwater struggle

One of the most important institutional and technical tools in Arizona's water management apparatus is the CAP, which resulted from decades of thought, planning, legal battles and political struggles, ultimately codified in the Colorado River Basin Project Act of 1968. The CAP was made possible only after various legal actions between Arizona and California provided greater clarity regarding the allocation of the waters of the Colorado River (*Cortinas et al., 2015*).

Over time, there has been growing emphasis toward more locally driven management, and away from large-scale federal action, such as the BOR's dam and diversion projects, and the array of aforementioned Supreme Court proceedings. This change in emphasis has important practical applications when considering the role that the federal government may be forced to play in the years following a shortage declaration. To illustrate this shift in water management, it is helpful to analyze the Groundwater Management Act (GMA), which marked a major transition in Arizona and western water management, because it was the result of clear political tensions leading to fundamental policy reform in Arizona (*Connall Jr., 1982*). It was not the result of direct courtroom action, which had been employed to resolve many conflicts in the past, although it was certainly influenced by *Farmer's Investment Company v. Bettwy*. Ultimately, water management in the 21st century has much more to do with managing recurring political conflicts. As Bruce Babbitt, former Governor of Arizona, wrote in 1986, "the second century of the Colorado River, now before us, will raise issues that no one even thought of back in 1922" (*Weatherford & Brown, 1986: xi*). His words still ring true today, thus it is increasingly important for scholars to look back into the richly sedimented history of the Colorado River to better understand the present, and look to the future.

Even though there are major issues with which to contend in terms of surface water, Western states face major concerns that have only recently been realized with respect to groundwater. The embedded issue for Arizona is that it was pumping groundwater at alarming rates, because it lacked major canal and delivery infrastructure from the Colorado River. In 1972, Lawrence McBride wrote an article in *Ecology Law Quarterly* entitled "Arizona's Coming Dilemma: Water Supply and Population Growth". McBride carefully examined demographic data and a variety of hydrologic data collected by the United States Geologic Survey and other agencies. His article sounded a significant alarm, because it provided a comprehensive analysis of the Basins and their populations. He described the method of water governance in Arizona as a "system of diversions because of the lack of surface water to meet demand" (*Mc Bride, 1972: 359*). At that time, the CAP was not yet finished for Tucson or Phoenix, and the challenges of evaporation, the possibility of desalinization plants, and diversions from Canada and Alaska were seriously considered. Meanwhile, groundwater recharge, which was eventually used to solve many problems in Arizona, was only mentioned in the context of Orange County, California. McBride concluded that, in Arizona, there was an *"impossibility of setting up a recharge program without changes to the law"*. In the end, it would take powerful leadership in Arizona and Washington, D.C. by Cecil Andrus, Bruce Babbitt and others, to begin to mitigate the years of groundwater overdraft (*Connall Jr., 1982; Ambrose & Lynn, 1986*). Clearly, many strides have been made since that time, although some disconnect still exists between water law and the science of hydrology. Some scholars (*Linton & Budds, 2014*) would go so far as to say that the constant obsession with the hydrologic cycle in matters such as law, which have profound social components, actually serves to obscure the human interaction with water, and disguise water beyond comprehension. However, adaptations for the better have been made, even if legal doctrines still hamper the Western United States (*Glennon, 2007*). Finally, despite Delphus Carpenter's efforts, a highly constructed and contentious system has been created, and its effects can be seen in today's methods of water management.

3.3 Arizona's water management tools: the GMA

Many of the tools and techniques that have allowed Arizona to effectively rebrand itself, from a state that was over-drafting water at a dangerous rate, to a leader in conservation and management techniques, have come about since the 1980's and 90's. Arizona had to take on somewhat of a "bunker mentality" (*Source: Interview, Kathryn Sorensen, Phoenix Water Services Department, April 2015*) in order to gain the stability and assuredness that it has today. As will be described, Arizona underwent a complete about face as it moved through the 1980's, beginning with the GMA and a new vision for its water resources.

Today, collaboration and negotiation are the preferred methods for resolving water conflicts. Court cases are often seen as expensive and time consuming, and are sometimes avoided because the decisions may contain undesirable outcomes for both parties. After a long history of depleting its groundwater, the state had to be summoned to action. As Arizona transformed itself, Bruce Babbitt emerged with the leadership and vision it required.

Bruce Babbitt: New Visions for Arizona and America

Bruce Babbitt is one of the foremost figures in Western environmental politics. Born into a ranching family in Flagstaff, Arizona in 1938, he rose to prominence with his election as Governor of Arizona in 1978, after serving as Attorney General of Arizona since 1975. He continued on as Governor until 1987 when he ran on the Democratic ballot for President. Babbitt also served as SOI under President Clinton from 1993 to 2001. As SOI, he left his most lasting impression on American politics, and he is considered to be one of the most successful to hold that position, because of his extensive conservation efforts through use of the Endangered Species and Antiquities Acts. Additionally, he became known for his ability to "reach bipartisan compromises on issues whenever possible" (*Leshy, 2001: 199*).

Babbitt maintained a strong commitment to the environment throughout his political career. His father was one of the founders of the Arizona Wildlife Federation, as well as the Arizona Game Protective Association. Following interests in the natural world, he received a degree in geology from the University of Notre Dame, and then moved on to the University of Newcastle, England, where he received an M.S. in geophysics, and finally to Harvard Law School before entering his political career. About the time that he became the SOI, Babbitt was twice considered for a position on the U.S. Supreme

Court by President Clinton (*O'Leary, 1998; Terrain.org 2006*). Throughout his career, he exhibited an uncanny ability to see the big picture. As *Leshy (2001: 201)* states: "Babbitt has been the most nationally focused of them all", (referring to the legacy of various SOI's). From the Flagstaff of his youth, his worldview was leavened by years at Notre Dame and Harvard, by graduate school in England, by much travel around the country and abroad, and by an inquisitive mind and voracious reading on many subjects".

Additionally, *Leshy (2001: 203)* goes on to say that it was Babbitt's ability to engage the details that was "a technique he had mastered soon after becoming Governor of Arizona, when he almost literally locked representatives of major water interests in his office for months while, under his strong direction, in 1980 they hammered out the first meaningful groundwater management law in the state's history". In this context, it is possible to see yet another example of strong leadership utilized to promote consensus to prevent future problems. It is a spirit that was also evident with Delphus Emory Carpenter during the initial years of work on the Colorado River Compact.

Today, Arizona's GMA is highly regarded for its success, as well as its ability to compensate for the damage that had been done to Arizona's aquifers by placing limits on the amount of groundwater that could be withdrawn from certain areas (*Connall Jr., 1982: 314*). It has been argued that *Farmers Investment Company v. Bettwy* was the event that set the actions of Babbitt and others in motion. *Bettwy* involved pecan growers in Arizona (South of Tucson) and the Anamax Mining Company. The copper mine needed water and was drilling wells in the Sahuarita-Continental Critical Groundwater Area, with the intention of then moving the mined water outside of the Critical Management Area, as was permitted by the Critical Groundwater Code of 1948. *Bettwy* proved to be a turning point, because it applied the reasonable use doctrine to prohibit the transportation of groundwater away from the land from which it was extracted. The mining company and major Arizona cities including Tucson, stood to lose a great deal from this legal decision because cities were transporting water from wells that were a great distance outside of their service areas (*Pearce, 2007: 42*).

As Governor of Arizona, Bruce Babbitt was able to recognize the potential implications that *Bettwy* had for cities. Rather than again going to the courts, he was able to successfully enact the GMA by utilizing the state legislature, which helped cities and allowed for new development opportunities in ways that were more secure than in the past. In this way, the GMA allowed for some transportation of groundwater to occur away from the basins and sub-basins from which the water originated, rendering the *Bettwy* decision irrelevant.

The GMA has had implications for Arizona beyond what anyone could have anticipated. It allowed Arizona to craft a unique model of water management, which was built around the production of consensus between interested parties, rather than lengthy litigation. Importantly, the GMA set the stage for a whole host of legal and institutional tools that would shape Arizona groundwater management for years to come. By instituting proactive measures to conserve water for the last twenty-five years, Arizona, and especially Tucson, appears well positioned for future shortages on the Colorado River.

3.4 Beyond Babbitt

By 1977, SOI Cecil Andrus made it clear that Arizona must demonstrate through legislative action that it could begin to withdraw groundwater in a healthy way; otherwise the CAP construction would not move forward. In 1980, Arizona responded to these problems with the GMA (*Connall Jr., 1982*). This law created Active Management

Areas (AMAs), which not only instituted an innovative style of managing groundwater based on basins, but also introduced limits to the expansion of irrigation and development. The cities in Arizona that were designated as being within an AMA had to demonstrate an Assured Water Supply for 100 years. According to A.R.S. 45–576 (J) (1),[2] assured water supply means "sufficient groundwater, surface water or effluent of adequate quality that is continuously available". Unfortunately, the system has sometimes been difficult to implement.

In 1993, the introduction of the Central Arizona Groundwater Replenishment Districts (CAGRD) gave real estate developers the chance to introduce subdivisions in places in which there were no direct connections to CAP water (*Valdez Diaz, 1996; Blomquist et al., 2001: 662*). These districts were given extended powers in 1999 with the aim of supporting real estate activities in Arizona (*Avery et al., 2007*). They function as water banking tools for developers in that those developers who join the CAGRD are allowed to pump water for their subdivisions, and in turn, the CAGRD purchases the water necessary to replace any groundwater withdrawn in excess of "safe-yield". This process allows developers to enroll their homeowners in the CAGRD, and the homeowners become responsible for paying groundwater withdrawal fees, which are seen as a line-item on property tax bills (*CAPc 2015*). Under this system developers must prove a "100 year water supply" to the Arizona Department of Water resources (ADWR), and obtain the right to sell lots in the subdivision. The developers become members of the CAGRD and purchase CAP water to satisfy the "safe yield" obligations to the Replenishment District. Problems with this system emerged in that an unevenly depleted aquifer was created at the site of subdivisions because that water was not directly injected into the more vulnerable hydrological zones from which it was pumped, causing a water deficit in the areas of the subdivisions, especially into the most vulnerable hydrologic zones from which it is being pumped (*Avery et al., 2007; Colby & Jacobs, 2007*). Finally this process allowed developers to essentially create a model of fragmentation in the urban landscape whereby subdivisions could be built outside of existing water networks.

The CAGRD is not to be confused with the Arizona Water Banking Authority (AWBA), which was created in 1996 as an additional underground storage mechanism (*Blomquist et al., 2001: 663*). In common parlance, AWBA is referred to as the "water bank". One difference between the CAGRD and the AWBA is that the AWBA deals exclusively with the state of Arizona's unused 2.8 MAF allotment, primarily supplied by the CAP. In this way, the state of Arizona is able to "use" the water, and prevent it from being used by California. At least initially, the CAGRD and the AWBA were seen as a success.

Another provision of the GMA was that the Arizona Department of Water Resources (ADWR) was created to replace the Arizona Water Commission (*Blomquist et al., 2001: 661–2; Colby and Jacobs, 2007*). The ADWR provided for a seven members board to be elected by Arizona voters, with the director appointed by the Governor. It was this quick and very deliberate action by the state of Arizona, especially Governor Babbitt, which caused the Department of the Interior (DOI), SOI, and BOR to make the CAP one of its top priorities through the 1980's (*Kupel, 2006; Colby & Jacobs, 2007*). Despite the strides made by Arizona in the 1980's and 90's,

---

2 Arizona Revised Statutes, see http://www.azleg.gov/ArizonaRevisedStatutes.asp.

more challenges lay ahead for Arizona including the delivery of CAP water in Tucson through Tucson's aging infrastructure which created a firestorm of anger in the local community in the early 1990's (*Kupel, 2006; Cortinas et al., 2015*). However, there is a commonly heard discourse about water that was made clear by the Director of the Arizona Department of Water Resources, Tom Buschatzke in a recent NPR interview when he gave the following response to questions comparing the drought in California to the drought in Arizona:

> The metropolitan parts of Arizona already have mandatory water conservation requirements in place. We also have stored a lot of water underground, so for a point in time when we see shortages, we've got over 3 million acre-feet of water. That's more than a year's (worth) water underground. We've definitely done things differently, we've made some different choices. I think Arizona is one of the better places you can be right now in the western United States (*LeClair, 2015*).

To be sure, Arizona has been forced to be proactive and it is much to its benefit as the drought that spreads throughout the West (albeit less severe in Arizona than California) continues. However, one must understand that Arizona's model of management has revolved around historical responses to the rival state of California over many years as has been illustrated. In fact, it may even be argued that California did not really "play the game," so to speak when the time to negotiate the 2007 Agreement came, which may be largely attributed to the fact that it has remained squarely in the field of power in water management in the West.

In adding new nuances to the highly imbricated structure of Colorado River Basin management and policies, however voluntary such measures may be, the observant participant in this play will undoubtedly encounter and be faced with notions of uncertainty. In a broad sense, this simply means that no one has an answer for the future. Each state can only be as well prepared as possible with planning strategies and iterations thereof. For Arizona, the 2007 Agreement provided a level of certainty in terms of amounts of flow to be received (or not), but also a level of uncertainty. What will the next agreement look like? Will there be another cooperative effort like the one in 2007?

According to the 2007 Shortage Sharing Agreement Guidelines, Arizona will face a reduction at the 1075 feet watermark in Lake Mead, and official CAP documents state the reduction will be 320,000 af. At a Lake Mead elevation of 1050 feet, Arizona faces a 400,000 af reduction. Finally at the Lake Mead level of 1025 ft., Arizona faces a 480,000 af reduction and some municipal and industrial users will be forced to cut back on consumption (*CAPa 2014; CAPb 2015*).

3.5 The shortage sharing agreement of 2007

The most pressing issue today is the ongoing 15-year drought, which the Basin states set out to manage in 2007 with the Shortage Sharing Agreement. As *Douglas Grant* writes in his article entitled "Collaborative Solutions to Colorado River Water Shortages" (*2008: 964*), the Colorado River *"supplies drinking water for over twenty-seven million people and irrigation water for over 3.5 million acres in seven western states. This vital resource has been gripped since 2000 by the worst drought in over a century of recordkeeping"*. The reservoirs of Lake Mead and Lake Powell are hovering near half full. However, the Shortage Sharing Agreement is further evidence of how managing water is a kind of strategic political management in which conflict and cooperation are both present. Considering that many states are quite apprehensive about federal involvement, and especially apprehensive about the SOI mandating an unfavorable model to manage a shortage in the Colorado River Basin, it may be surprising to learn that in 2005, SOI Gale Norton (Colorado) provided the initial impetus by requesting the BOR to provide "guidelines" under which Lakes Mead and Powell could operate in times of official shortage (*Grant, 2008: 964*). For a cooperative, and admittedly temporary, solution to be reached however, the structure for the agreement did not result from litigation, but rather came through the National Environmental Policy Act (NEPA) with the direction of the SOI. In fact, without the direction of the force (i.e., SOI Gale Norton) external to the state authorities, the agreement may never have been reached, and the process may not have begun (*Grant, 2008*). In this way the determining authorities were quite concentrated, which allowed for collaboration to take place. This is not at all implausible considering that part of the Colorado River Basin Project Act of 1968 indicates that the SOI must consult the Basin states to establish long range plans for reservoirs. Clearly, the legal structure of the Basin gives rise to the contemporary model of water management.

The 2007 agreement may not have fallen into the traps of the past for a number of reasons. As the Record of Decision (ROD) states, the process created "voluntary" reductions in flows and voluntary signing of the forbearance agreements, and the federal government was present to "facilitate and not dictate" a solution that was not intended to be "permanent" (*ROD, 2007*). Therefore, the stakeholders came to the table with somewhat less of a burden to make a long lasting, hard and fast type of decision that may have plagued collaboration in the past. The final agreement allowed for flexibility in Basin operations and it seems that the process in reaching agreement was also somewhat flexible given NEPA constraints on the autonomy of the players (*Table 1*). Additionally, the number of states necessarily involved in such an agreement is daunting, but the number of individuals actually negotiating with and representing the following different entities within the Basin, must also be considered.

In this instance, the Department of the Interior (DOI) served as the mediator, which allowed for a certain degree of constrained autonomy of the negotiators (*ROD, 2007*). Some of the parties who were privy to this negotiation would simply have one member present, or a few members, and perhaps assistants. Interestingly, there are clear inequalities of representation that immediately appear when considering this membership. The Upper Basin has very minimal representation from the individual states, and historically, has not used all of its allotted flow. However, from the earliest days, California has been able to secure the money, and therefore the infrastructure, to control even more water than it was officially allotted (*Resiner, 1986; Glennon & Kavkewitz, 2013*). Arizona and Nevada have very specific representation, but it is not nearly as extensive as that of California.

One criticism of the 2007 Agreement could be that it only buys Arizona slightly more time before it starts to take shortages. It also appears that while other Basin states have been forced to deal with new legal rules for extracting groundwater, taking shortages, conservation measures, and (in general) real concerns about future

Table 1 Entities involved in interstate negotiation.

Representing Arizona:
Arizona Department of Water Resources (ADWR)
Central Arizona Project (CAP)
Yuma Irrigation District
Lake Havasu City

Representing California:
Governor's Board of Directors
Municipal Water District of Los Angeles (MWD)
Imperial Irrigation District
Coachella Irrigation District
Palo Verde Irrigation District
City of Needles

Representing Nevada:
Colorado River Commission of Nevada
Southern Nevada Water Authority

Representing the Upper Basin States:
Executive Director of the Upper Basin States
Upper Basin Colorado River Commission
City of Denver Representative
Bureau of Reclamation
DOI Solicitor from Washington D.C.

Source: Interview, Tom Buschatzke, ADWR, March 2015.

water, California has not really had to play the game in the same manner, which is to say, altering the future of Californian society and lifestyle.

Eventually, the seven states then provided their own proposal, and the BOR received public comment. Finally, on December 13, 2007, the SOI issued a final set of guidelines, including operational elements, based on the proposal made by the Basin states in April 2007 (*Grant, 2008: 965*). The four key components of the agreement were that (1) the states agreed to coordinate the operation of Lakes Mead and Powell, (2) the Lower Basin states would develop a set of operating guidelines for times of declared shortage, (3) guidelines established in 2001, which were devised with surplus in mind, would be modified, and (4) the delivery plan would create an intentionally created surplus to allow more water to be stored in Lakes Mead and Powell to forestall a declaration of shortage (*Grant, 2008*). Arizona will still take the brunt of a shortage in the initial years under this scenario, but the main focus is to ensure that enough water is in Lakes Mead and Powell to prevent a shortage declaration for as long as possible. Through Arizona's extensive groundwater recharge, storage and recovery programs, Arizona has been preparing for these challenging times for many years (*Grant, 2008; Glennon & Kavkewitz, 2013*).

The 2007 Shortage Sharing Agreement is another example of the complexity involved in studying the relationship between water and society. The values, institutional and conflictual history, instruments and agencies of each state across the Colorado River Basin (and Mexico), were on full display (*USBOR, 2010*).

4 CONCLUSION: CHANGING TIDES IN WATER MANAGEMENT

Throughout this history, it can be seen that the model of managing water in the West has been a consistent movement away from what may be considered more traditional forms of conflict, such as litigation and court action. Consequently, novel institutional tools have developed within the Western states. The larger model of management of the Colorado River Basin's water resources leaves Arizona with much to lose from long-term drought conditions. Because Arizona cannot afford to repeat the past, it has created a unique water management model for itself that stresses cooperation and consensus over conflict.

In terms of studying water management, much of the history has involved extensive research into the hydrologic cycle, or issues of who/what governs and at what level, be it local, state or federal. However, *"a notable development has been the increasing recognition that it is not just society's relationship with water that is at stake, but the social nature of water itself"* (Linton & Budds, 2014: 170). This social nature can clearly be seen in the aforementioned examples. Many more academic and newspaper articles as well as op-eds, will discuss the drought crisis in the West.

But, how is today's situation different, and what framework of investigation can be applied to it? The current drought is important, not only because it is the most severe in recent history, but because it presents a chance to break with the past. A consensus has been built allowing for the sharing of water that is largely unequal and has created clear social and political divisions, but it has also created a norm of collaboration, and a need to produce consensus. Clearly, water management does not only involve managing flows, it also involves managing trust, people, and political power and even the forces of domination, which may often be implicit and charged with a particularized and historical energy.

Hence, a struggle in policymaking remains; however, it has taken a novel form, that of struggling to reach consensus. As time passes, the struggle becomes less about taming the waters of the Colorado River, as was done in the first part of the century, and more about working within the confines of the established model of management in the Basin, and also simultaneously, the model within each state.

Many people are asking what the future will hold for Arizona. To be sure, it will face the brunt of a shortage in the initial years following such a declaration. It is unclear what role the SOI will play in the Colorado River Basin, and what ongoing efforts to manage the drought will mean for Arizona as the populations of California and Arizona continue to grow, and the drought becomes more widely felt. Many of these same concerns about population growth, and the role of water in the development of economies, are the same as those that the framers of the Compact struggled with in the first part of the 20th century. Although no state wants to be forced into a situation in which the courtroom must be revisited, the future remains uncertain.

Finally, it seems that conflict and cooperation will persist in many forms, perhaps in constant fluctuation. In 1893, Emile Durkheim wrote that "the greater part of our relations with others is of a contractual nature" (*Durkheim, 1893: 213*), before adding that "everything is not contractual in the contract". Therefore, while legal or contractual legitimization may provide the necessary coherence to water laws, the uncertain and unforeseen future might constitute the ultimate problem when charting the convoluted course of the Law of the River. With this in mind, the issue of the Western U.S. drought takes on new meaning. Just as many scientific articles, and much popular press yet to be written, will seek to document and explain what the drought is and what it means, it may be equally important to ask, what would the West be without its laws on the Colorado River? The issues surrounding a drought, spanning from how to deal with depleting aquifers and decreasing Colorado River flows to choosing which fields to fallow, or what to do about desalination, call into question the actions involving more than 100 years of history. As has been shown, it is a history of hard fought battles for and against dams and canals, of struggles over drinking water, and ultimately over the existence of Western American culture. Water management, and the policies that allow the water to be managed, continue to be of great importance as states craft innovative ways to manage crisis, but also to prevent crisis from reaching their doorsteps.

REFERENCES

Ambrose, W.A. and Lynn, P. (1986). Groundwater Recharge: Enhancing Arizona's Aquifers. *Journal of the American Water Works Association*, 78(10): 85–90.

Avery, C., Consoli, C., Glennon, R. and Megdal, S. (2007). Good Intentions, Unintended Consequences: The Central Arizona Groundwater Replenishment District. *Arizona Law Review*, 49(2): 339–359.

Blomquist, W., Heikkila, T. and Schlager, E. (2001). Institutions and conjunctive water management among three western states. *Natural Resources Journal*, 41(3): 653–683.

Colby, B.G. and Jacobs, K.L. (Eds.). (2007). *Arizona Water Policy: Management Innovations in an Urbanizing Arid Region*. Washington: RFF Press.

Connall Jr, D.D. (1982). History of the Arizona Groundwater Management Act. *Arizona State Law Journal*, 2: 313–343.

Cortinas, J., Coeurdray, M. and Poupeau, F. (2015) Du Reclamation act de 1902 aux megaprojets fédéraux – Une perspective socio-historique sur la genèse des politiques hydriques aux USA : le cas de l'Ouest étasunien, *Cuadernos de Desarrollo Rural*, 12(76): 135–153.

Durkheim, E. (1893). *1964. The Division of Labor in Society*. New York: Free Press of Glencoe.

Feller, J.M. (2007). Adjudication That Ate Arizona Water Law. *Arizona Law Review*, 49(2): 405–440.

Getches, D.H. (2009). *Water Law in a Nutshell* (4th ed.). St. Paul, MN: Thomson/West.

Glennon, R.J. (2012). *Water Follies: Groundwater Pumping and the Fate of America's Fresh Waters*. Washington: Island Press.

Glennon, R.J. and Culp, P.W. (2001). Last Green Lagoon: How and Why the Bush Administration Should Save the Colorado River Delta. *The Ecology Law Quarterly*, 28(4): 903–992.

Glennon, R. and Kavkewitz, J. (2013). Smashing Victory: Was Arizona v. California a Victory for the State of Arizona. *Arizona Journal of Environmental Law & Policy*, 4(1): 1–38.

Grant, D.L. (2008). Collaborative Solutions to Colorado River Water Shortages: The Basin States' Proposal and Beyond. *Nevada Law Journal*, 8(3): 964–93.

Kupel, D.E. (2006). *Fuel for Growth: Water and Arizona's Urban Environment.* Tucson: University of Arizona Press.

LeClair, A. (2015, March 18). As California Sets New Water Restrictions, Arizona Resources Dwindle. *Arizona Public Media.* Retrieved March 19, 2015, from https://www.azpm.org/p/crawler-stories/2015/3/18/59276-could-new-california-water-restrictions-become-arizonas-future/.

Leshy, J.D. (2001). The Babbitt Legacy at the Department of the Interior: A Preliminary View. *Environmental Law,* 31(2): 199–228.

Linton, J. and Budds, J. (2014). The Hydrosocial Cycle: Defining and Mobilizing a Relational-Dialectical Approach to Water. *Geoforum,* 57: 170–180.

MacDonnell, L.J., Getches, D.H. and Hugenberg, W.C. (1995). The Law of the Colorado River: Coping With Severe Sustained Drought. *Journal of the American Water Resources Association,* 31(5): 825–836.

McBride, L. (1972). Arizona's Coming Dilemma: Water Supply and Population Growth. *Ecology Law Quarterly,* 2(2): 357–384.

Megdal, S., Nadeau, J. and Tom, T. (2011). The Forgotten Sector: Arizona Water Law and the Environment. *Arizona Journal of Environmental Law & Policy,* 1(2): 243–293.

Meyers, C.J. (1966). The Colorado River. *Stanford Law Review,* 19(1): 1–75.

Mostert, E. (1998). A Framework for Conflict Resolution. *Water International,* 23(4): 206–215.

O'Leary, B. (1998). Bruce Babbitt. Retrieved April 8, 2015, from http://www.washingtonpost.com/wp-srv/politics/govt/admin/babbitt.htm.

Patashnik, J. (2014). Arizona v. California and the Equitable Apportionment of Interstate Waterways. *Arizona Law Review,* 56(1): 1–51.

Pearce, Michael J. (2007). Balancing Competing Interests: The History of State and Federal Water Laws, In B.G. Colby and K.L. Jacobs (Eds.). *Arizona Water Policy: Management Innovations in an Urbanizing, Arid Region.* Washington: RFF Press.

Reisner, M. (1986). *Cadillac Desert: The American West and its Disappearing Water.* New York: Penguin Books.

Record of Decision [ROD] (2007). Colorado River Interim Guidelines for Lower basin Shortages and the Coordinated Operations for Lake Powell and Lake Mead. Retrieved from http://www.usbr.gov/lc/region/programs/strategies/RecordofDecision.pdf.

Stegner, W. (1953). *Beyond the Hundredth Meridian: John Wesley Powell and the Second Opening of the West.* New York: Houghton Mifflin.

Tarlock, A.D. (2001). Future of Prior Appropriation in the New West. *Natural Resources Journal,* 41: 769–793.

Terrain.org. (2006, March 22). Interview with Bruce Babbitt – Terrain.org: A Journal of the Built Natural Environments. Retrieved April 7, 2015 from http://terrain.org/2006/interviews/interview-with-bruce-babbitt/.

Thompson, B.H., Leshy, J.D. and Abrams, R.H. (2013). *Legal Control of Water Resources: Cases and Materials* (5th ed.). St. Paul, MN: West Pub. Co./Thomson Reuters.

Trelease, F.J. (1960). Federal Limitations on State Water Law. *Buffalo Law Review,* 10(3): 399–426.

Tyler, D. (1997). Delphus Emory Carpenter and the Colorado River Compact of 1922. *University of Denver Water Law Review,* 1(2): 228–274.

Tyler, D. (1998). The Silver Fox of the Rockies: Delphus Emory Carpenter and the Colorado River Compact. *New Mexico Historical Review,* 73(1): 25–43.

Valdez Diaz, C. (1996, April 22). City studies water supply options. *Tucson Citizen.* Retrieved (October 15, 2014) from http://tucsoncitizen.com/morgue2/1996/04/22/39002-city-studies-water-supply-options/.

Weatherford, G.D. and Brown, F.L. (1986). *New Courses for the Colorado River.* Albuquerque: University of New Mexico Press.

Worster, D. (1992). *Under Western Skies: Nature and History in the American West*. Oxford: Oxford University Press.

Zeitoun, M. and Mirumachi, N. (2008). Transboundary Water Interaction I: Reconsidering Conflict and Cooperation. *International Environmental Agreements: Politics, Law and Economics*, 8(4): 297–316.

DOCUMENTS, CASES AND LEGAL REFERENCES

Arizona v. California, 283 U.S. 423, 51 S. Ct. 522, 75 L. Ed. 1154 (1931).
Arizona v. California, 373 U.S. 546, 83 S. Ct. 1468, 10 L. Ed. 2d 542 (1963).
Arizona v. California, 460 U.S. 605, 103 S. Ct. 1382, 75 L. Ed. 2d 318 (1983).
Boulder Canyon Project Act, 43 U.S.C. §§ 617 et seq., (1928).
CAPa (2014). Strategic Initiatives and Public Policy: Colorado River Shortage. *Central Arizona Project*. Retrieved March 1, 2015, from http://www.cap-az.com/documents/planning/Shortage_Issue_Brief.pdf.
CAPb (2014). Colorado River Shortage. *Central Arizona Project*. Retrieved March 1, 2015, from http://www.cap-az.com/documents/planning/Shortage_Issue_Brief.pdf.
CAPc (2015). Property Tax Q&A. *Central Arizona Project*. Retrieved July 26, 2015, from http://www.cap-az.com/departments/finance/property-taxes.
Farmers Investment Company v. Bettwy, 558 P.2d 14 (Ariz. 1976).
Hurley v. Abbott, 259 F. Supp. 669 D. Ariz. (1966).
Kansas v. Colorado, 206 U.S. 46, 27 S. Ct. 655, 51 L. Ed. 956, (1907).
Sims Ely to Carpenter, 18 April 1944, box 37, Carpenter Papers, NCWCD.
U.S. Department of the Interior (2007). Final Environmental Impact Statement: Colorado River Interim Guidelines for Lower Basin Shortages and Coordinated Operations for Lakes Powell and Mead. Bureau of Reclamation. Boulder City, Nevada.
USBOR (2010). Field Hearing: Collaboration on the Colorado River. Retrieved (March 19, 2015) from http://www.usbr.gov/newsroom/testimony/detail.cfm?RecordID = 1622.
Winters v. United States, 207 U.S. 564, 28 S. Ct. 207, 52 L. Ed. 340 (1908).
Wyoming v. Colorado, 259 U.S. 419, 42 S. Ct. 552, 66 L. Ed. 999, (1922).

Chapter 5

Water for a new America: The policy coalitions of the Central Arizona Project (Part 1)

Joan Cortinas
UMI iGLOBES, CNRS/University of Arizona, USA

Murielle Coeurdray
UMI iGLOBES, CNRS/University of Arizona, USA

Franck Poupeau
UMI iGLOBES, CNRS/University of Arizona, USA

Brian O'Neill
Water Resources Research Center, UMI-iGLOBES, University of Arizona, USA

INTRODUCTION: HOW TO FACE THE DROUGHT

The Central Arizona Project (CAP) is made up of 336 miles of aqueducts, tunnels, pipelines, and pumping stations designed to transport water from the Colorado River to southern Arizona. Proposed in the 1960s, the CAP was promoted as a means of fighting the effects of past and future drought and helping to bring an end to transborder conflicts based on unequal water distribution between the states (Arizona, California, Nevada, etc.). Drought was already a much-feared factor in the 19th century:

> "When all the rivers are used, when all the creeks in the ravines, when all the brooks, when all the springs are used, when all the reservoirs along the streams are used, when all the canyon waters are taken up, when all the artesian waters are taken up, and when all the wells are sunk or dug, there is still not sufficient water to irrigate all this arid region." (*Powell, 1893*).

And today, as the drought situation worsens, water managers are still trying to find a way to keep water reservoirs high in order to avoid future water shortage in the West. This continuing attitude towards regulating water deficit stress might seem surprising: one might think that the construction of federal mega-irrigation infrastructures along with huge water reservoirs would have provided the ultimate technological solution to cope with drought cycles in the West. However, by looking back to the genesis of water policies, it is clear that the management of water resources has always been complex, with numerous levels of decision making – municipal, regional, state, inter-states and federal. Further, the tensions and articulations between levels have not only served to fashion water policies, they have carried forward a certain idea regarding how the economic development of the West should proceed.

It is, therefore, difficult to understand what is at issue in such a region (regarding water management) without taking into account the historical and social forces that have contributed to the construction of the technical system for managing water and the CAP. Specifically, these forces include the approaches implemented by economic, political and administrative teams who formed opportunistic coalitions to advance their interests and their vision of the world, and thereby promoted their view of the role that should be played by the federal government in the regulation of resources in general and the development of the American West in particular. An examination of the social origins of these water policies reveals that many of the contemporary questions addressed by water managers and decision makers are not new. Indeed, they touch upon old conflicts between Arizona and California, as well as on the recurrent issue of drought.

The scientific perspective developed in this chapter, is based on a social history of water policies, and diverges from the current paradigms of political studies on water governance. We show that it cannot convincingly be claimed, as did *Elinor Ostrom (1990)*, that everything might be negotiated and resolved at the local level by local actors. One has therefore to examine the social and historical logic underpinning issues that are presented as being new, by paying attention to the social conflicts that have arisen as a consequence of water policies in the American West. From this perspective, it is possible to defend the hypothesis that the current management of water and drought in the American West is conducted via a decentralized and multi-level institutional structure corresponding to an absence of national planning in terms of water use, an instrumental vision of water as a vector of economic growth, and a political field heavily influenced by local and regional forces.

To develop this hypothesis, a distinction will be made between three phases of water policy in the western United States (U.S.). The first, which began in the late 19th century and lasted until the 1920s, corresponds to the genesis of the federal policy, whose advocates and structuring effects will be identified. The second phase encompasses the fifty years between 1920 and 1970 and is articulated around the battle between Arizona and California for the water of the Colorado River, the legal resolution of which led to the elaboration of the CAP as the main source of supply for Arizona. This phase makes it possible to shift the focus from the federal to the regional level and to throw light on the third phase, exemplified by the conflict over the quality of water that started in the city of Tucson in 1992. The second and the third phase will be presented in the following chapter. This chapter will present the first phase of the historical process 1890–1929, and will highlight the basis of a long term ongoing process for the making of American water policy. This research based on secondary information and data, on the archives of the protagonists of the conflicts, on the gray literature associated with them, and on interviews with water managers from Arizona constitutes the first part of a larger survey of disputes about water, the logic that underpins them, and their social effects.

I A NEW ELDORADO

1.1 Irrigation projects and economic crisis

The first federal water policy was introduced in the US in the form of the Reclamation Act passed by Congress in 1902. This legislation was the result of a project, promoted and steered by the economic powers of the West in conjunction with business lawyers well-connected to economic circles and Washington politicians, the objective of which was to bring to the foreground problems concerning the distribution of water. Those same economic actors had, a few years earlier, fought against Powell's power and influence. In 1890, they introduced a new strategy designed to address two new series of constraints, on the one hand, the economic structure of the West (as constituted in the 1860s), and on the other, the effects of climatic conditions (notably drought) on the way in which that structure functioned.

By the end of the 19th century, most of the land in the West was in the hands of railway companies, banks and other large landowners. However, most of the land was sold to farmers in the first decades of the 20th century (*Jacobs, 1998*). This period witnessed the transformation of subsistence agriculture into commercial agriculture, in which the exchange value of agricultural products was the basis of the existence of producers. These producers were strongly oriented toward the "market", taking enormous financial risks to mechanize their farms, while the railway companies charged sometimes abusive prices for transporting goods (*Pisani, 1984*). The third structuring element of the economy of the West was its speculative character: land changed hands on a regular basis, a phenomenon promoted not only by large landowners, but also by farmers, who viewed their land as a commodity (*Hofstadter, 1955: 43–54*).

This speculative dimension of the economic structure of the West reached its apogee between 1870 and 1887, during which period the land prices increased exponentially. The bubble burst in 1887 when the existing economic structure collapsed due to the drought of the winter of the same year (the drought affected a third of the West). The decline in the price of farms destroyed the confidence underpinning the speculative market, land prices dropped in an unprecedented manner, people left the country for the city in droves, and many farmers found themselves unable to repay their debts. This explains why the region's economic powers (banks, railway companies, large-scale manufacturers, and suppliers of products consumed by farmers) began to promote federal irrigation projects. The railway companies in particular supported these projects with a view to increasing the value of their land and increasing rates of traffic. Aware that mining, despite its profitability, would not be sufficient to ensure their continued expansion, they viewed new arrivals as being the most effective way to transform the West into a new economic Eldorado (*Worster, 2001: 483*).

San Francisco's financial leaders saw in these irrigation projects the chance to develop a planned economy protected from economic crises and climatic fluctuations (*Pisani, 1984: 296*). Other industrial producers and suppliers that were part of this coalition of interests included the National Board of Trade, the National Businessmen's League and the National Association of Manufacturers. They used the services of two highly regarded business lawyers that were based in the West and had close connections with the region's business milieu and political elites. In 1897, the first, George H. Maxwell, became the founder and leader of the National Irrigation Association, an institution that lobbied Congress on behalf of the advocates of federal irrigation projects. The second, Francis G. Newlands, a Congressman since 1893, would use his position to promote a federal irrigation policy.

> **George H. Maxwell, a reformist in the field of water**
>
> George Hebard Maxwell is a reformist who played a key role in the Progressive Era by generating a substantial number of ideas that were taken up by politicians. Reformists of this kind were professional men from the urban middle class: local business leaders, entrepreneurs and professionals outraged at the economic power wielded by new economic elites emerged in the East of the USA and linked to the flourishing industrial sector. These men invested in a broad range of different areas, including the replacement of sub-standard housing, the regulation of working conditions, and social insurance. Their reforms were designed to regenerate an America that they thought of as suffering from moral decadence. Certain of these reformers, including Maxwell, were to attain important social roles in this struggle for their political ideal.
>
> Maxwell was born in Sonoma, California in June 1860. After being educated in public schools in Sonoma, San Francisco and San Mateo, he became a lawyer in Arizona in 1882. He specialized in disputes concerning water management in California, and was particularly sensitive to the problems of small farmers in regard to major land owners. In 1896, he helped to develop federal irrigation projects in the American West, before going on to join the National Irrigation Congress, where he promoted federal, rather than state and private projects. In 1899, he set up his own organization, the National Irrigation Association. At that point, he abandoned his legal career to focus on promoting legislation favoring a national water policy. He achieved his goals in 1902 when Congress approved the Reclamation Act, a document he helped to draw up.
>
> From this date on, his professional career was informed by two main commitments. On the one hand, he was actively involved in managing the water projects that emerged in the wake of the Reclamation Act. He was the Executive Director of the Pittsburg Flood Commission from 1908 to 1911, Executive Director of the Louisiana Reclamation Commission from 1912 to 1913, and member of the Ohio State Water Conservation Board from 1931 to 1942. On the other hand, he promoted the reformist ideal close to his heart by encouraging the development of a West made up of small communities of farmers functioning as an alternative to saturated urban centers that were a source of social problems. In 1907, he became a founding member and the Executive Director of the American Homecraft Society, an organization that defended the interests of small farmers. He promoted a large number of projects based on the ideal of homecraft in Arizona, Massachusetts, Minnesota and Indiana. However, like many of the reformers of the time, Maxwell's political influence waned and died as the Progressive Era came to an end in the 1920s.

Maxwell shared the ideas of middle class urban reformers involved in crusades aimed at solving the issues of deteriorating living standards and values in industrial cities – notably the decline of the family, the spread of disease, and urban overpopulation. They believed that this situation, if left unaddressed, could lead to a proletarian revolution, or at the very least to social unrest (*Hofstadter, 1955; Topalov, 1988*). Maxwell's grand idea was a new America of small landowners based on the concept of "homecraft" developed by the Cadbury brothers (*Lovett, 2000*): an America where each family would own its own home in a healthy environment and in which workers would be motivated by contact with the land and the development of the family farm. In Maxwell's view, this notion of an America (or more precisely of the West) as an Eden for small landowners offered a solution to the problems of joblessness in the major cities, whose revolutionary potential was feared by the economic and political elites of the period. *"We believe that, as a Nation, we should be less absorbed in Making Money, and should pay more heed to raising up and training men who will be Law-Abiding Citizens: that the welfare of our workers is of more consequence than the mere accumulation of wealth and that stability of national character and of social and business conditions is of greater importance to the people of this country as a whole than any other one question that is now before them"* (Maxwell, in *Pisani 2002: 28*).

Newlands, on the other hand, was less interested in defending the ideas of a reformist movement than in promoting an alternative approach to irrigation, an alternative that he had already tested. In the late 1880s, he had promoted the Truckee Irrigation Project, a private initiative that had not succeeded for a number of reasons, among them a failure to ensure agreement between the divergent interests of multiple land owners (*Hays, 1999: 12*). Further, for Newlands, the only way to develop the economy of his state (Nevada) was through agriculture and cattle rearing. But in the end, like Maxwell, his approach led to the emergence of a coalition of economic groups and institutions that contributed to the development of water policy.

1.2 The Reclamation Act of 1902

The lobbying campaign led by the coalition between these successful lawyers and the large landowners of the West received a favorable hearing in Congress: the Reclamation Act of 1902 led to the setting up of the Reclamation Service which, in 1923, became the Bureau of Reclamation. The transformation of a problem previously defined by this coalition of Western elites into a public problem, or, more specifically, into a recognized problem taken into account in the elaboration of public policy, was linked to two factors. The first, and most important, concerned the bureaucratic arena. The early part of the 20th century, which has been defined as the dawn of political capitalism (*Kolko, 1963*), witnessed the emergence of a strong alliance between the economic power of Big Business and the political power of Congress and the federal government. This alliance was designed to confront the "dangers" that could put them both in peril. *"There were disturbing groupings ever since the end of the Civil War: agrarian discontent, violence and strikes, a Populist movement, the rise of a Socialist party that, seemed, for a time, to have an unlimited growth potential. [...] The political capitalism of the progressive Era was designed to meet those potential threats, as well as the immediate expression of democratic discontent in the States. National progressivism was able to shot-circuit state progressivism, to hold nascent radicalism in check by funding the illusion of its leaders – leaders who could not tell the difference between federal regulation of business and federal regulation for business"* (*Kolko, 1963: 285*).

In the West, the sudden expansion of populism (that had first emerged in 1880) threatened not just the economic elites, but also the established two parties, the Republicans and the Democrats (*Lash, 1991*). This party-movement was successful in the Western states, where farmers experienced enormous difficulties due to the instability of prices of agricultural products, high rates of debt linked to the purchase of their land, and transport problems that prevented them from effectively distributing their merchandise. These farmers and ranchers, who were seriously in debt and whose products and livestock were losing value on the market, gave their support to the party most likely to defend their interests against those of the banks and railway companies, accused by farmers of being the source of all their ills (*Hofstadter, 1955: 64*). But beyond the local economic power directly targeted by the propositions of populists, it was the hegemony of the Republican and Democrat parties that was being contested. To diffuse the threat of the populist movement, the two parties adapted a catchall strategy with a view to appropriating the issues on which its popularity was based (*Hofstadter, 1955: 94–130*). An alliance was struck between the financers of Maxwell's campaign and the political elites of Congress and the federal government, with a view to introducing the Reclamation Act, designed to mollify the populists of the West by limiting the size of government-funded irrigation water to farms of a maximum of 160 acres.

The dominant interests of the political sphere thus aligned with those of the economic power brokers of the West, who were seeking to increase land values after the land price bubble burst in 1890. The idea was to promote irrigation projects that could only be funded by the federal government. Federal intervention in the West would have the effect of shoring up the political and economic power brokers of the region who, thanks to the gradual incorporation of Western states (Idaho, Montana, North Dakota, Washington and Wyoming) into the Federation in the 1880s, wielded an ever-increasing degree of political power.

The content of the law that gave birth to the Reclamation Service in 1902 was based on reformist ideals that favored small farmers. By law, irrigation projects run by the Reclamation Service were linked to farms that, in order to promote the development of smallholdings, were limited in size to 160 acres per person or 340 acres per married couple. Further, the law included measures designed to preclude speculation on land designated for irrigation projects (*Hays, 1999: 12*). A water management system was also established by the users of these irrigation projects. The system was made up of districts tasked with setting up associations to regulate irrigation within the geographical limits of water basins. However, the idea of small farmers was not debated in the draft law, which demonstrates that it was not considered to be the most important political issue at play. There was a consensus among the participants about the need to develop a model of economic expansion capable of resolving the economic crisis that had ravaged the country. But within this consensus there was disagreement between those who thought that water policy was key to economic expansion and those who believed that a different approach, including (notably) the conquest of overseas markets, was the best option. *Worster (1985: 163)* describes debates in the Senate and Congress in the following terms: "*What they really were wrangling over was the wisdom of the traditional American policy of economic expansion and its future direction in the new twentieth century. Did the country need more farmland in production? Was the westward movement now outmoded? What impact did expansion have on older settled regions? Was it wiser to expand overseas or at home?*"

The law proposed by Newlands made it possible to meet the needs of economic expansion by convincing a majority of people that a water policy based on the irrigation of the West was, in the long-term, the only viable solution: *"Newland's bill passed, whether it was constitutional or not, whether there was a national need for it or not, because it promised to augment American wealth and muscle"* (*Worster, 1985: 165*). The dominant coalition that presided over the implementation of the major water projects can thus be understood in terms of convergent interests and shared beliefs: the development of the West had to support the wealth of Eastern cities (*Cronon, 1992*), but no one imagined the ecological consequences of the growth of agroindustry.

1.3 Engineers and the failure of the America of small farmers

It quickly became clear that the water policy of the first two decades of the 20th century had failed to usher in the new America of small agricultural landowners described in the promulgation of the Reclamation Act. This failure occurred at several different levels. First, in the period 1900–1920, very few projects were developed and the number of hectares irrigated was small (*Pisani, 2000*). Second, after small holders and small farmers abandoned their land *en masse*, the West came to be largely dominated by large-scale agri-business (*Pisani, 1984*). These two setbacks were associated, in particular, with a lack of political links among Reclamation Service directors in the West. In fact, their profile as civil engineers was unsuitable to the region's economic and social structures, and they found it difficult to deal with local elites, speculation associated with the sale of land, and the requirement to take local needs into account.

The Reclamation Service was directed by Frederick H. Newell between 1902 and 1915, and by Arthur Powell Davis between 1915 and 1923. As engineers who had spent their entire careers working for the US Geological Survey, their professional specialization was the construction of reservoirs (*Carpenter, 2001: 331*). They were both "New Engineers", products of the boom in the profession that occurred between 1880 and 1920. Their training was based on a vision of the discipline that emphasized a mastery of nature and technological development to the detriment of the political, social and economic aspects of public projects. These "New Engineers" considered technical expertise based on the laws of nature as a tool for getting around economic interests and class conflicts. They believed that science was above politics and business, and, as such, beyond discussion (*Pisani, 2002: 24*). They therefore focused entirely on the laws of physics and paid little or no heed to the problems that the beneficiaries of irrigation projects would encounter prior to project completion, and failed to spend time establishing the kind of links with local power brokers required to ensure that the projects were successful. Unlike Maxwell and Newlands, Newell and Davis shared a vision that focused strictly on the construction of dams and reservoirs; that vision marginalized soil analysis and drainage, and failed to provide support for new agriculturalists and farmers in land irrigated by federal projects (*Pisani, 1984; Carpenter, 2001*).

Frederick H. Newell and the *New Engineers*

Frederick H. Newell, head of the Reclamation Service from 1902 to 1915, was one of the New Engineers who emerged during the boom in the profession. Between 1880 and 1920, a substantial number of engineering schools were founded, leading to a dramatic rise in graduates. Trained in the belief that technology, of which they regarded themselves as the masters *par excellence*, would enable them to apply natural laws to governing the world more effectively, they thought of themselves as being above politics. Their professional ethic was based on the notions of organization, efficiency, rationality and expertise, and not on democratic ideals. Their technical knowledge earned them recognition from their peers and the professional prestige required for brilliant careers in public service, in a period in which reformist ideas informed policies concerning the management of natural resources. Nevertheless, Newell shared the ideas of the advocates of the Reclamation Act, according to whom small communities of farmers had to be developed in the West in order to regenerate an America whose decline had manifested itself in the social revolts of the 1890s. All that had to be done was to complete useful engineering projects and introduce planning for infrastructure to transport water to the communities of this regenerated America. His attachment to the Jeffersonian ideal can be linked to his social trajectory.

Newell was born in 1862 in Bradford, a poor, rural town in Pennsylvania. He went to public schools and, thanks to his impressive academic results (and to the fact that he was able to live in his grandfather's house), he was accepted by the Massachusetts Institute of Technology in 1880. In 1885, he graduated with an engineering degree, and in 1888 he met John Wesley Powell, who advocated a water policy based on small irrigation projects managed by consumers in small farming communities in the West. Powell hired him to work on the study he was conducting on potential irrigation projects, thus giving him a place in the most prestigious geological institution of the time, the US Geological Survey. Belonging to this institution opened doors to other scientific establishments and provided Newell with political contacts that eventually enabled him to become the head of the Reclamation Service in 1902. He joined the Great Basin Lunch Mess, a think tank made up of the leading experts of the time in the field of natural resources. He later joined the National Geographic Society, the American Geographical Society and the American Forestry Association. Thanks to his contacts, he became President Roosevelt's advisor on natural resources. Like many people whose reforming activities reached their apogee during the Progressive Era, Newell's professional career trajectory began to trend downwards in the late 1920s. During this time and until the end of his life, in 1931, he spent an increasing amount of time teaching and conducting research in universities.

Furthermore, knowledge required to ensure the success of irrigation projects in the West (soil analysis, planning and the building of canals) was part of the remit of the Department of Agriculture and the Army Corps of Engineers. However, cooperation between the two organizations was reduced to a strict minimum due to a conflict opposing the men heading them, and, more broadly, the engineers of the Army Corps and those of the Reclamation Service (*Carpenter, 2001: 335*). Moreover, few of these engineers were from the West. This meant that the majority of them were unfamiliar with the region and its soil characteristics. Although irrigation systems were an obligatory subject, the reality was a long way from the specialist areas of the universities from which they had graduated (*Carpenter, 2001: 335*).

Irrigation projects were confronted with other social obstacles associated with the position of engineers within both the states bureaucracy and the federal administrations. These engineers belonged to federal agencies that enjoyed a large degree of political autonomy. Between 1902 and 1914, the Reclamation Service was not obliged to answer either to Congress or to the President of the United States; it had its own budget which it could use as it saw fit, with only the Secretary of the Interior to review it (*Carpenter, 2001*). During this period, the problems deriving from the directors of the Reclamation Service's conception of the role of the organization became more serious due to the agency's administrative organization. The Reclamation Service – later known as the Bureau of Reclamation – was part of the Department of the Interior, which at the time had twelve divisions tasked with overseeing the Service's work and dealing with certain aspects of its work. In regard to legal questions, all decisions were validated by the assistant of the Department's Advocate General; all expenditures were overseen by the finance section. The size of the administration and the way in which tasks within it were organized meant that it took the Reclamation Service an inordinate amount of time to implement its projects; the Service was thus slow to react to problems in the field. To this form of organization, based on an extreme division of tasks, should be added the hierarchical aspect; the Reclamation Service had to wait for the signature of the Department's lawyer, which often took months. For example, such bureaucratic delays deprived the Nevada farmers of water for an entire season (*Carpenter, 2001: 333*).

However, the Reclamation Service had to face an unforeseen increase in the cost of the first reclamation projects. Due to poor calculations about the cost of drainage, the quality of water supplied to irrigated properties, and the yield of irrigated land, they underestimated the potential capacities of new farmers to reimburse the Service for the cost of their properties. This prompted Congressmen to question the competence of this federal agency, and its autonomy ceased in 1914 when Congress took control of its budget. Furthermore, the method applied to calculating the value of land to be purchased for the development of projects left too much room for negotiation with landowners. The speculative attitudes of those landowners caused a considerable increase in the cost of land purchased by the Service. All these unforeseen costs would have had less of an impact if the Reclamation Service had been able to recuperate its outlay in the form of bank drafts paid by the beneficiaries of its irrigation projects. However, in the first two decades of the 20th century, most of the occupants of irrigated plots of land vacated them without settling their debts. The increase in the cost of the construction led to an increase in the price of the projects, which had a negative effect on incoming farmers thereby making it even more difficult for them to succeed.

Another problem that made the job of the Reclamation Service difficult during this period was the territorial political divide between the federal government and individual states and local centers of power. Individual states were unhappy at the idea of what they saw as their water being used to irrigate land in other territories: "*An even more troublesome problem confronted the Reclamation Service: water in one state often could most efficiently irrigate lands in another. Yet the transfer could rarely be accomplished. In Nevada, for example, Newlands and others waged an unsuccessful campaign to obtain water from Lake Tahoe in California to irrigate lands in the lower Truckee Valley in Nevada. Newlands had hoped that the federal government could plan for full development of interstate streams by retaining the freedom to locate reservoirs and irrigable lands irrespective of state lines. Yet, the Reclamation Service met great resistance from local people who wanted to use the water in their own state and complained of federal interference with state rights"* (*Hays, 1999: 18*).

In spite of these issues, the First World War gave second wind to the dream of a new America of small farmers. Migration from the country to the city increased (the number of urban employees rose thanks to the war industry) while, at the same time, there was a need to provide jobs for veterans. Contemporary fears over the "Yellow Peril" in California also contributed. However, one of the projects of the Reclamation Service in its second youth, the Yuma Project, suffered from the same ills as most of the old projects. Of the 173 farms set up by the Service in the Yuma area, 99 were abandoned between 1910 and 1913. The Klamath Project, met the same end, largely because it was located on the border with Oregon in a region far away from agricultural markets. Similarly, the Orland Project was also confronted with problems associated with speculation, the lack of experience of new farmers, and technical problems regarding the distribution of water. These problems led to contracts between the Service and recently installed farmers being broken, causing serious disputes over conditions of payment for farm properties. The most visible consequence of this state of affairs was that farms were abandoned and debts left unpaid.

During this initial period, very few projects were completed. Indeed, the impact of the Reclamation Service as promoter of a new America in the West can legitimately be called into question. In 1920, only one major project was actually completed, the Arizona Salt River Project. Some reached 80–90% of their capacity, while others such as the Newlands, Klamath and Shoshone barely reached 50% (*Reisner, 1985*). Even with the impetus provided by the First World War, fewer than half of the 60,000 government project farms were irrigated; only 3% of the public land in those projects was cultivated, while private land went unused. In 1925, a sixth of all government farms were vacant, and less than 10% of irrigated land in the West was exploited by the federal government (*Pisani, 2002: 293*).

The cumulative effect of these factors was to usher in a new period, which began in 1928 and became truly established in 1929 with the advent of the Great Depression. The era of irrigation projects, based on a law that attempted to promote the interests of smallholders, had come to an end. The alliance between economic and political institutions managed by engineers who believed in the power of science as the primary instrument for running the country was confronted by new political forces affecting the way in which the capitalist system was organized in the Western United States.

2 THE GENESIS OF THE CAP AS ONE OF THE MEGA-PROJECTS OF THE FEDERAL GOVERNMENT

2.1 A remedy to the Depression

From the late 1920s, and to a greater degree from 1933 with the arrival of the Roosevelt administration, the federal government's policy on water entered a new phase in terms of objectives, resources and beneficiaries. The idea was no longer to use irrigation projects to create a new America of small farmers, but instead to help America clamber out of the Depression. Beyond the need to serve irrigation projects, mega-water projects based on the construction of large dams were also designed to generate electricity and control river flows. This period was also characterized by an increase in the number of water projects. From an average annual budget of $9 million/year between 1902 and 1933, the average budget for Reclamation Service/Bureau of Reclamation's projects increased to $52 million/year between 1933 and 1940. Between 1902 and 1928, the agency completed 36 irrigation projects (reservoirs, lakes, canals, dams), as against 228 in the following 30 years (*Reisner, 1985: 165*). Further, this period was characterized by the rise of agribusiness and the growth of cities, who were the leading beneficiaries of the water projects resulting from the water policies designed to favor small farmers.

Above all, this period marked a new phase in capitalism in which the focus was to re-establish the national economy following the collapse of the financial system and quelling social tensions with the aid of a federal program of investment in which major projects played a key role. In this quest for new foundations for economic growth, water was an important factor: "*Hoover argued [that the federal government] should not merely regulate business and referee economic disputes; it should direct, promote, and sustain economic growth. It should integrate and rationalize the nation's economy, making the marketplace as orderly and predictable as possible. Without water management, there could be no planned economy. And without a planned economy, the United States would remain vulnerable to boom and bust cycles*" (*Pisani, 2002: 244*). While to overcome the economic and social consequences of the Depression the federal government was obliged to act at the level of individual states, the American economy of the 1930s was much more interdependent than it had been at the turn of the century (*Pisani, 2002: 252*), so that crises and natural disasters in one state had an impact on the economies of other states. The nature of the new initiatives, which principally took the form of large hydroelectric plants (that also controlled the flow of the great American rivers), enabled several states to benefit from a single project.

These transformations had repercussions in Congress with, in particular, the approval of the first two laws of the new period: the Flood Control Act and the Boulder Dam Bill. These two laws were passed in 1928 after having been blocked in Congress for several years. The Boulder Dam Bill had been presented in 1922 by Phil Swing, Congressman for Imperial Valley in California and by California Senator Hiram Johnson. This proposition was refused by Congress in 1922, 1924 and 1926, largely due to the opposition of the states of the northern basin of the Colorado River that did not share the same interests. At the same time, the states of the Northwest Pacific supported a different project to build a dam in the Columbia Valley. Similarly, the states of the Midwest and the South had no particular interest in the Boulder Dam. "*The old mercantilist view that one state or region's gain was, inevitably, another's loss remained powerful during the 1920s*" (*Pisani, 2002: 259*). This view was mirrored in the irrigation projects of the original Reclamation Service, which promoted initiatives focusing on a specific territory within a single state, an approach that exacerbated tensions with other states, provoking their opposition in Congress. On the other hand, the new policy of the late 1920s focused on the large-scale production of electricity using hydroelectric power sources that were not confined to a single state.

Congress's approval of the Boulder Dam project in 1928 was linked to that of the Flood Control Act, passed in the wake of the 1927 floods caused by the Mississippi breaking its banks. These floods caused massive damage along the entire course of the river and provoked economic and social disaster in the states of the Midwest and the

South composing its watershed. A policy designed to control the flow of rivers then emerged as the only way to protect these states from further catastrophe. Representatives promoting the Boulder Dam decided to negotiate with the states affected by the Flood Control Act, offering their support for the law in exchange for a vote in favor of the Boulder Dam Act. This episode marked the emergence of much broader coalitions for water policies than had been the case with earlier irrigation projects. It was thanks to this association between representatives from the Midwest, the South and the West that the Flood Control Act was passed.

In regard to the Boulder Dam, the representatives of California needed the votes of the states of the Colorado River's northern basin. Electricity constituted a key element of the water policies of the period. The Boulder Dam was a hydroelectric project designed to generate energy for a large part of the West. During the 1920s, private electricity companies began to lobby against the Boulder Dam in order to avoid a situation in which most electricity was produced and controlled by government agencies. These companies formed a powerful trust, just as the railway companies had done at the turn of the century. The states that had formerly opposed the Boulder Dam understood that a vote against the project would mean that a key resource like electricity could fall under the control of private interests, implying a loss of political control over future economic development in their territories. It was this shared interest between the different states of the West, created by the emergence of electricity as an object of new water policies, which made it possible to move beyond oppositions based on particular interests, and led to the states of the West casting a majority vote for the Boulder Dam Act in Congress in December 1928 (*Pisani, 2002: 253–262*).

This change in approach to water policy was also made possible by non-political factors, particularly by transformations in American agriculture. Little publicly owned land remained for the creation of new communities; agriculture demanded increasingly more knowledge and equipment and therefore more investment; cities, meanwhile, became more attractive in terms of living conditions. On the one hand, the difficulties of the 1920s (the drought of 1925 and the Depression) widened the gap between large landowners, who were able to ride out such problems, and smallholders, who were forced to give up their land to those landowners in the West. On the other hand, the increase in the Californian urban population that started in 1900 encouraged the development of agri-business. The state's urban population grew by 89% from 1900 to 1910, by 58.5% from 1910 to 1920, and by 78.8% in the 1920s (*Pisani, 1984: 452*).

2.2 Changing coalitions

Until 1914, decision-making in the sphere of water projects was shared by the Department of the Interior, its technical agency (the Reclamation Service), and the White House. Presidents could leapfrog Congress and green light the federal budgets required to develop the West. This coalition can be explained in regard to the fact that Congress, strongly represented by the elites of the East at a time when the states of the West were constructing their legitimacy, saw no interest in voting through budgets for projects that did not directly affect their constituencies. A series of water projects was therefore authorized, primarily to the advantage of California (which had achieved statehood in 1850). This situation created tensions with other (younger) states in the

West, including Arizona (founded in 1912), that fought to protect the control of the use of their natural resources against the federal government, while at the same time aspiring to economic development. From the 1930s, the more the delegations from the states of the West organized themselves and gained influence, the more decision-making powers they acquired. The battle to obtain federal funding took place in the Congressional arena, in the House of Representatives and the Senate, in which representatives of the states of the East and the West engaged in debates about the wisdom of initiating major projects such as the construction of dams. Some Arizona elites understood this and, to promote the development of water projects, attempted to develop a united front instead of systematically opposing the federal government.

The growing influence of Congress was accompanied by bureaucratic change. In the early years of the century, the Reclamation Service's irrigation projects still focused on precise territories within single states, an approach that exacerbated state interests in Congress. The new policy of the late 1920s, which was continued with the New Deal and on into the post-War period by focusing, beyond irrigation systems, on the large-scale production of electricity and controlling the flow of rivers, generated effects far beyond the territorial power structure, encompassing inter-state relations within the federal government. This situation presupposed compromise and consensus in Congress. The change in policy also had an impact on bureaucratic practices which, in order to provide a framework for multipurpose regional projects, needed, in order to attain legitimacy, to enlarge their zones of influence to the national level.

Other transformations of a sociological order should also be taken into account. Between the 1940s and the 1960s, the Bureau of Reclamation was not only a breeding ground for engineers, but also employed a considerable number of assistants who provided a link with Congressmen. Some of those assistants were promoted to the level of directors within the Bureau of Reclamation. During this period, the Bureau was no longer able to count on the support of the Secretary of the Interior and the White House in the promotion of its technical projects, since it was Washington politicians who now held decision-making powers. But to their technical resources, the Bureau added political support, courting politicians less because of their party affiliation, and more because they shared a common interest, namely the development of the West.

3 CONCLUSION

This research has shown that Western United States water policy in the end of the 19th century and the beginning of the 20th century has been implemented by evolving coalitions consisting of local economic powers, political elite, reformers and new engineers. Water policy has been built to promote economic development and to avoid social protest. In that sense, water policy has been strongly influenced by what historians have called "the age of reform" (*Hofstadter, 1955; Gusfield, 1963; Topalov, 1988*) – a period characterized by the emergence of "organizations and movements devoted to improving mankind, changing social conditions, and reforming the character of human beings" (*Gusfield, 1986: 53*). The promotion of this "water for a New America" has failed because of the vision promoted by the new engineers, who preferred science and technology as being to politics and economy. Reformers like Maxwell and new engineers like Newell had a clear vision: build small owners communities in the West through hydraulic engineering. This conception of water policy is linked to their social trajectory, which constitutes a key element towards understanding water policies, its success, and failures. Finally, water policies at the beginning of the 20th century were developed at the federal level and has come to shape the water policy of today. The dominant Western American policy and management paradigm, which assumed a position to water infrastructure megaprojects that went largely unopposed until the 1970's, was essential for economic growth. What would the West be without its megaprojects and without the federal government's involvement in the interest of Western growth and expansion? Indeed, megaprojects have carried enormous amounts of water to the West, supporting big cities and agribusiness. The West's undeniable dependence on water continues to have a strong impact on decision-making related to water management, but also on the implementation of water policy throughout the 20th century, as will be illustrated in the following text (see *infra* chapter 6).

REFERENCES

Carpenter, D.P. (2001). *The Forging of Bureaucratic Autonomy: Reputations, Networks, and Policy Innovation in Executive Agencies, 1862–1928*. Princeton, N.J./Oxford: Princeton University Press.

Gusfield, J.R. (1963). *Symbolic crusade: Status politics and the American Temperance movement*. Urbana and Chicago: University of Illinois Press.

Hays, S.P. (1999). *Conservation and the Gospel of Efficiency*. Pittsburgh: University of Pittsburgh Press.

Hofstadter, R. (1955). *The Age of Reform*. New York: Vintage Books.

Jacobs, H. (1998). *Who Owns America? Social Conflicts about Property Rights*. Madison, Wis.: University of Wisconsin Press.

Kolko, G. (1963). *The Triumph of Conservatism. A Re-interpretation of American History, 1900–1916*. New York: Free Press.

Lash, C. (1991). *The True and Only Heaven. Progress and Its Critics*. New York/London: W.W. Norton and Company, Inc. Traduction française: (2002) *Le seul et vrai paradis. Une histoire de l'idéologie du progrès et de ses critiques*, Paris: Climats.

Lovett, L. (2000). Land Reclamation as Family Reclamation: The Family Ideal in George Maxwell's Reclamation and Resettlement Campaigns, 1897–1933. *Social Politics*, 7(1): 80–100.

Ostrom, E. (1990). *Governing the Commons: The Evolution of Institutions for Collective Action*. Cambridge University Press.

Pisani, D.J. (1982). State vs. Nation: Federal Reclamation and Water Rights in the Progressive Era. *Pacific Historical Review*, 51(3): 265–282.

Pisani, D.J. (1984). *From the Family Farm to Agribusiness: The Irrigation Crusade in California and the West, 1850–1931*. Berkeley: University of California Press.

Pisani, D.J. (2002). *Water and American Government: The reclamation bureau, national water policy, and the west, 1902–1935*. Berkeley: University of California Press.

Reisner, M. (1985). *Cadillac Desert: The American West and its Disappearing Water*. New York: Penguin Books.

Topalov, Ch. (1988). *Naissance de l'urbanisme moderne et réforme de l'habitat populaire aux États-Unis, 1900–1940*. Rapport au Plan Urbain. Paris: CSU.

Worster, D. (1985). *Rivers of Empire: Water, Aridity, and the Growth of the American West*. New York: Pantheon Books.

Worster, D. (2001). *A River Running West: The Life of John Wesley Powell*. New York: Oxford University Press.

Chapter 6

Sharing the Colorado River: The policy coalitions of the Central Arizona Project (Part 2)

Murielle Coeurdray
UMI iGLOBES, CNRS/University of Arizona, USA

Joan Cortinas
UMI iGLOBES, CNRS/University of Arizona, USA

Brian O'Neill
Water Resources Research Center, UMI-iGLOBES, University of Arizona, USA

Franck Poupeau
UMI iGLOBES, CNRS/University of Arizona, USA

INTRODUCTION: ARIZONA VERSUS CALIFORNIA

This chapter focuses on the last phases of the socio-history of western water policy that have been described in the previous chapter. In the aftermath of the golden age of megaprojects, the Central Arizona Project (CAP) is the culmination of two social components of the New Deal era; a great belief in the engineering skills to build up big water infrastructures and the politicking practices for territorial sovereignty over water in the West (*Willey & Gottlieb, 1982*). These prerequisites for success, which are key to understand how the idea of Central Arizona Project came up, have deepened in the context of the increased economic development of the West affecting the post Second World War period. Furthermore, the Central Arizona Project mirrors a transitory phase towards a much decentralized and multi-layered decision-making process in water issues. Lastly, choosing to come back to the history of the CAP enables to better appreciate what is going on in the current context of looming water shortage in the West. In addition to be designed as a technical response to an excessive groundwater pumping, the Central Arizona Project is also the product of specific and temporary alliances between elites from the political, economic and administrative spheres, which have structured and still structure social tensions with regard to the existing drought situation.

In Arizona, support for the CAP had its origins in a struggle between individual states (anxious to develop their economies) for a share of the Colorado River. The period of the "*New Deal*" enabled the expansion of the economy in California, notably thanks to the policy of building dams to irrigate agricultural land. The completion of the Hoover Dam on the Colorado River in 1936 consecrated the legitimacy of the federal government's approach to the development of the state. However, the American West neither begins nor ends at the Californian border, and other states had a major interest in controlling and exploiting the river that *Reisner (1986)* called the "*American Nile*". The Arizona elites envied California's economic prosperity, while at the same time accusing it of overexploiting the Colorado River (made possible by the Imperial Valley's senior water rights on the river). Their struggle with California over access to the Colorado River manifested itself in various ways depending on whether the elites concerned were state leaders or advocates of Arizona's interests in Washington. When, in the 1930s, against the backdrop of the New Deal, the Department of the Interior decided to sign a contract with the Metropolitan Water District (a primarily urban water distributor in California) to build the Parker Dam in Arizona, the state governor Benjamin Baker Moeur (informed federal officers that if Arizona's rights in the matter were not clearly defined he would oppose the deal (*August, 1999*). In effect, he considered that a dam could not be built to divert the Colorado River without the consent of Arizona. When the project was approved, Governor Moeur declared martial law on the site, bringing in the Arizona National Guard and accusing the federal government of calling Arizona's sovereignty into question. Since any negotiations with the state of Arizona were doomed to fail, the federal government took its case for building the Parker Dam to the Supreme Court. The dispute took an unexpected turn when the Supreme Court ruled in favor of Arizona, ruling that the federal government had failed to demonstrate that the dam had been authorized. Nonetheless, shortly after the verdict, Congress authorized the dam's construction.

This conflict between Arizona and the federal government is not just one episode among others. Indeed, on many occasions in the 1930s Arizona used the Supreme Court as an arbiter to impose its sovereignty and its legal rights to the Colorado River. Believing that it had been poorly treated in regard to water resources (*Hundley, 1975: 289*), Arizona refused to ratify the Colorado River Compact, designed to regulate water distribution between the river's Upper and Lower States. However, the Supreme Court declined to uphold Arizona's complaints about California, preferring instead to emphasize the rights of the federal government to enforce water policy in the watersheds of the American West. In the post-World War II context, the conflict between Arizona and California was still there but took a different turn. Shaping powerful coalitions that would not only serve their interests but also would meet the logic underpinning the federal government's approach, was the main aim of some Arizona decision makers, who searched for being united over a project – in this case, the Central Arizona Project (CAP). In the next two sections, this article will see how these coalitions were built, have met stiff opposition, changed facing the rise of environmental perspectives in the 1970s, and finally resulting in political compromise with California and a consensus with the federal government: when the CAP arrived in Tucson, the conflict over the CAP shifted from interstate controversies to local tensions, deriving from coalitions of local elites being pitted against each other for the very viability of the project. In addition to a technical standpoint, this approach to coalitions formation thus enables to highlight the social dimension of a project, leading to make visible the multi-layered decision-making process as well as the significant role of the multi-positioning of professionals dealing with water issues.

1 A NEW COALITION FOR A NEW PROJECT

1.1 The CAP association

After the Second World War, the legacy of the New Deal was apparent in the way in which natural resources were managed. The development of the West was still envisaged in terms of the construction of more and more dams; the generation of electricity was considered a sure fire bet in that it had helped to produce the ships and planes that guaranteed victory against Germany, while the creation of irrigation systems helped agri-business. It was in this post-New Deal, post-War context (which was favorable to large-scale water projects) that the CAP, initiated by the State of Arizona, took shape. Indeed, at the time, Arizona was experiencing unprecedented demographic growth, a fact that encouraged the state's elites to find solutions to problems associated with the seasonal migration of significant numbers of "Snow Birds" who consumed water and used air-conditioning in a region in which aquifers, still the main source of water, were gradually drying up due to the needs of agriculture.

To ensure legitimacy of the CAP, the elites of the State of Arizona sought federal funding, which required authorization from Congress. After the ratification of the Colorado River Compact by Arizona in 1944, concrete initiatives were taken to develop a political consensus, an approach that contrasted sharply with the conflict-ridden period of the 1930s. In 1946, the CAP Association (CAPA) was created by local decision-makers (farmers, bankers, lawyers, companies working in the general interest) who shared the belief that water from the Colorado River fundamentally important to the future of Arizona's economy. From the outset, CAPA was linked closely to the Arizona Congressional Delegation, notably the legislative activities of the two Democratic Senators, Carl Hayden and Ernest McFarland, who were in favor of the CAP.

The CAPA was involved in setting up hearings on early legislation to authorize the CAP sponsored by two Arizona senators, Carl Hayden and Ernest McFarland to the Senate Subcommittee on Irrigation and Reclamation. In 1948, CAPA supported the creation of a state agency, the Arizona Interstate Stream Commission, the aim of which was to advocate in Congress and in the courts on behalf of the state's claims to access to the Colorado River (*Mann, 1963: 128*). Wayne Akin, President of the CAPA, was appointed President of the Commission by Governor Osborn. This appeared not to be a neutral choice in that, through CAPA, Akin could count on the advice of the most influential local decision-makers in the state (*Johnson, 1977*). Moreover, since graduating in the 1940s, he had been a member of a group called the "*League of 14*". Made up of two members from each of the seven states of the Colorado River Basin, the League's objective was to bring together decision-makers in the water sector to discuss shared problems and, where possible, to resolve their differences.

This initiative was perceived as having the potential to contribute to building a federal consensus around the CAP project. Furthermore, Wayne Akin could also count on Charles Carson, legal advisor to Governor Osborn and the Arizona Interstate Commission, and also a member of the Board of the Phoenix Chamber of Commerce, who had sought Akin out when the CAPA was first set up, and who later supported the legislative efforts of Congressmen Hayden and McFarland. This form of a lobbying organization is best understood if we bear in mind that in the West, political participation was not structured exclusively around the political parties (which were considered weak), but also focused on interest groups that attempted to influence public policy in their own particular fields (*Thomas, 1991: 165*). In the case of the promotion of the CAP in Congress, the CAPA interest group and a handful of Arizona politicians decided to put their differences to one side and work together, thus strengthening their influence to oppose California, which was both strongly represented and highly organized.

Carl Hayden, Western lawmaker: from the Salt River Project to the Central Arizona Project

The use and distribution of water is central to the public career of Senator Carl Hayden. Born in Arizona in 1877, Hayden enjoyed a 57 year career as a Representative and Senator. After spending ten years as a local politician, he represented the new state of Arizona, first in the House of Representatives (1912 to 1927) and then in the Senate (1927 to 1969). Although famous for the part he played in developing the American West by means of the CAP, Carl Hayden was a product of the legendary conquest of the "frontier".

Hayden's father, Charles Trumbull Hayden, was born in Connecticut on April 4, 1825. He was one of the pioneers who left the East to make their fortune out West. While working the region to the north of Salt River Valley as a manufactured product salesman, Hayden saw the potential of this arid and hostile territory. His ambition was to transform the Salt River Valley into a canal-irrigated agricultural empire. In the mid-1870s, the irrigation community – based on the south bank of the Salt River – in which he lived, assumed a pioneering role thanks to a variety of rapidly expanding farms (grain fields, mills, orchards, etc.). As well as this family success, Hayden recalls that, at the time, his parents' farm was repeatedly hit by drought and flooding. The farmers and businessmen of the Salt River Valley asked the government for aid to halt the flooding and provide water storage facilities. These environmental problems occurred in the context of the irrigation movement of the 1880s, led by its promoters, Maxwell and Smith, who organized meetings dedicated to the "conquest of arid America", which Charles Hayden attended. Distributing tracts and magazines at conferences, Charles Hayden initiated the young Carl into the problems of his community and other territories of the West. His family spent considerable sums of money in the cause of preserving the rights of these territories, winning and losing cases along the way.

During his time at California's Stanford University, one of the West's pioneering academic institutions, a career in law and politics with a focus on water issues seemed an obvious choice to Carl: "*I want to make water law a specialty not only because it is a new and open field where the prizes are large to the winner, but also because through it I can have a greater power for good and evil than at any other branch of the law. I know that the law of water is not taught in schools nor found in books, but that is all the more reason why it will be so valuable when known*". (…) "*I have no fear of not getting along in this world. Just let me train rightly for the right thing and the result is not in doubt. I am going into politics – I shall make honest water laws and see that they are honestly executed*" (August, 1999: 24). The project was somewhat delayed by the death of his father: he spent a number of years successfully running the family business. However, he did not give up politics, which he considered useful in solving local problems. Later, he gradually left his business interests to one side, starting out in politics at the local level, where he immediately sought aid from Washington to develop water distribution networks. He later started campaigning in state, and later federal elections. His Washington ambitions can be explained by a series of factors: Carl Hayden and other local leaders, aware of the economic potential of Roosevelt's new federal policy, encouraged the development of water projects with a view to enticing people to move to /farm in arid areas. On March 4, 1903, the Salt River Project, for which he was the spokesman in Washington, was the first of 26 projects authorized by the Department of the Interior in the first decade of the national irrigation program. This successful experience was the point of departure for his political campaigns for election to Congress, where he dedicated the best part of his political career to issues concerning the development of water distribution in the West.

During the fourteen years he spent in the House of Representatives, Carl Hayden witnessed a transformation in American society. When he entered Congress in 1912, America still defined itself in terms of Jefferson's agrarian ideal of small communities, and decentralization. By the time Hayden became a Senator in 1927, the Progressive Era was tracing out a new America that was more urban, more centralized, more industrialized and more secular than ever before. The development of the Colorado River became a major issue in the 1920s, dominating the agenda of the politicians of the American Southwest. Hayden was already involved, taking part in negotiations about the Colorado River Compact in 1922, and remaining active in the field up to and including the passage of the Colorado River Basin Project Act of 1968, which, among other things, authorized the CAP. With the development of the Colorado River, Hayden was confronted by other issues. Instead of advocating for an irrigation project centered on a river in a single state, as had been the case of the Salt River Project, he was responsible for resolving the problem of how to share a river crossing seven states of the Southwestern United States and a part of Mexico. Hayden also had to resist the All American Canal project promoted by California and, along with it, the advocates of the Imperial Valley. While he supported a regional conception of the development of the river, he opposed California's desire to obtain the exclusive right to access water from the Colorado River with the help of the federal government with the sole objective of increasing the prosperity of the Imperial Valley: "*you are now coming to Congress asking that an extraordinary thing be done by the passage of his legislation and Congress must look to the development not only of the Imperial Valley, but the Colorado River Valley as a whole, and that can only be fully developed by storage*" (ibid.: 76).

Carl Hayden also fought against leaders in Arizona, particularly Governors Hunt and Moeur, and Senator Fred Colter, advocates of a state-based approach to rights concerning the development of the Colorado River. In 1923, promoting a federal approach, Hayden announced his support for the ratification of the Colorado River Compact. "*Any fair-minded person must conclude that Arizona alone cannot undertake the development of the great river without the consent of the United States, and without understanding with the other states of the Colorado River Basin, all of which leads to the conclusion that sooner or later the Colorado River Compact must be approved by the State of Arizona*" (ibid.: 92). According to Hayden; the main partner in the development of the Colorado River was the federal government. In the 1940s, a change of governor in Arizona combined with increasing urbanization within the state, created a new situation that, in turn, led to the emergence of a new approach to water policy. Senator Hayden introduced the first legislation for the CAP, a process which brought him up against Californian interests and alternative projects promoted by Arizona. Nevertheless, legislative negotiations about how water from the Colorado River was to be shared culminated in the authorization of federal funding for the CAP, crowning the Senator's political career.

In the late 1950s, the Arizona's economy had reached a turning point. With demographic expansion concentrated in the "Sun Corridor" (Phoenix and Tucson), agriculture was no longer the main source of wealth. Indeed, there was a proliferation of new jobs in other industries (*Sheridan, 2012*). In 1961, this context forced CAPA to adopt a new approach to promoting the CAP, which recognized the demands for water from industry and the cities (*Johnson, 1977*). At the same time, the Arizona Interstate Stream Commission requested the Bureau of Reclamation's local agencies to re-evaluate water use in areas outside the distribution zone originally designated for the CAP, including, notably, an extension of the aqueduct toward Tucson. As part of the quest for national unity in the face of Congress, the aim of the new approach was to demonstrate that the CAP would serve growing urban populations in Tucson and Phoenix, thus providing a tactical advantage in terms of undermining the validity of California's argument, according to which the CAP was essentially intended for agricultural purposes. In 1966, the members of the Arizona Interstate Stream Commission and the CAPA joined a task force designed to support the efforts of the Arizona Delegation to Congress, a process that culminated in the authorization of the CAP in 1968.

Having exhausted all legal possibilities, the State of Arizona had practically no chance of satisfactorily settling its differences with California in the courts (*Hundley, 1975: 299*). Thus, Arizona had to prevail in the political arena. To have a chance of making their voice heard in Congress and successfully defending the CAP in the face of political opposition from California, the Arizona politicians and lobbyists had to present a united front. Within Arizona itself, water policy went through a radical change with the election of Sidney Osborn (Democrat) as Governor in 1940. Osborn signed the Colorado River Compact in 1944, validating a regional vision of water distribution in spite of strong opposition from the state's utilities, as well as from the Arizona Highline Reclamation Association (an organization promoting the interests of the state and irrigation projects in each of the state's counties and districts, whose president was the Democrat, Fred Tuttle Colter), and from two other Senators and six Congressmen (*Hundley, 1975: 299*). In so doing, Osborn distanced himself from the kind of conflicts that characterized the Arizona political scene in the preceding decades, in particular the tradition of systematic opposition established by the preceding Democratic governors, George W. P. Hunt and Benjamin Baker Moeur.

This change in policy also reflected a growing awareness of the potential consequences of long-term population growth in Arizona, particularly in and around Phoenix and Tucson. Not only did this growth prefigure Arizona's urban development in the following decades, it also posed the question of water supply in a state that had already experienced water and energy crises in the droughts of the 1930s and 1940s. Governor Osborn's new political orientation, represented in Washington by Senator Hayden, was also part of the effort to develop the consensus needed by the State of Arizona to pass the CAP in Congress (*August, 1999*). However, in the 1960s, the political and legal disputes over the CAP led to the formation of new coalitions among the nation's senators.

1.2 Legislative compromises and environmental pressures

Between the United States Supreme Court's 1963 decision in favor of Arizona (in regard to sharing the Colorado River with California) and the confirmation of that ruling by Congress in 1968, the federal administrators and representatives of the states of the Colorado River Basin began to elaborate a regional plan for the development of the West. While debates in Congress focused primarily on the CAP, a new project, the Pacific Southwest Water Plan (PSWP) was promoted in 1963 by the new Secretary of the Interior, Stewart Udall, appointed to the post two years previously by President John F. Kennedy, to whom he was close (*Johnson, 2002*). Eschewing a concept focused on state borders and, instead, promoting a regional approach to water needs, the objective was to unite the interests of Arizona and California without passively accepting the energy policy of the moment; the proposition, presented by Morris Udall, Stewart's half-brother, included the construction of two giant dams (Bridge Canyon and Marble Canyon) near the Grand Canyon National Park.

Democrat Congressman for Tucson, Morris Udall, a lawyer by training and profession, emerged as the mediator of opposing interests. For Senator Carl Hayden, the main advocate of the CAP, the PSWP was a competing and contradictory legislative initiative (*August, 1991*). The Senator regarded the PSWP as a stalling tactic on the part of California designed to delay the authorization of the CAP. Paul Fannin (Republican), elected Governor in 1958, also thought of the PSWP as "*a conspiracy against Arizona born in California*". He defended Senator Hayden's approach, which consisted of promoting the idea of the CAP in a separate legislative proposition. This vision was rejected by Morris Udall, who was favorable to the more regional approach of the CAP. However, in Arizona, the reputedly influential *Arizona Republic* condemned Udall's future political career in Arizona by accusing him of being beholden to the California water lobby represented by James Carr, the Under Secretary of the Department of the Interior.

The administration's preference for a regional approach to developing the Colorado River (in the form of the PSWP) enabled Senator Hayden to gather support. In 1966, the Arizona Task Force supported the legislative effort by uniting the main sources of expertise on water of the time (the Arizona Interstate Steam Commission, the Arizona Public Service, the CAP Association and the Salt River Project). Moreover, the CAP had the support of powerful allies in the Senate: Senator Henry Jackson (Democrat, Washington), head of the Senate Interior Committee, and Senator Clinton Anderson (Democrat, New Mexico), President of the Power and Reclamation Subcommittee of the Interior Committee. In their struggle in favor of the CAP, these Democratic senators supported Carl Hayden against his Californian Republican opponents, Thomas Kuchel, member of the Senate Committee on Insular and Interior Affairs, and Claire Engel, who had helped torpedo the CAP project in the House of Representatives in 1951. In the end, the real battle for the CAP was the one waged in the House Interior Committee and the Irrigation and Reclamation Subcommittee. The latter was directed by Wayne Aspinall (Democrat, Colorado), who was very wary of California (and Arizona) and what he saw as their expansionist aims, and who was concerned that the Upper Basin States, especially Colorado, might not obtain their share of the water allocated to them under the Colorado River Compact. Hayden also clashed with John Saylor (Republican, Pennsylvania), a member of the House Interior Committee. Saylor, a conservationist, opposed the construction of dams included in the CAP; he was opposed to the continuous development of public sector electricity generation schemes, and his position earned him the respect of those of an ecological persuasion.

In the end, the Colorado River Basin Act of 1968 authorized the projects supported by the representatives of different states, or a single state, either completely, as in the case of the CAP, or more partially, as in the case of the Pacific Southern Water Plan. This result, produced by a compromise between the various forces at play, was based on "pork barrel politics", according to which Congressional allocation of public funds served the interests of the legislators' constituencies rather than the national interest. At issue was not only the economic development of the West, but also political careers and reputations. This meant that everyone had to take into account the needs of their respective constituencies in terms of new projects and the allocation of sections of the Colorado River. While, for example, California lost the legal battle, it nevertheless obtained a guarantee of 4.4 million acre feet of water from the Colorado River in case of drought, which later turned out to be more than useful in countering the water crisis of the 2000s.

To hope to be able to ensure that their needs were met, coalitions had to find a point of agreement. They also had to take into account external pressures deriving, in particular, from environmentalists fundamentally opposed to the construction of new dams that may have flooded one of the jewels of the West, the Grand Canyon, and favorable to alternative energy sources. In the 1970s, the rise in the number of environmentalist votes changed decision-making processes; authorization for major water projects was based not only on economic viability and technical reliability, but also increasingly reflected a new dynamic involving the reconciliation between environmental protection and economic growth. Furthermore, a new legislative framework, the National Environmental Policy Act, had also been passed in 1970. It is certainly not by chance that, during this period, a number of dam construction projects were abandoned in favor of projects based on alternative energy sources which coincided with the emergence of a new generation of bureaucrats and politicians more focused on adapting to the emergence of values different from those of their predecessors, who had been motivated exclusively by economic considerations. And it is also probably not by chance that during this period, Congress's authorization of federal funding for major water projects was accompanied by the requirement that the states meet certain legislative imperatives, notably with respect to the protection of their natural resources.

Up until this point, Congress and the Bureau of Reclamation had championed the development of the West. The need to incorporate environmental concerns was expressed by a new generation of Arizona elites, including Stewart Udall, then Secretary of the Interior. They were doubtless helped in this effort by their legal background and their experience as lawyers. By suggesting that, in order to support the CAP, the Bureau of Reclamation should become a shareholder in a coal-powered electricity plant, Stewart Udall managed to echo the concerns of the environmentalist movement (to which he had dedicated a book, *The Quiet Crisis*, in 1963), while at the same time promoting the economic ambitions of the elites of the West. Finally, on September 30, 1968, the Colorado River Basin Act was passed by Congress and signed by President Lyndon Johnson. Among other projects supported by Congress, it authorized the construction of the CAP. Due to the authorization of the CAP, tensions between the states over the allocation of Colorado River water had been politically neutralized, and the balance of power between the fierce advocates of economic development, the promoters of dams, and the inveterate defenders of the environment had reached a new equilibrium.

> **Stewart Udall, between the pioneer spirit and the environmental cause**
>
> Stewart Udall was born in Arizona on January 31, 1920. He spent his childhood in the Mormon community of St. Johns, growing up in a semi-arid milieu in which water management was part of basic education and irrigation seen as a way of life and a scientific principle. The Mormons were known at the time for their ability to transform previously arid areas into fertile agricultural land by building dams and small canals. Beyond his familiarity with water issues, the sense of justice inculcated by his mother, and the public service ethos associated with his father, who had been Chief Justice of the Arizona Supreme Court, also shaped his interests. Over the course of his professional career, Udall attempted to reappraise the heritage of the pioneers, reconciling it with the law and the democratic ideal.
>
> After graduating with a degree from the University of Arizona in law in 1948, Stewart set up a law firm in Tucson with his brother Morris to fight against segregation in the city. Along with his legal activism, he harbored political ambitions. He was elected Vice President of the Central Committee of the Democratic Party, then Treasurer of the Legal Aid Association of Pima County. He was also elected three times to the House of Representatives in Washington D.C. When John F. Kennedy announced that he was running for President, Stewart was one of his most fervent supporters, recruiting delegates from Arizona to help him. This link with Kennedy helped him gain a promotion in 1961 to Secretary of the Federal Department of the Interior, managing 65,000 employees and a budget of 800 million dollars, allocated to regional development and water management. During the eight years that he occupied the post, he bore witness to the end of the Golden Age of water projects and the emergence of the environmental movement: "*I began with the idea that dams were probably a good thing.... I presumed that if anyone, the Corps of Engineers, the Bureau of Reclamation, wanted to build a dam, it was a good thing. I ended up thinking that we ought to be highly skeptical of any dams*" (Johnson, 2002: 31). This point of view accurately reflects the attitude he took in negotiations over the CAP between 1963 and 1968: while ensuring the adoption by Congress of the Colorado River Basin Compact authorizing the federal funding of the latest major water project, he also campaigned for alternative sources of energy to dams, thus incorporating the environmentalist cause into federal water policy for the first time.

In public debates in the 1960s, the most controversial CAP legislation concerned dams. The idea that an alternative source of energy was required to pump CAP water to central Arizona provoked a reappraisal of approaches to the project. These environmental concerns were promoted by the Sierra Club, whose president, David Brower had made a name for himself in the 1950s for his anti-dam campaigns in the Grand Canyon and his promotion of coal as an alternative source of energy for transporting water. Moreover, the need to find an alternative to dams was made more urgent by the fact that, at the time, pressure for environmental issues to be taken into account in water projects was growing. This trend was given concrete form by President Johnson, whose administration set up the National Environmental Protection Agency in 1970 which forced federal agencies to evaluate the environmental impacts of federal water projects. Having won the legislative battle in 1968, the advocates of the CAP had to take into consideration this new institutional framework, which allowed social groups that had previously been excluded from the decision-making process to express their concerns, and which obliged federal agencies to take notice of them. Winning battles in Congress no longer guaranteed local support (*Ingram, 1990*): it was probable that the electors concerned no longer systematically represented the interests of the Eastern allies of the federal agencies.

From the 1970s, water projects in the West were targeted by Congressmen from the East and Midwest, who criticized them on the grounds of regional favoritism and a lack of economic efficiency. At the same time, those projects were increasingly unpopular with local people. President Carter attempted to shut down the CAP project, whose promoters were forced to adjust their position. To ensure that the CAP was not eliminated from the federal budget, the Senators from Arizona had to abandon construction work on the most controversial dams. Meanwhile, federal agencies were told to find alternative approaches to funding the storage of local water and providing flood control for the project. In addition, the implementation of the CAP was conditioned on Arizona's reforming its laws in regard to the management of groundwater.

While the period 1940–1960 had been structured around stable coalitions between a few key Congressmen from the West, the Bureau of Reclamation, and the interests of farmers and promoters of local development who shared a desire to see the West prosper, the 70's saw the emergence of new sources of influence and coalitions, based on an idea of politics defined by a respect for formal rules and the protection of the environment at a time when the American West, now increasingly urbanized, was confronted with water shortages resulting from its economic success.

2 DELIVERING THE CAP

2.1 Regulating groundwater: federal government and new local leaders

While the CAP was associated with the struggle between two states, Arizona and California, over the Colorado River, the project was finally approved due to its positioning within the balance of power between federal authorities and local elites concerning the regulation of groundwater. It is probable that the CAP obtained federal funding because the dominant idea among the advocates of the project was to build a consensus over the need for the project in the name of well-understood interests, namely the economic prosperity of the state, guaranteed by the development or reinforcement of agriculture, industry and tourism.

In Arizona, the territorial constitution of 1864 (the Howell Code), which declared that surface water was public property, made absolutely no mention of groundwater (*Mann, 1963*). In the 1930s, due to the growing efficiency of pumping mechanisms, the increase in the price of cotton, and cheap electricity, groundwater was increasingly used for agricultural purposes. Its extensive use became a feature of the political debate. But it was above all the federal CAP project that triggered legislative action in Arizona between 1948 and 1980. The 32 years between the promulgation of two sets of laws regarding the CAP can be interpreted as the transition from one coalition to another, a transition rendered possible by the emergence of a new political configuration.

In the 1940s, the Department of the Interior had declared that the CAP would only be approved if the State of Arizona committed to a legislative plan to restrict agricultural irrigation methods involving the pumping of groundwater. Under the aegis of Governor Osborn, the Groundwater Code was passed in Arizona in 1948. The Code forbade the expansion of agriculture that was irrigated using groundwater without resolving existing problems regarding the pumping of water or the quantity of water allocated to landowners, or resolving the issue of the overexploitation of groundwater. In spite of federal pressure, the successive governors of Arizona were confronted by strong opposition within their state from mining companies, farmers and city governments which were unhappy about rigorous legislation and who claimed exceptional rights to administer their own water. This coalition had every chance of success in a context in which the Arizona Supreme Court swayed between contradictory decisions (groundwater as public property *versus* rules governing the use of groundwater according to the needs of landowners). Legislators were constrained to a "policing role", which created a legislative impasse because of a lack of political unity regarding the issue. Although he had supported the CAP as a Congressmen a few years before, Governor McFarland (Democrat), elected in 1954, introduced no legislation on the subject (*Mann, 1963*), as if he were somehow echoing dominant interests that had little time for federal grievances.

However, the more support the CAP gained in Washington, the more the local decision-makers were obliged to legislate the use of groundwater. This development threatened the very existence of the project. In the context of the 1970s, increasing criticism of major projects, especially in regard to their impact on the environment and their prohibitive cost, was enough to call federal funding into question. Moreover, the continuing urbanization of Phoenix and Tucson exacerbated competition for water among farmers, mining companies and cities. A coalition of new local elites encouraged the State of Arizona to change its policy on groundwater, which was the *sine qua non* for the CAP to benefit from federal funding. This change was, doubtless, also rendered possible by the fact that states in the federal system were becoming increasingly influential in terms of controlling and implementing water policies at a time when federal funding was increasingly hard to obtain as policies announcing the end of the Welfare State were introduced. These structural transformations affecting political power in Washington, where the beliefs of the dominant coalition were undermined by the environmentalist cause and the all-conquering advances made by neoliberalism, created the preconditions for a realignment of existing coalitions within the field of a local power configuration that was itself characterized by new urban paradigm.

The Critical Groundwater Code, promulgated in 1948 under the aegis of Governor Osborn, did not address the issue of supplying the two cities with water. In the late 1960s and early 1970s, differences persisted in approaches to the question. While the pumping of groundwater was generally limited to the area covered by a single supplier, the resource was sometimes pumped beyond these parameters to urban areas for both industrial and domestic uses. This kind of water distribution created tensions that were resolved between 1969 and 1974 in a series of legal cases brought against the City of Tucson and decided by the Supreme Court of Arizona. Tucson was also forbidden to transport groundwater from wells in the Avra and Alter Valleys, designated as critical areas. For the Arizona Supreme Court, property rights in groundwater were linked exclusively to rights of use. Those rights were limited to "reasonable use" and were not associated with ownership of the resource.

The issue was not limited to Tucson, and other disagreements between local farmers and mining companies soon emerged. In 1976, the Farmers Investment Company (FICO), a large pecan producer located in the Santa Cruz Valley to the south of Tucson, brought a case against the copper mining company Anamax that was ultimately decided by the Arizona Supreme Court. This case has often been considered the trigger for the Groundwater Management Act of 1980. In effect, the problem was that several mining companies to the south of Tucson pumped water from an area considered as critical (Sahuarita-Continental Critical Groundwater Area), and transported it to their own sites outside that perimeter. FICO, which owned nearly 7,000 acres of farmland, located in that area, took the view that the mining companies were not respecting the law and had broken rules governing the use of water when it transport the water beyond authorized areas. The mining companies argued, on the contrary, that it was necessary to precisely define the area from which the water was pumped, since the water they used came from the same aquifer as the water used by the farmers.

The mining companies also claimed that pumping groundwater did not cause aquifers to diminish as long as the water was used and replaced. The City of Tucson also got involved in the controversy, claiming that FICO and the mining companies were polluting the groundwater in the basin from which the city extracted most of its water. The mining companies countered that, contrary to law, the City of Tucson transported groundwater a long way from its basin of origin. Citing rights of use, the Arizona Supreme Court found in favor of FICO and against the mining companies, but nevertheless ruled that it was incumbent on the legislature to define rights based on economic interests and decide whether it was in the interest of the state to encourage mining to the detriment of agriculture. It was based on this ruling that the mining companies and the City of Tucson formed a coalition against the farmers with a view to encouraging the legislature to undertake a legal reform of regulations governing the transportation and use of groundwater.

In the past, the mining companies and FICO had often clashed in Tucson over water management, but it seemed that interests had evolved (*Connall, 1982*). Mining companies and City Hall both wanted to change the law governing the transportation of groundwater. For example, the monopolistic position of farmers (who consumed 89% of water supplies) encouraged City Hall to take a conservationist stance. On the other hand, farmers criticized the mining firms and the cities as being openly hostile to agriculture. During the arduous negotiations, organized with a view to resolving these problems, each party was represented by a different lawyer: agriculture by Jon Kyl (the Salt River Project's lawyer), Mark Wilmer (lawyer for FICO), and Brock Ellis (a lawyer from Phoenix). The mining companies were represented by James Johnson and James Bush (from Phoenix), and Thomas Chandler (from Tucson). Furthermore, the spokesmen for the cities were Jack DeBolske, Director of the League of Arizona Cities and Towns, and Bill Stevens, a Phoenix lawyer.

Mark Wilmer, lawyer for Arizona against California

Mark Bernard Wilmer was born in July 1903 in Wisconsin. Son of a farmer, he grew up in the small town of East Troy, essentially a community of farmers, dairy workers and shopkeepers. Attracted to literature and to the new, increasingly urban America, the embryonic growth of which he saw every morning on his way to Burlington College, he developed a genuine desire for mobility, encouraged by his father, who wanted him to continue his university studies. In 1926, he gained a place at the College of Law at the Georgetown University in Washington D.C. Two years after graduating in 1929, he qualified for the State Bar of Arizona in May 1931.

In the early days of his career in Phoenix, his first legal case concerned disputes between the Roosevelt Water Conservation District and the Salt River Project. He did not yet know that water would become an important part of his legal career. Indeed, the complex problematic of regional rights, going beyond the merely local framework, placed him at the center of the political, environmental and legal history of the West in the 20th century. At the time, the dispute between Arizona and California was only just beginning, and he was still to establish a professional reputation. In the 1930s, his legal portfolio was essentially made up of criminal cases. He was a litigator. The skills he displayed in the local courts impressed the region's more politically-minded judges. He was invited by the County Attorney of Maricopa County and the Attorney General of the State of Arizona to help them in their work. During this period, he met Frank L. Snell, a well-known, well established attorney, a graduate of Kansas School of Law, based in downtown Phoenix, with whom he set up the Snell and Wilmer Law Firm.

Between 1940 and 1950, Snell & Wilmer became one of the largest law firms in the American West. Later, it provided a wide range of services and expertise and considerable resources to its clients, with 400 lawyers in six offices in Arizona, California, Nevada, Colorado and Utah. Due to its size, the firm helped to shape Arizona's political and economic agenda. And when Governor McFarland of Arizona, at a time when the legal battle against California seemed to have been lost, went in search of the best litigator in the state, the Association of the Bar of Arizona and the legal community as a whole recommended Mark Wilmer. In 1957, at the height of a remarkable legal career, Wilmer accepted the challenge of defending Arizona's case against California, which he did until the Supreme Court delivered its decision in 1963. By accepting the case, Wilmer entered a world defined by water (or the lack of it), as well as by culture and tradition. He familiarized himself with the legal case, the rulings of the courts, and cross-border conflicts, imbibing as much information on the state's most valuable resource as he could and defining his relationship with those against whom he applied his defensive strategy on behalf of Arizona against California.

The legal heritage of Spain and Mexico that gives primacy to *prior appropriation* (see *supra* chapter 4: 46) was not exclusively applicable to Arizona and did not only govern local water supplies. Wilmer quickly realized that this legal doctrine was an issue for the states of the West who coveted the Colorado River, leading to serious disputes, notably between Arizona and California, whose explosive growth at the beginning of the 20th century constituted a regional threat in terms of rights to the Colorado River. He discovered the degree to which California was capable of influencing the water policies of the West; indeed, the Imperial Irrigation District was one of Wilmer's most powerful legal adversaries. In 1911, Phil Swing, Imperial Valley's lawyer, who later converted his legal influence into a seat in the House of Representatives, along with the real estate developer, Mark Rose, set up the Imperial Irrigation District (IID) which lobbied Congress and the Secretary of the Interior to develop water projects (notably, the All American Canal). This was part of a race to acquire as many prior rights to the Colorado River as possible for California. The coalition between the (IID) and the federal department set alarm bells ringing in the Arizona congressional delegation and alerted other states with an interest in the river. Meanwhile, the City of Los Angeles attempted to secure increasing volumes of water to fulfill needs associated with its rapidly expanding population. Like Arizona, the Upper Basin States – Wyoming, Utah, Colorado and New Mexico – saw Southern California, with its urban centers and agricultural users, as a threat to their future development. Prior appropriation had to be fought.

As Mark Wilmer discovered in the Congress archives, the tensions between California and the other states of the West were at the origin of legal actions, which led to the promulgation of the Colorado River Compact of 1922, and the Boulder Canyon Act of 1928. During this period, the Governors of Arizona were strongly opposed to this legislation, which they regarded as being overly favorable to California. They took legal action defending state rights to the water of the Colorado River against the federal government and the Golden State. Forty years later, however, Wilmer used the legislation as a guideline in his legal strategy in favor of Arizona at the Supreme Court. In spite of the fact that it sanctioned the construction of the Hoover Dam, the legislation abrogated the principle of *prior appropriation* and provided an annual allocation of water from the Lower Colorado River Basin as follows: California, 4.4 million acre-feet; Arizona, 2.8 million acre-feet; Nevada, 300, 000 acre-feet.

The *Law of the River* included legislation passed by Congress. The laws, exclusively concerned with the distribution of water from the Colorado River, excluding its tributaries, provided the keystone of Mark Wilmer's legal defense system. Wilmer attempted to convince the Supreme Court by using arguments derived from legislative history and previous debates. California, which denied Arizona any legal right over the Colorado River, took the position that the desert state already benefitted from water from its tributaries (the Gila River and the Salt River). California's lawyer, Northcutt "Mike" Ely based his case on legal history, citing the Supreme Court's decision of January 5, 1922, on a dispute between Wyoming and Colorado concerning a tributary of the Colorado River applying the priority rule to rivers running through two states. Finally, on June 3, 1963, the Supreme Court validated Mark Wilmer's strategy: "*We are persuaded by the legislative history as a whole that the Act was not intended to give California any claim to share in the tributary waters of the Lower Basin States. […] Where Congress has exercised its constitutional power over waters courts have no power to substitute their own notions of "equitable apportionment" for the apportionment chosen by Congress*" (August, 1999: 89).

Legislative activity was encouraged above all by debates about the CAP taking place at the same time in Washington. In effect, the environmental impacts of the project were increasingly being called into question, and the existence of the project itself was threatened by the Carter administration on the grounds of costs. To ensure that it was not deprived of all federal funding, the government posed a precondition (relayed by the Secretary of the Interior of the time) that the State of Arizona had to introduce a new legal framework for regulating the management of groundwater. Faced with an impasse in negotiations at the local level, the Democratic Governor of the time,

Bruce Babbitt, who was from the West (see *supra* chapter 4: 53–54), assisted by the Republican Senator, Stan Turley, introduced a new dynamic by serving as a mediator between the cities, the mining companies and farmers, a commitment that was doubtless linked to his university career (he had graduated from Harvard Law School) and professional background (as a former Attorney General), combining law and politics, both of which were useful in reconciling a respect for the law while at the same time taking local issues into account.

These negotiations were based on a new belief in the need to seriously legislate groundwater in order to guarantee that the CAP was financed. They culminated in the promulgation, in 1980, of the Groundwater Management Act, which limited rights to use groundwater in four areas (Phoenix, Tucson, Prescott and Pinal County), referred to as "Active Management Areas".[1] In the end, the CAP did not lose its federal funding. It should be noted that realty promoters, the Chamber of Commerce, and private water companies were not invited to the negotiating table (*Connall, 1982*), even though the urban development of the region – particularly the sale of properties – appeared to be a decisive factor.

Consensus was actively sought at the local level: FICO provided its support; decision-makers in Tucson attempted to maintain a belief in the necessity of the project – City Hall voted on funding for the CAP Association on several occasions – and the Chamber of Commerce lobbied in favor of the CAP (*Dames and Moore Inc., 1995*). However, at the same time, two agrarian economists from the University of Arizona broke rank by rendering public the fact that farmers in Arizona were financially unable to buy CAP water. While they were later forced to leave their university positions, their actions nevertheless marked the emergence of local protest against the CAP project.

2.2 Urban development, conflicts over water and the realignment of coalitions

The arrival of the CAP in Tucson illustrates the fact that if a city in Arizona wanted water, it had to buy it from water transport infrastructure projects and not only pump groundwater. The event marked a new stage in water policy: *"the farmers got established in the central part of the state because of the Salt River Project. The cities grew up in the middle of the farmland. The real estate interests, the money people – they are all in Phoenix and Scottsdale and Tucson. They didn't want to move. So we're going to move the river to them. At any cost."* (*Reisner, 1986: 305*). Evidently, the urban factor was not new. From the 1920s onward, the spatial expansion of American cities had been promoted by coalitions of developers keen to boost the real estate market, helped in this not only by the growth of the automotive industry, but also by policies designed to subsidize consumption and by federal support for social housing (*Gonzales, 2009; Wiley and Gottlieb, 1982*). What changed was the political influence of the cities on the coalitions formed around water issues. Responsibility for water distribution, which had initially been guaranteed by private companies in the

---

1 Active Management Areas & Irrigation non-expansion areas: http://www.azwater.gov/AzDWR/WaterManagement/AMAs/

late 19th century passed, in the 1920s, to public administrations run by city mayors whose goal was to extend the network and guarantee quality. After the Second World War, the growing needs of the cities meant that other sources of supply were required, a situation that relied on federal aid. Thus, in 1946, while, at a time when groundwater was increasingly being used to supply expanding cities – mainly Phoenix and Tucson (*Kupel, 2003: 153*), the Central Arizona Project Association was founded. To the degree that drawing off water destined for agricultural purposes, especially in regard to the Salt River Project serving Phoenix, was a delicate strategy, local politicians became fervent advocates of the CAP, in spite of federal norms and pressure exerted by environmentalist movements.

The Groundwater Management Act of 1980 (GMA) reorganized the distribution of water in several zones referred to as Active Management Areas (AMA), not only instituting an innovative style of managing groundwater based on basins, but also introducing limits to the expansion of irrigation. However, the system was difficult to implement: each city located in one of the zones had to use CAP water for a period of 100 years. Later, the introduction of the Central Arizona Groundwater Replenishment Districts (CAGRD) gave real estate developers the chance to introduce subdivisions in places in which there were no direct connections to CAP water (*Valdez Diaz, 1996; Blomquist et al., 2001: 662*). Another measure introduced by the Groundwater Management Act of 1980 was the creation of the Arizona Department of Water Resources (ADWR) to replace the Arizona Water Commission. This ensemble of institutional and economic measures, introduced by the State of Arizona and its governor, Bruce Babbitt, encouraged the Department of the Interior and the Bureau of Reclamation to make the CAP one of its priorities in the 1980s (*Colby & Jacobs, 2007*). These new approaches to water management help to explain how the arrival of CAP water in Tucson marked a displacement of conflicts at the city level and why it did so much to reshape local political coalitions.

In 1989, construction work started on a CAP processing plant in Tucson, as well as on the Clearwell Reservoir, designed to stock water treated with a mixture of ozone and chloramines, a process considered by some as at best experimental, but by others as "progressive" (*Kupel, 2003: 193*). CAP water was finally delivered on October 5, 1992, but numerous problems came to light over the next few years. In the first few months, leaks were detected in the pipes, and the press reported complaints against the "water mafias" and corrupt bureaucrats who swamped their fellow citizens with toxic substances from the Colorado River. Among the opponents of the CAP water delivered by Tucson Water throughout the 1990s, Richard Wiersma was the spokesman of the Citizens Voice to Restore and Replenish Quality Water. "*In 1992, disaster struck Tucson. The water utility began serving chemically treated CAP water to half its customers, and the switchboards lit up with complaints almost immediately. This brown, foul tasting, highly corrosive water destroyed plumbing and appliances. It killed pets and plants, and caused rashes and allergic reactions in people*" (*Wiersma, 1995: 1*).

When CAP water arrived in Tucson, it destroyed pipes and flooded private homes. In 1995, 20,500 complaints were made against Tucson Water due to leaks (*Dames & Moore Inc., 1995: xvii*). Popular trust in the city's leaders was rapidly eroded and strong opposition emerged. For example, Bob Beaudry became one of the most vehement activists and critics of the use Tucson Water made of CAP water. Moreover, the water's brown color, which was due to a slightly acid PH and corroded pipes (*Dames &*

Moore Inc., 1995: vii), added to the local people's dissatisfaction with to the political leaders. It should be noted that, at the time, Tucson's infrastructure system was one of the oldest in the state (*Dames & Moore Inc., 1995; Kupel, 2006; Colby & Jacobs, 2007*). The lack of maintenance of this infrastructure was at the root of most of the damage caused by CAP water and had triggered a high degree of mistrust of the state and the federal government, a mistrust which informed their oppositional strategies.

Between October 1992 and October 1993, Tucson City Council was led by the Democrat Molly McKasson, of Ward 6. It was in City Hall that a number of coalitions promoting water quality in Tucson emerged, among them Citizens for Water Protection, headed by Molly McKasson; Citizens Voice to Restore and Replenish Quality Water, whose president was Richard Wiersma; as well as lobbies headed by Ed Moore, the Republican Supervisor of Pima County, Bob Beaudry, the Tucson entrepreneur and car salesman, and Gerald "Jerry" Juliani, who promoted the idea of holding a referendum on water quality. These coalitions acted against a Pro-Development Coalition centered on Jim Click, the local car salesman, Chuck Freitas, Director of the Safe and Sensible Water Committee, and the Southern Arizona Leadership Council. These campaigns in favor of water quality in Tucson resulted in a number of decisions being taken by City Hall in October 1993, including the withdrawal of CAP water from the city's pipes, and limiting CAP water to the west of the city, since the majority of complaints had been made by residents of the east (*Editors, 1995; Pitman, 1997; Kupel, 2003*), who were excluded from the supply zone. In this context, Mike Tubbs, the head of Tucson Water, resigned.

By 1994, most of the administrative agencies had become aware of the situation in Tucson. Reports indicate that most difficulties caused by CAP water derived from obsolete infrastructure, mostly in the east of the city. CAP water was entirely dissolving the solid material (*Kupel, 2003*) produced by corrosion in the pipes, thus causing major damage. Anxious to keep the federal government out of the dispute by respecting provisions of the Safe Drinking Water Act (SDWA) and other legislation in the field, the Arizona Department of Environmental Quality (ADEQ), an administrative agency authorized by the Environmental Protection Agency to apply the SDWA, levied a fine of 400,000 dollars against Tucson Water in November 1994 on the grounds of having failed to respect drinking water quality regulations by testing wells inadequately. In a declaration made after the fine was announced, the ADEQ attempted to reassure those concerned, claiming that the CAP situation was under control (*Newman 1994*). Furthermore, Tucson Water adjusted its prices in an attempt to compensate for the fact that the water was brown and contaminated (*Moore, 1995; Kupel, 2003: 194*). Tucson Water, then directed by John S. Jones, was forced to stop using CAP water due to maintenance issues associated with draining the canal (*Newman, 1994: 1; Kupel, 2003: 194*). Then, still in 1994, Tucson Water appointed a new president and vice president.

If such institutional dysfunctions occurred, it is because the majestic Colorado River was simply too powerful for Tucson's aging infrastructure. The temporary closure of the CAP enabled the coalitions headed by Molly McKasson and Richard Wiersma to lobby the city government on behalf of the Citizens Water Protection Initiative (Proposition 200). The proposition, which achieved a 57% majority among the city's voters (*Chesnick, 1999: 4; Kupel 2003: 194*), was voted through the legislative system in the form of the Water Consumer Protection Act (WCPA). The Act outlawed chloramines

> **Molly McKasson and the Water Consumer Protection Coalition**
>
> In Molly McKasson's family, an interest in local politics was cultivated from one generation to the next. In an interview with the Pima County Oral History Project, Molly recalled that her mother had become politically active as a protestor of the Vietnam War. She encouraged her daughter to remain faithful to her values, but always to be willing to change her opinions in line with any new knowledge she acquired. In the 1990s, Molly McKasson focused on improving the quality of life of Tucson's citizens by promoting initiatives like the WCPA. Furthermore, thanks to the political and social satires she delivered in Tucson theaters, and by attending political rallies in Phoenix, she was able to better define her beliefs and make important political contacts.
>
> In 1991, five different associations in the city supported Molly McKasson in her campaign to become councilor for Ward 6. At the time, she was in the process of emerging as a public figure in the theater and writing articles for the local papers. The fact that she was not a career politician was one of the factors that enabled her to gain support. As McKasson recalled in her interview with the Pima County Oral History Project, *"in 1993, the pipes broke and a firestorm of anger was created and people felt ignored. Growth was on the front burner and quality of life on the back"* (McKasson, 2011). She claimed that development and planning initiatives did not take long-term concerns into account. In Tucson, greater emphasis was placed on growth and urban development than on increasing living standards.

and ozone, which Tucson Water had previously added to the supply. Mention was made of the fact that, unless the company could ensure that the quality of its water was as high as that of groundwater, the law would be subjected to another vote in five years (*Valdez Diaz, 1996; Kupel, 2003*). In the end, Tucson City Hall proved to be the main link between local people and government bureaucracy. From the outset, Michael Brown, Mayor of Tucson, Molly McKasson, city councilor, and Michael Tubbs, head of Tucson Water, organized task forces to study the CAP and examine social responses to it (*Dames & Moore Inc., 1995: 10*). In fact, in 1994, the City of Tucson hired a consulting firm (Dames & Moore Inc.) to find solutions to the city's water problems.

In 1994, McKasson and Wiersma began to collect signatures supporting the Water Consumer Protection Act initiative in view of the 1995 vote in City Hall. Among the opponents of the CAP, Gerald "Jerry" Juliani can be considered as "the Pro-Prop 200 Group's researcher". From 1993 to 1999, he wrote a number of articles for local papers arguing that protective measures should be introduced on behalf of water consumers, as did other members of the Water Consumer Protection Coalition. Between 1997 and 1999, he was spokesman for the Pure Water Coalition (*Nitzel, 1999: 1*). On the other side of the fence, a Pro-Development/Pro-CAP coalition began to take shape in 1997. It included the Southern Arizona Leadership Council (SALC), a group criticized for being "a Republican aristocracy made up exclusively of white men", but which represented local business leaders anxious to defend strong values and contribute to the development of urban areas. The SALC made substantial contributions to causes that they believed would aid economic growth and urban development, including 1997's Proposition 200, which advocated a return to CAP water for Tucson. Thus, Bob Beaudry and Jim Click, rival car salesmen, clashed throughout the 1990s over water quality and CAP-based legislation, using their financial resources to support their initiatives until a decision was finally taken in late 1999 to reinstate CAP water. From the beginning, Jim Click played a central role in the Tucson dispute. He started out his career in Los Angeles, working at a car dealership before buying his own in Tucson. He was a generous donor to numerous political organizations, citizen movements, the University of Arizona, and various causes involving health and welfare (*Press release, 2014*).

In 1997, Proposition 200 was approved by a 59% majority. The SALC then became actively involved in supporting Proposition 201 (1999). Jim Click came out against the Coalition for Adequate Water Supply, making a contribution of 30,000 dollars. Another local entrepreneur, Karl Eller, president of the Circle K convenience market chain, made total contributions of 25,000 dollars to the 1997 and 1999 campaigns (*Beaudry, 1999*). Closely involved in the world of business and politics in the state for decades, he played a pivotal role in encouraging the Phoenix Suns NBA franchise to come to Arizona. Meanwhile, the University of Arizona recognized his efforts by appending his name to the School of Management and the name of his wife to the campus-based Dance Studio (*Wang, 2012*). It is reported that the Pro-Development Coalition spent over 1 million dollars bringing CAP water to Tucson (*Davis, 1999: 3*). In 1999, Tucson Water's "Ambassador Program" was designed to instill new faith in the CAP. A number of homes on East Fourth Street, between Craycroft Road and Wilmot Road, were supplied with a mixture of groundwater and CAP water (*Chesnick, 1999*). The Pro-Development Coalition distributed bottles of this mixture at football games and in shopping malls to enable the people of Tucson to form their own opinions.

According to the SALC, the 1999 vote proved a great success with the election of Bob Walkup, who defeated Molly McKasson to become Mayor of Tucson. This was one of the "clear signs" that strong leadership was emerging in Tucson's business community (*Southern Arizona Leadership Council, 2014*). The SALC stated that this this did not represent "*a business community that seeks unfettered growth and development ... but a business community that, for once, saw that it needed to be involved in creating Tucson's future.*" The new mayor, Bob Walkup, a retired aerospace executive, declared that "*water is our No. 1 problem ... the vitality of our community demands a solid, long-term water policy and everybody knows this*" (*Davis, 1993: 3*). In 1999, Mitch Basefsky, now spokesman for the CAP, but previously spokesman for Tucson Water, added that "*the majority in this community were willing to accept CAP water. They understood this was an alternate resource that we had to make use of in order to sustain our environment*" (*Chesnick, 1999: 2*). The CAP was supposed to solve the problem of the overexploitation of groundwater and the risk of shortages that it involved (*McKasson & Devine, 1998*). Instead, it seems to have generated additional tensions between economic leaders, citizen organizations, local politicians and Tucson Water.

3 CONCLUSION: COALITIONS AND DECISION MAKING

The history of water policies draws many lessons for a sociological analysis with regard to environmental issues. First of all, history shows us that water policies implemented in the West are the product of specific and temporary coalitions between economic elites and elites from the political and administrative spheres, who regard any water use project as the engine for economic development. Second, this conception of water use has to do with a territorial structure of political power in which state interests are crucial to shaping the very type of projects likely to be implemented and selecting their location. Third, the way in which water policy has been structured is specific in a sense that its organization is part of a multi-layer institutional framework, from which policy responsibilities are divided into different management bodies, ranging from a federal level to a municipal one. Furthermore the high cost of water megaprojects such as the CAP tends to dictate not only the shape of management policies but also the strategies to be adopted in the face of the lasting drought situation affecting the American West for many years. In fact, local authorities that bear the burden of paying the building cost of such a megaproject have no choice but to honor that debt through a water intensive sale to users who depend on surface water. Indeed, increasing the amount of water sold brings broader benefits for the federal reimbursement, which pushes local decision makers towards leading change or innovation through a headlong rush into unbridled urban sprawl and economic growth.

REFERENCES

August, J.L. (1999). *Vision in the Desert. Carl Hayden and Hydropolitics in the American Southwest*. Fort Worth: TCU Press.

August, J.L. (2007). *Dividing Western Waters. Mark Wilmer and Arizona v California*. Fort Worth: TCU Press.

Chesnick, J. (1999, October 25). Special Report: Prop. 200. *Tucson Citizen*. Retrieved from http://tucsoncitizen.com/morgue2/1999/10/25/227787-special-report-prop-200/.

Beaudry, B. (1999, October 18). Yes On Prop 200. *Tucson Weekly*. Retrieved from http://www.tucsonweekly.com/tucson/yes-on-prop-200/Content?oid=1065474.

Blomquist, W., Heikkila, T. and Schlager, E. (2001). Institutions and conjunctive water management among three western states. *Natural Resources Journal*, 41(3): 653–683.

Colby, B.G. and Jacobs, K.L. (Eds.). (2007). *Arizona Water Policy. Management Innovations in an Urbanizing, Arid Region*. Washington: RFF Press.

Connall Jr, D.D. (1982). History of the Arizona Groundwater Management Act. *Arizona State Law Journal*, 2: 313–343.

Davis, T. (1999, October 25). Water Starts Fires in Tucson Election. High Country News. Retrieved from https://www.hcn.org/issues/166/5359.

Dames and Moore Inc. (1995). CAP Use Study for Water Quality: A Review of CAP-Related Decisions from 1965 to present. Retrieved on October 3, 2014 from http://www.savethesantacruzaquifer.info/CAPQualityReport.pdf.

Hundley, N. (1975). *Water and the West. The Colorado River Compact and the Politics of Water in the American West*. Berkeley: University of California Press.

Ingram, H. (1990). *Water Politics. Continuity and Change*. Albuquerque: University of New Mexico Press.

Johnson, J.W. (2002). *Arizona Politicians. The Noble and the Notorious*. Tucson: University of Arizona Press.

Johnson, R. (1977). *Central Arizona Project (1918–1968)*. Tucson: University of Arizona Press.

Mann, D.E. (1963). *The Politics of Water in Arizona*. Tucson: University of Arizona Press.

McKasson, M. and Devine, D. (1998, June 25–July 1). Water Log: Think those two votes on direct delivery of CAP water meant something? Think again-CAP water is almost certainly headed back to your tap. *Tucson Weekly*. Retrieved from http://www.tucsonweekly.com/tw/06-25-98/feat.htm.

Newman, H. (1994, June 28). City to settle water suit. *Tucson Citizen*. Retrieved from http://tucsoncitizen.com/morgue2/1994/06/28/180143-city-to-settle-water-suit/.

Press release (2014, January 23). Andy Tobin Raises More Than $232,000. *Sonoran Alliance*. Retrieved on March 6, 2015 from http://sonoranalliance.com/tag/jim-click/.

Reisner, M. (1986). *Cadillac Desert. The American West and its Disappearing Water*. New York: Penguin Books.

Sheridan, T.E. (2012). *Arizona. A History*. Tucson: University of Arizona Press.

Thomas, C.S. (Ed.). (1991). *Politics and Public Policy in the Contemporary American West*. Albuquerque: University of New Mexico Press.

Udall, S.L. (1963). *The Quiet Crisis*. New York: Rinehart and Winston.

Valdez Diaz, C. (1996, April 22). City studies water supply options. *Tucson Citizen*. Retrieved from http://tucsoncitizen.com/morgue2/1996/04/22/39002-city-studies-water-supply-options/.

Wang, A. (2012, January 30) Arizona's been good to billboard entrepreneur Karl Eller. *AZcentral.com*. Retrieved from http://www.azcentral.com/business/news/articles/2011/11/28/20111128arizona-been-good-billboard-entrepreneur-karl-eller.html.

Wiersma, R. (1995, November 2–8). Vote yes on Prop 200. *Tucson Citizen*. Retrieved from http://www.tucsonweekly.com/tw/11-02-95/cover.htm.

Willey, P. and Gottlieb, R. (1982). *Empires in the Sun. The Rise of the New American West*. Tucson: University of Arizona Press.

Chapter 7

The making of water policy in the American southwest: Environmental sociology and its tools

Franck Poupeau
UMI iGLOBES, CNRS/University of Arizona, USA

Murielle Coeurdray
UMI iGLOBES, CNRS/University of Arizona, USA

Joan Cortinas
UMI iGLOBES, CNRS/University of Arizona, USA

Brian O'Neill
Water Resources Research Center, UMI-iGLOBES, University of Arizona, USA

> "This was the American Dream: a sanctuary on earth for individual man: a condition in which he could be free not only of the old established closed-corporation hierarchies of arbitrary power which had oppressed him as a mass, but free of that mass into which the hierarchies of church and state had compressed and held him individually thrilled and individually impotent."
>
> —William Faulkner (1965: 62)

MANUFACTURING THE DESERT

Since the 1990's, the environment has become a subject of central importance in humanities and social sciences – interest in political ecology and environmental ethics has grown dramatically in both the United States and Europe (*Hache, 2012; Kalaora & Vassopoulos, 2014*), environmental history has become a vast field of study, at once structured and diversified (*Lochet & Quenet, 2009*), political science is investigating approaches to the governance of resources (*Olstrom, 1990*), and "instruments" have been implemented by institutions responsible for the environment (*Lascoumes, 1994; Lascoumes & Le Galès, 2004*). In particular, the field of environmental sociology has made important advances, focusing particularly on issues related to environmental justice in regards to the most vulnerable social groups, e.g., migrants (*Gagnon, 2008; Park & Pillow, 2002*). Whereas research was previously the monopoly of federal agencies and Californian "think tanks," an increasing number of researchers began to explore the impacts of the ongoing ecological transitions, and possible adaptation to these rapid changes.

Of particular interest for field of an environmental sociology is the western United States, a region whose economic development was historically based on the exploitation of its natural resources (*Pincetl, 2011*), and which now faces the combined effects of climate change and urban sprawl. First, the management of environmental issues in this region raises the question of the modes of domination associated with the unequal distribution of natural resources and their transformation into basic services via major technological systems. This transformation has been made possible by the construction of contemporary cities and metropolitan areas along with their "networks of power" (*Tarr & Dupuy, 1988*). In effect, the region has witnessed the development of massive infrastructures (dams, aqueducts, etc.) designed to encourage the large-scale irrigation of agricultural land and the expansion of urbanized regions, especially since the Federal Reclamation Act of 1902. And, beginning in the 1930s, due to public policies developed in the urban East (particularly programs developed by federal agencies), the installation of technical systems gradually transformed the West into the breadbasket of the East (*Cronon, 1992*), and subsequently into an autonomous center of economic development (*Leslie, 2005*).

Second, this process of construction was not without its ups and downs, and conflicts over the distribution of water among valleys, cities, states, and professional sectors were commonplace. Centralized decision-making encountered powerful collective resistance, ranging from community-based protest to the environmental movement and legal and administrative action. Importantly, the *"water wars"* that began in the late 19th century present a large field of research that has been explored with great fervor by environmental historians, geographers and sociologists such as *Reisner (1985), Worster (1985, 1992), Pisani (1984, 2002), Gottlieb & Fitzsimmons (1991), Walton (1993), Espeland (1998), Erie (2006)* and *Gottlieb (2007)*.

Third, the aforementioned studies on conflicts about water provide not only an image of the way in which the environment is regulated in the western United States, but also, above all, of the American social sciences themselves. Indeed, this body of research emphasizes power relations and acts of resistance, and demonstrates that the West cannot be considered an empty, arid tourist destination in which pioneers searched for a new frontier in order to realize the "American Dream". Instead, it reveals that the scarcity of water is at once a resource and a problem or, more precisely, the product of multiple appropriations by groups – social organizations, local decision-makers, and federal elites – struggling for control of both water and power.

Contrary to the enchanted vision of the "American Dream," it is of course possible to question, as did Marc Reisner in *Cadillac Desert (1985: 481)*, whether it was really necessary to build such large dams (as opposed to off-stream reservoirs), or to irrigate relatively unfertile land instead of developing a less extensive form of agriculture. Nevertheless – as indicated by the title of the last chapter "A Civilization, if You Can Keep It" – this vision of a forthcoming environmental, and even civilizational, catastrophe, leaves unaddressed the question of the logic underlying the approaches at play as one considers the field of water management and policy. Reisner's narrative of the various companies involved in the field delivers a somewhat absurd vision that is devoid of meaning or, more precisely, of motivation. Politicians, administrators, engineers and landowners fighting for or against major projects are either cast as cynics corrupted by power and money, or as "political idiots" who fail to understand the implications of their actions, namely the construction of infrastructure and the kind of economic development enabled by it. They are portrayed either as traitors to the "American Dream" or as fanatics of technological modernization.

The narrative provided by *Cadillac Desert* does indeed illustrate the brute, and indeed brutal force of economics. However, it provides little understanding about how the physical construction and management of rivers and dams is also a symbolic construct that goes well, beyond mere personal interest, to involve a collective struggle over the principles informing a legitimate *vision and division* of the world and its development. The account provides, therefore, an interpretative framework that integrates various levels of analysis, notably relations of domination at the intersection of local powers, and the fields of economics and bureaucracy, that need further elaboration with a view to reconstituting the chronology of individual environmental events, and to illuminating their underlying social logic.

This chapter does not present a well-defined and identifiable corpus, and still less a "school" (*Topalov, 2003*). Instead, it provides a kind of "enlarged case study" (*Burawoy, 2009*), that highlights the analytical principles applicable to a sociology of the environment: a sociology that takes into account the multiple layers of institutions and instruments which fashion the environment as a "natural problem." And since water regulation cannot be reduced to a problem of institutional management alone, it also takes into account the ways in which conflicts are inscribed in spaces of power, and in which local power centers and multiple economic forces are confronted with a bureaucratic field that is at once centralized and decentralized.

I WATER IN THE DESERT

Few regions are associated with images as idealized as those linked to the desert landscapes of the American West. The region's spectacular landscapes constitute the incarnation of the Wilderness considered as an ideology structuring the relationship between North Americans and their natural environment (*Nash, 1967*) – even if the vision of nature of the first pioneers was surely not so romantic. These landscapes represent a "frontier," an external territory that society attempts to circumscribe by introducing parks and conservation policies. While the American West is a desert, it is nevertheless a desert that is "inhabited," or, in other words, "prey" to a growing urban expansion and that is now encountering the effects of climate change. As Edward Abbey wrote in *Desert Solitaire*, one of the founding works of political ecology: "*Water water water… There is no shortage of water in the desert but exactly the right amount, a perfect ratio of water to rock, of water to sand, insuring that wide, free, open, generous spacing among plants and animals, homes and towns and cities, which makes the arid West so different from other part of the nation. There is no lack of water here, unless you try to establish a city where no city should be…*" (Abbey, 1968: 130).

The ultimate site of ecological nostalgia and conservationism (*Jacoby, 2003*), the West is also the birthplace of American anthropology in the 19th century, as unknown lands were explored and cartographic missions organized, an interest in local native populations developed (*Fowler, 2010*). Information gleaned in the various exploratory missions to the Colorado River, the Green River and the Grand Canyon undertaken by John Wesley Powell around 1879 was later used in the planning of major hydraulic projects in the region on the basis of a form of science piloted by the federal government (*Stegner, 1953; Worster, 1992*). Whereas the nation's land management was led by a cadastral approach, which did not recognize topography or resources, Powell's suggestions about how to manage water in the West was based on the concept of the watershed (*Pincetl, 2011*). In the following decade, Powell was appointed director of not only the US Geological Survey, but also the Bureau of Ethnology at the Smithsonian Institution in Washington DC, where he developed, before Stuart Hall, an interest in Native populations, but where his position was somewhat weakened by the maneuvers of Senators defending "local interests" against federal projects. He was probably the first to define the West as the "arid region" of the United States at a time when the area still represented a "frontier" and a dream of freedom.

This "Paradox of the American West," to quote the historian *Donald Worster (1985, 2001)*, raises a series of questions concerning the impact on natural resources of aridity and the use made of the resources, as well as on ways of living and the institutions regulating economic and social life. How did people who had taken possession of the land adapt to ecological conditions? Achieving that promised freedom was consistent with a desire to transform nature, inherited from the colonial project over the Americas:

> "Those five centuries of learning about the Colorado (actually most of the learning was concentrated in about five decades) seem almost never to have raised in people's minds a simple question: What changes in society would be required to master the Colorado's course? Only Powell gave the matter much consideration; he recommended that both big government and big business stay away from the river, that ordinary settlers be encouraged to organize themselves into 'cooperative commonwealths' and set about in their own way to make use of the watershed. Had his suggestion been followed, development would have been confined to the smaller side streams and upcountry valleys, for the main river was too strong for any local group of settlers, with limited capital and expertise, to harness" (*Worster, 1985: 67*).

In just a few sentences, Worster paints a convincing picture: the West could have been conquered by organized and autonomous communities of smallholders that incarnated the "American Dream." However, in the end, it was the work of the federal government and of powerful economic forces that developed the West when by taking advantage of the rules of the 1902 Federal Reclamation Act: settlers could farm no more than 120 acres with federally subsidized water (340 acres for a couple), and above that amount farmers had to either sell that land at pre-water prices or pay the full cost of developed water.

In fact, it is generally acknowledged that the major projects of the 20th century were not managed by self-organized smallholders, for the simple reason that they did not have the capacity to finance such infrastructure, particularly the construction of the various dams along the 2,300 km course of the Colorado River. Framed by the Colorado River Compact of 1920 (see *supra* chapter 4: 48sq), these projects were designed to regulate the flow of a river running through the arid and semi-arid areas of several states (Colorado, Utah, Nevada, Arizona, California) before flowing, reduced virtually to the size of a stream, into the Gulf of Mexico. In 1991, the completion of the Central Arizona Project Canal (begun in the 1960s to meet the irrigation needs of the agriculture of southern and central Arizona), also contributed to meeting the growing urban demand of the Sun Corridor (Phoenix, Tucson). In the end, the development of the Colorado River both enabled California – notably the Imperial Valley and Los Angeles (*Leslie, 2005*) – to expand throughout the 20th century, and also made possible the expansion of the entire area to the west of the Rockies (*Summitt, 2013*), a region less desert-like than at first appearance and, in reality, largely structured by agricultural irrigation. In any case, as with small farmers, agribusiness did not on its own have the resources required to build such infrastructure. Toward the end of the 19th century, large-scale farmers had to call upon the federal government to mobilize and finance the ideal of the "American Dream" of the frontier and the development of smallholdings to achieve its ends. In so doing, however, they provoked a number of conflicts over rights to the use of water (*Pisani, 2002*).

In a semi-arid environment, irrigation helps to dramatically reduce the uncertainties linked to natural factors. Nevertheless, a high degree of institutional regulation is still required to resolve conflicts over access to water among different groups of farmers. The political ideal of autonomous community organizations managing water supply, presented as the most effective solution in terms of guaranteeing economic development, cooperation and social justice, echoed Jeffersonian ideology, according to which agricultural smallholders constitute the heart of American democracy – as if the civic virtues allotted to them, supported by the pastoral narrative of their communion with nature, enabled them to assume the role of regulator. With growing urbanization affecting the desert lands of the West, and with the New Deal period destined to provide a remedy for the Depression, approaches to economic development were linked to the federal government, which became the manager and main contractor of major water projects at the national scale (*Pisani, 1984*). The predominant role played by the driving forces of economic development (large farmers, San Francisco investors, and the Los Angeles elites) undermined the democratic hopes of the previous period (*Gottlieb & Fitzsimmons, 1999*). Although now at a different scale, conflicts over the use of water persisted. Disputes now involved not small farmers, but various states of the West in fights against California (and the project supported by the Bureau of Reclamation) for a share of the Colorado River in order to boost their economic prosperity.

The region is, therefore, more than a natural framework colonized by man, as in the semi-mythical stories about "how the West was won." Indeed, it forms a complex socio-ecological system, marked by the presence of social organizations and institutions actively working to transform and regulate nature (*Ostrom, 2009*), a process marked by the "outgrowths" represented by the cities of Las Vegas, the tourist destination located in the Mojave Desert (*Nies, 2014*), and Phoenix, an "unsustainable" city in the Sonora Desert (*Schipper, 2008; Ross, 2011*). The West is located at the crossroads of a series of needs and desires: the conquest of new demographic spaces and pressures; the influence of federal bureaucracies and local authorities; demand for agricultural activity to guarantee food supply and exports and the absence of sufficient water to support that activity. It is, therefore, all the more difficult to reduce the development of the US desert to an ineluctable process of population increase and economic development, especially in that the "American Dream" was used to legitimize a wide variety of approaches, including economic modernization through agribusiness., Therefore, the symbolic construction of the desert, nourished by various narratives concerning the conquest of the West – ranging from paeans to "modernization" to catastrophist critiques of that same process – that should, first and foremost, be analyzed from a historical and sociological perspective.

2 HYDRAULIC SOCIETY AND THE AMERICAN EMPIRE

Worster's work is often presented as a simple transposition of the perspective of the despotic states of Asia developed by Wittfogel to the vast landscapes of North America. According to Wittfogel, agricultural irrigation would lead to a centralization of political power, due to the organization of labor and capital required to build the infrastructure. But Worster's interest is broader in scope; in fact, his focus is on an original definition of the hydraulic society and the power structures that developed in the American West: "*a social order founded on the intensive management of water. That regime did not evolve in isolation from the industrial system, of course, but all the same it was a distinctive emergent, reflecting the geography and arid climate of the state*" (*Worster, 1992: 55*). His emphasis is less on the omnipresence of a centralized state and its approaches to water management, and more on the impact of relationships of power traversing all the social groups concerned.

Worster outlines the implications of the social and ecological transformation of California, which enabled it to strike a balance between the ideals of the West (freedom, democracy, individualism) and its mythical history (the saga of men and women leaving civilization to dig out with their own hands a means of subsistence from the bowels of nature, the story of liberation from the East, of tradition and control, the tales of cowboys and other intrepid adventurers). Veering away from a conception of the West as a simple colony of the East – as described by *Cronon (1992)* – he suggests that the concept most suitable to defining California is that of empire. This empire emerged in the 19th century with the transformation of a desert region into a verdant, prosperous territory: the introduction of major hydraulic projects oriented a fast-developing California toward the export of agro-industrial products. This empire was structured around a politico-bureaucratic elite from the metropolises of the eastern U.S. and wealthy local farmers belonging to the growing agro-industry, motivated by a shared desire to transform an arid desert into a fertile oasis and a source of profit.

Within this framework, the manipulation of rivers led not only to the introduction of new models of human interaction, but also of new relationships of domination. In the American West, the approach to water management characteristic of the capitalist state was accompanied by the emergence of a social order within which power and influence were held at once by a private sector dominated by wealthy farmers, and a public sector composed of bureaucratic planners and representatives of politicians, supporting one another in regard to the creation and control of water resources. This created a hierarchical society in which workers were used as instruments of environmental manipulation while, at the same time, rivers became a means of control over the workers. In such a society, water is no longer thought of as a biological necessity, as it would be in the kind of subsistence economy characteristic of traditional societies; water becomes a commodity, the value of which is assessed as a function of what it is capable of representing (an irrigated parcel of land, kilowatts of energy) or producing (bales of cotton or truck loads of oranges).

To transform a desert into an area of economic production, the capitalist state thus accorded a place of honor to the engineers, whose technologies made it possible to build gigantic dams. The main protagonist of the development of water infrastructure in an arid environment, the state appears less as a body echoing the voice of the people, and more as the product of an elite motivated by a spirit of conquest and power. In this regard, Worster partakes of the ferocious critique developed by Arthur Maas in *Muddy Waters* (1951). Professor of Political Science at Harvard, Director of the Harvard Water Program from 1955 to 1965, and consultant to a number of public sector environmental agencies, Maas denounced the unbridled power of a federal elite, the US Army Corps of Engineers that he defined as an insubordinate, exclusive "clique," with little concern for public well-being. Linked to the senators of the West and the economic forces that they represented, this lobby, one of the most powerful in Washington, composed of around two hundred officers from the upper echelons of society (controlling nearly 50,000 engineers on the ground) was responsible, according to Maas, for preventing the needs of the region from sufficiently being taken into account.

In the early 20th century, the expansion of the American West was threatened by a problematic agricultural situation due not only to repeated droughts, but also to a lack of sufficient capital for constructing the irrigation systems needed by small, family-run farms. The conquest of the desert demanded large-scale irrigation that could not be managed by unaided small farmers, which in turn required a new mode of control and organization. Regional politicians and entrepreneurs thus turned to Washington. They appealed primarily to a federal agency set up in 1902 – the Bureau of Reclamation – which promoted the idea that exploiting the wild rivers of the West with a view to boosting local economic prosperity and carving out a place in the wider world should be built around the expertise of administrator-engineers, the only people capable, due to their technological expertise, of implementing a large-scale irrigation system. The question therefore arose as to who was to control the system.

> Some saw in the conquest of the desert an opportunity to create a technological democracy: the natural characteristics of the region presupposed a form of social organization based on watersheds (eventually a system divided into districts was realized in California, while an altogether different model was taken up by its rival, Arizona) regulated on an associative basis by a local corpus of entrepreneurs and politicians; access to irrigable plots of land would be limited in order to preclude cattle barons from making free/private use of public domains, or forestry companies from selling land off cheap, while non-irrigable land would be managed internally, without the intervention of professional bodies from outside the American West. The setting up of a district following the natural topography of the local watershed would thus rely upon/challenge the capacity of citizens to achieve self-determination by decentralizing sources of authority and power. It would express defiance in regard to the federal government and capitalists and, at the same time, a quasi-absolute faith in technocrats and administrator-engineers. For others, however, the new social order announced by the introduction of large-scale irrigation systems was tantamount to the creation of an industrial empire based on agricultural land. The transformation of the desert into productive farmland presupposed access to a huge amount of capital which regional administrations were unable to provide without help from external sources. Consequently, a large-scale water management system could not be envisaged without federal aid. Thus, in the early 20th century, "federalization" of the development of the West reflected a questioning of the modalities of economic expansion policy.

According to Worster, the main reason the centralizing elites promoted an irrigation program to boost wealth was that they represented a new generation of administrator-engineers who enjoyed a powerful position within government, who had little belief in democracy as a social remedy, and who defended the idea that only a technological elite could contribute to national expansion. In this regard, Worster disagrees with those historians who see in the legal and institutional concretization of this conception (the Federal Reclamation Act of 1902), a victory for ordinary people, who were provided with the formal right to own land and accrue wealth. The federal decision to make land available represented, above all, an opportunity for speculators to get rich quick by purchasing plots of land, not with the intention of farming, but with a view to selling them as soon as access to water had been guaranteed. Indeed, that decision also reinforced the position of previously established landowners[1] (Hiltzik, 2010). In fact, the conquest of the West was structured around a relationship between a national politico-administrative elite and a regional economic elite. It cannot be maintained that the West was developed by an omnipotent federal government, as in Wittfogel's analysis; Worster suggests, on the contrary, that the effects of a composite field of power, made up of coalitions established at multiple administrative levels are to be sought in the sphere of the control of resources.

California provides a fine example of a transformation from a rural region to an agricultural-industrial complex through the emergence of a new form of domination based on a parallel accumulation and combination of two types of resources, namely land assets owned by wealthy farmers, and the technical expertise of federal officers. Up until the 1930s, agro-industry farmers represented a kind of economic avant-garde, employing around 100,000 workers, selling fruit, vegetables and cotton for export, and exerting immense influence on local and national policies (Pisani, 1984). These private proprietors in the Californian agricultural sector controlled access to land in Imperial Valley in the south of the state and, further north, in the Great Central Valley. The management of irrigation, organized around quasi-governmental districts originally introduced in order to regulate access to property and, therefore, to irrigable plots of land, was not really democratic: few small farmers had access to decision-making processes, which were largely dominated by influential representatives of the private sector elected to the board of directors, and by expert managers suspicious of the community-based participation of small farmers in water planning on the grounds that it would, or so they supposed, endanger the maximization of technological and economic efficiency. Water was, therefore, managed for the private interests of large agro-industrial irrigators, a factor that created tensions between water districts in California.

According to Worster, in the early 20th century, the agricultural elite which ran the Great Central Valley was increasingly dependent on and vulnerable to a resource that was escaping its control. The agricultural development of the area presupposed irrigation on a large scale – new regional alliances were required. The Bureau of Reclamation was thus destined to become an indispensable partner in the development of the West's industrial agriculture sector, which was in need of its expertise. Its influence did not derive from its role as an agent of social reform, but from local irrigation agencies motivated by the logic of the market. Consequently, it functioned as a kind of "service bureaucracy," that served more to perpetuate a well-established elite than to

---

1 Although the 1902 Newlands Reclamation Act was to provide water to small farmers, it allowed the development of the large scale irrigation systems to be captured by large scale interests who manipulated land disposal laws and rules (Pincetl, 2011).

eliminate poverty in the region. Beginning in the 1940s, the emerging hydraulic society was the result of a convergence of instrumental forces: farmers harvested a profusion of cereals, fruits and vegetables, with no end in sight other than accumulation; federal technicians, indefatigable in their reorganization of natural watersheds, rendered reasonable what they thought of as previously having been irrational; the state of California, which wanted to develop its own water projects to irrigate the Central Valley, faced an economic depression that led it to ask the Bureau of Reclamation to fund the construction of water infrastructures (the Central Valley Project was established in the 1950s). Thanks to this convergence, the desert was transformed into a source of wealth.

From the 1960s, the development of large-scale hydraulic engineering ($7 billion was invested in irrigation), along with the wealth generated in the region, notably by agribusiness, meant that the West was on an economic high. But Worster asks questions about the power structure associated with this triumph. Was the region a model of democracy, as the advocates of its development maintained? Was it a society in which power and profit were widely distributed? Or was it, as in the case of more or less all the earliest hydraulic societies, characterized by a hierarchical power system rife with social inequality? Was authority concentrated and centralized in the hands of a few individuals to the point that small communities were reduced to a state of impotence?

The main contribution of Worster's research is to demonstrate that an emphasis on the opposition between market and state is not the most effective approach to describing the power structure in the American West, where capitalism had to adapt to specific ecological conditions. In arid and semi-arid environments (the Central Valley is not a desert) in which water was scarce, the private sector found it hard to prosper on its own and called upon the state in order to overcome the problem and pursue accumulation of wealth by exploiting every river of the region. After the Second World War, this power system gradually established itself, accompanying regional industrial transitions and encouraging the creation of future desert metropolises with unprecedented water infrastructure.

But, in the America of the 1970s, and until the election of President Reagan, the legitimacy of this empire was increasingly called into question: social and ecological militants disapproved of the engineering triumphs, which were the objects of the neoliberal critiques of the role of the state in land regulation and, consequently, in the field of access to water. The state was charged with achieving contradictory goals: promoting the accumulation of private wealth through the augmentation of the availability of a scarce resource (water), while maintaining social harmony via its equitable distribution. The question of justice in its various forms was raised more and more frequently: protests against the introduction of new dams; legal and political debates on limitations on land allocations seen as an attack on the freedom to do business or, inversely, as a form of democracy prejudicial to economic growth; strikes led by agricultural workers against agribusiness; claims made by native populations to water rights dating from the 19th century.

This dissension revealed that the hydraulic society was in fact fragmented and fragile. Alongside technological feats and the accumulation of profit, it had also generated growing levels of inequality: the elaboration and scope of irrigated agriculture was accompanied by the exploitation of low-income, working immigrants; the Bureau of Reclamation did not hold to the legal requirement to deliver water to farms of 160 acres only (or 340 in the case of a married couple), ensuring the Jeffersonian ideal of access to property and prosperity possible for the greatest number (particularly for the urban working class of the East). Public criticism of major hydraulic projects, considered economically unjustifiable and harmful to ecological systems, highlighted the need for an alternative movement focusing on the conservation of nature. These criticisms also marked the emergence of a new struggle concerning the empire's legitimacy, a struggle not only about the validity of modes of administration, but also about the principles used to define that legitimacy and, consequently, the legitimacy of the domination of the water sector by a coalition of agribusiness and the federal administration.

3 COLLECTIVE MOBILIZATIONS AND THE FIELD OF POWER

While the American West was conquered or purchased in order to be incorporated into the geopolitical architecture of a "foreign nation," it would be simplistic to see in this process of expansion as the systematic reflection of an uncontested domination on the part of a state bureaucracy informed by an ethos of instrumental rationality. At certain times and in certain places, the process of political and economic incorporation generated resistance movements, as evidenced in Walton's study of the battles for water in Owens Valley over a period of 130 years.

Located in the east of California, Owens Valley is a rural, arid and mountainous area which, over the course of its history, has been the object of a multitude of desires due to its natural resources, notable among which is water. And yet, while the federal government and the City of Los Angeles looked to control the area, they were obliged, from the outset, to confront the resistance of the local community. By placing this community's revolt in the context of the state's process of economic growth and modernization, Walton provides a different, but complementary perspective to Worster's: both, in effect, attempt to establish a correlation between the management of natural resources and the involvement of the federal government in the regional economic development of the American West.

> From 1860 until its transformation in the 1890s, Owens Valley was commercially isolated. The area was characterized by an economy based on subsistence farming carried out by colonists struggling for survival in a region traditionally populated by the Piaute Indians. While the federal government encouraged the setting up of colonies, it provided a bare minimum of security and land. Infrastructure (roads, canals, etc.) was entirely funded by the colonists, who developed cooperative organizational approaches to irrigation, mechanisms of popular justice, and engendered community values. Over the course of time, these values and approaches provided the foundations of a rebellion to the growing involvement of the public authorities in the economic development of the region: they were considered as threaten-

ing the autonomy of independent farmers and local traders by subjecting them to external competition that dictated prices, monopolized distribution and imposed foreign standards.

At the turn of the 20th century, Owens Valley suffered the direct consequences of a change in national policy. Attitudes towards federal intervention started to evolve. The federal government began to place an emphasis on urbanization, the bureaucratic management of natural resources, and regional incorporation. Initially, Owens Valley was happy to be absorbed into a process of development oriented toward large-scale irrigation projects designed to guarantee agricultural efficiency by the Bureau of Reclamation. However, when the federal project was abandoned in favor of Los Angeles in order to support local expansion,[2] local people organized protests demanding the restoration of federal involvement. Traditional picnics, a symbol of social cohesion, were transformed into political rallies leading to the creation of a legal organization representing Owens Valley (the irrigation district) which sought support from California's public institutions (the governor, politicians) and brought the dispute before the state courts. In fact, when requests from representatives of civil society for negotiations with the Valley's public authorities were rejected, traditional methods of "popular justice" were adopted by citizens: commandos sabotaged the aqueducts. By finally neutralizing the community revolt, Los Angeles achieved control over water and the territory, but it nevertheless had little control over local residents. According to Walton, dissidence persisted, limiting the power of Los Angeles to that of an owner-employer, rather than that of an occupying authority.

In the early 1930s, the New Deal, designed to rejuvenate the nation's economy, made its effects felt in rural areas, including Owens Valley. The policy of conservation became an essential factor in efforts to climb out of the economic recession by delivering employment and repairing land damaged by drought, erosion and flooding. In the American West, the federal agencies were reinforced: the Bureau of Reclamation extended its influence by means of new federal projects that subsidized the development of agribusiness. The very idea that the Department of the Interior, which was, at the time, responsible for managing those agencies, should be reorganized or supervised in the interests of the conservation of water as a resource, was met by widespread opposition among Congressmen. According to Walton, it was well known, although rarely explicitly acknowledged, that federal water agencies were a major source of both "pork barrel" projects for Congressmen and of subsidized water projects designed to promote local interests. This conception of the management of the resource gave rise to the over-exploitation of groundwater reservoirs, which, in turn, provoked opposition from Owens Valley inhabitants.

The 1960s and 1970s marked a turning point in the history of the resistance of Owens Valley. The Welfare State became ever more closely involved in environmental conservation. This new political orientation, combined with Los Angeles's need to protect rest and relaxation areas in order to attract ever-growing numbers of seasonal migrants, transformed the Owens Valley into a new tourist and governmental service economy. Owens Valley's resistance, rather than capitulation to the desires of the metropolis, was rendered possible and legitimized by changes on the national political scene, with Congress introducing new legislation (The National Environmental Policy Act, or NEPA, 1970) and creating a new legal instrument (the Environmental Impact Statement) demanding that all federal agencies assess the environmental impacts of their programs. Owens Valley was able to find the arms to fight Los Angeles's expansionist political programs that, once again, Owen Valley residents considered to threaten its communities and natural environment. As in the water

---

[2] For more details on the history of Los Angeles and how the city acquired water rights: (Gottlieb, 2007).

wars of the 1920s, traditional citizen committees were set up in Owens Valley to circulate petitions and put pressure on the federal government and its representatives; legal action was taken, etc. The resistance of the 1970s was, according to Walton, characterized by the fact that it was a legal battle based on state and federal legislation mainly conducted within the courts by means of lawsuits supported by social movements. With NEPA, the policy of extended state responsibility for environmental protection introduced a new source of legitimacy for community-based protests that brought the dispute before state courts with a view to protecting the environment. As well as ousting the Los Angeles Water Department and laying claim to full self-determination, the citizens, with the support of county government, won most of their legal battles and forced Los Angeles into an agreement: the water still flowed from the Owens Valley to LA, but in reduced amounts in order to preserve the biodiversity of the lake in Owen Valley.

The success of the environmental movement of the 1970s and 1980s was also the result of a combination of cultural values, militant strategies, and the incorporation of environmentalism into the politics and laws of the state. According to Walton, the community-based revolt that had preceded it had created a culture encouraging the emergence of political conditions that the following generation could use in its own circumstances to transform the opportunity of opposition into a local reality. For, since the 19th century, civil society had traditionally provided local services and organized water management by means of cooperative bodies responsible for irrigation, and this tradition was actively restored in a modern political struggle in favor of community-based autonomy, environmental conservation, and ideals of shared economic prosperity.

In terms of theoretical contributions, Walton's longitudinal analysis of the specific case of the water wars between Owens Valley and Los Angeles makes it possible to create a model describing relations between the state and collective action in terms of efforts to protect the environment. Counter to theories of rational choice, mobilization of resources, and moral economy, it links collective action to its origins in cultural beliefs about legitimate rights and the contexts of state authority. A decisive step was taken when collective action, no longer limiting itself to community-based resistance, became a social movement fighting to protect the environment within that same field of power. From Walton's point of view, the relationship between different forms of the state and collective action is reciprocal: the state does not only transform the orientations of an environmental movement by creating a new legal framework; within the federal politico-administrative system, the environmental movement can also exploit contradictions between different levels of state power with the objective of enlarging the public arena of legitimate social demands.

4　INTERNAL STRUGGLES WITHIN THE FIELD OF POWER

At the end of *Rivers of Empire*, Worster suggests that, in the context of the expansion of the environmental movement, the American hydraulic society generated a form of internal opposition that was not founded on class relations but, instead, on rival approaches to valuing nature. In her book, *The Struggle for Water*, the sociologist Wendy Nelson Espeland explores this hypothesis by focusing on the decision-making process leading to the decision to abandon a project to build a dam in the 1980s.

She analyzes the ways in which the political controversy over the Orme Dam became a "contest over the terms of rationality," which called into question the frontiers of traditional water sector bureaucracy. Protests culminated in the institutionalization of new administrative practices oriented toward the resolution of environmental conflicts.

The Central Arizona Project (CAP), built by the Bureau of Reclamation, consisted of the introduction of a complex system of dams, electricity plants, pumping stations and aqueducts designed to divert water from the Colorado River to irrigate Central and Southern Arizona (see *supra* chapter 6). The Orme Dam was part of this architecture. Originally thought of in the 1930s, gradually formalized in the 1960s, the project came in for increasing criticism from Washington because of its cost in the 1970s. It was also the object of increasing opposition from the Yavapai, an Indian community that would have been relocated if Congress approved the dam. When Jimmy Carter became President, the CAP found itself on a black list of 19 hydraulic projects that the White House intended to delete from the federal budget. But in Arizona, due to the vital need for additional water supplies, the President's approach caused a political crisis. Carter agreed to fund a study – the Central Arizona Water Control Study (CAWCS) – on the alternatives to the construction of the Orme Dam. Undertaken by the Bureau of Reclamation, the results were made official in the form of an environmental impact report. This process led, a few years later, James Watt, Secretary of the Interior in the Reagan administration, to validate an alternative to the Orme Dam and, in so doing, ratify an approach to evaluating environmental projects irreducible to the instrumental rationality that had, up until that point, dominated water policies.

> In 1976, the Bureau of Reclamation's first EIS (Environmental Impact Statement) on the dam was published, stirring up lively opposition. Written by engineers in favor of the dam, the report failed to mention the forced relocation of the Yavapai community. In response to the scandal that ensued and the political debate concerning the CAP, the CAWACS was set up with a view to restoring the Bureau of Reclamation's credibility. Experts with a new profile (the New Guard) – environmental specialists with a background in the social and natural sciences – were recruited. This new experts were to create and expand a new decision-making process based on procedural rationality designed to make the Bureau of Reclamation more democratic, an approach that triggered the hostility and incomprehension of the traditional engineers (the Old Guard).
>
> According to NEPA, the job of assessing the social and environmental effects of alternatives to the Orme Dam required not only traditional economic and technical data, but also the inclusion of data from archeological, historical, anthropological, sociological, biological and meteorological analyses. Efforts were also made to document and incorporate the opinions and values of the public. How could such disparate information be reconciled? How could it be guaranteed that social and environmental questions would be included in the agency's initial decision-making process? How could abstractions such as "social well-being," "environmental quality," and "efficiency" be evaluated? According to Espeland, in order to answer these questions, members of the New Guard turned to science, adopting rational decision-making models derived from economics and cognitive psychology. They therefore introduced Public Values Assessment (PVA), a procedure designed to measure the public's attitudes to and preferences for different projects and to enable citizens previously excluded from the debate about the dam to participate in the

decision-making process. The Yavapai community, intent on defending the territory on which the dam was to be built, was thus consulted and involved.

The CAWCS became an arena in which to accomplish in a practical sense what numerous decision theorists advocated theoretically and to change the decision-making routines employed by engineers within the organization. For the approach taken by the New Guard at the time militated against, on the one hand, the Bureau of Reclamation's traditional technology, namely, the planning and promotion of dams and, on the other, with the local political lobby which arrived at a decision to approve the dam, rather than exploring alternatives that took external opinions into account. Meanwhile, due to its marginal position, the New Guard deployed a strategy of inclusion that put instrumental rationality at the service of democracy by creating procedures authorizing more people to take part in the decision-making process, while improving the quality of information used to render decisions. It believed firmly in the potential of science as the source of neutral, specialized knowledge for resolving disputes and improving the decision-making process.

The engineers of the Old Guard were, due to their status as an enlightened elite, used to politicians and members of the public giving them the authority to make all decisions concerning the irrigation and development of the West. Its members were wary of the new legal framework (NEPA) and were markedly unwilling to abandon a system that had enabled so many important decisions and generated so many major projects. They had dedicated over twenty years to promoting the CAP and the Orme Dam, focusing on finding technical solutions to social problems (flooding in urban areas, irrigating arid land). In this sense, they had followed in the footsteps of their predecessors from the era of progressive policies and the federal irrigation movement that had culminated in the setting up of the Bureau of Reclamation and the deployment of an ideology of professionalism underwritten by a vision of nature as exploited through the implementation of a useful, monumental architectural aesthetic taking the form of gigantic dams. According to Espeland, for the Old Guard, abandoning the Orme Dam was tantamount to admitting that their careers had been a failure, that their technological ideals were no longer the object of consensus, that their agency was a relic from the past, and that, in sum, their power had been eroded. What was at stake for them in the struggle over the Orme Dam was, more than anything else, the preservation of their glorious past.

The "outmoded past" in which trust was placed in the technology of the engineers and in which substantial budgets were allocated to ever-more expensive water projects was fading away. While, previously, the dam would have been the expression of a technical solution to an irrigation problem, it now became the object of a referendum on water development. Furthermore, the affirmation of the validity of a native community vision revealed itself as incompatible with an instrumental rationalism applied to modernity and economic development. The Old Guard found itself out of step with, on the one hand, the new ideals informing their agency, characterized by an emphasis on science and a greater distance from networks of influence, and on the other, with a more open attitude toward a public that, at *a priori*, did not take their side. And when an alternative to the Orme Dam was finally chosen by means of the decision-making procedures implemented by the New Guard and communicated to the Department of the Interior, they were unable to see it as a triumph for the kind of rational planning to which they had dedicated their professional careers. The members of the Old Guard rejected the legitimacy of this alternative, but having been publicly discredited by the publication of the first report on the project for the dam, they were unable to bring a halt to the process and did not offer anything in exchange – they were of the opinion that the project was the best and most rational solution to the technical problems that they were attempting to solve in order to irrigate the Arizona desert.

Last, the professional identity of the Old Guard, their loyalty to the Bureau of Reclamation, and the value and time they dedicated to their work explained their support of the CAP and the Orme Dam, a position coherent with the traditional ethos of the agency, which was at once hard to quantify and ill-suited to conflict resolution. However, the New Guard took a different view of the decision-making process. The elaboration of an alternative to the dam provided an opportunity to promote a vision of the public service in which the government was more receptive to citizens. Similarly, participating in the decision-making process enabled the Yavapai to express their rights and to lay claim to a cultural attachment to the land that could not be reduced to a monetary evaluation. In deciding not to build the dam, the political authorities introduced the possibility of creating and participating in a new form of democracy, rooted in a procedural rationality deriving from a plurality of opinions. Consequently, the controversy over the Orme Dam in the new NEPA legislative framework led to the views of the public at large being taken into account, enabling a reappraisal of the Bureau of Reclamation's decision-making procedures and a redefinition of its *raison d'être*.

It may seem strange to focus on decision-making processes regarding a dam that was never built. But Espeland's study provides an understanding of the rivalries between different approaches to evaluating nature, and their links with the social and professional groups that defended them in the bureaucratic field. Unlike the technical vision of decision-making procedures exclusively controlled by scientific experts designed to improve the proposals of bureaucrats and politicians, transforming perceptions about nature into quantifiable data designed to reflect public choice in the water sector is far from a neutral act. According to Espeland, this process of transposition modifies the categories used by individuals to represent what is important to them. Changing ways of manufacturing public decisions serves to redefine the terms of the debate and set the parameters of what can and cannot be said. Evaluating nature becomes a political process in the sense that interests, values and preferences are at issue and have practical effects on the strategies adopted, as well as on professional identities. Marginalized within the Bureau of Reclamation, it was in the interests of the New Guard, which was seeking professional legitimacy, to ensure that instrumental rationality became institutionalized, that the agency's network expanded, and that mechanisms of participation were set up. In becoming a part of the decision-making process, the Yavapai, skeptical about the legitimacy of the procedures of the New Guard, had to express their cultural differences, their demonetized relationship with the land. Organizational interests and the construction of identity can also play a role in the environmental decision-making process.

Espeland takes a similar position to Walton when she asserts that it is difficult to find a satisfactory analytical approach using theories of rational choice that properly describes environmental decisions that integrate rationalities based not on efficiency but, instead, on moral values (public participation, social justice), multiple interests, and collective identities. Far from being consensual, theories of rational choice trigger conflicts over the most legitimate definition of approaches to evaluating and thinking about nature. In order to fully grasp the meaning and scope of environmental decisions, Espeland, taking a leaf out of the books of Worster and Walton, examines the origins of those decisions and the historical context of their production, thereby reconstituting their institutional trajectory. The case of the Orme Dam confirms how the emergence of a new legal framework (NEPA) perturbed organizational routines, thus calling into question the hegemony of a professional culture that seemed natural and universal by institutionalizing a new way of thinking about nature that integrated environmental and social concerns in a decision-making process that was at once bureaucratic and political.

5 ANALYTICAL TOOLS FOR AN ENVIRONMENTAL SOCIOLOGY

This sociological perspective on the genesis and the historical development of water policy in the western United States shows that water projects cannot be considered simply as matters of technical ability or the products of rational decision making, but should be regarded as deeply embedded in long-standing struggles for control of both water and power. Any struggle of this kind deserves to be highlighted: this is the starting point as well as the bedrock for a better understanding of what social relationships are involved in water projects and to what extent tensions among various groups with different vested interests play a part in the design of these water projects over time.

Accordingly, water policies cannot be reduced to an expression of external economic forces. In this regard, Donald Worster's analysis of hydraulic society raises another important point for an environmental sociology: claiming and pushing for big infrastructure on behalf of a "new America" implies the building of a strong coalition between federal administration (with a leading role for engineers) and local elites (based on agroindustrial companies). However, the way in which water policies are implemented cannot be expressed as a mere instrument of domination through which the population plays the role of passive puppets. As shown by John Walton, the collective capacity to resist such a domination means that water management should be understood as a space of struggle or a field (with reference to Pierre Bourdieu's theory) wherein dominant groups must constantly refine and demonstrate the legitimacy of their models of management in the face of the continuing involvement of other stakeholders in local issues. From that perspective, water megaprojects require another model of water rights to share the resources among various states; and engineers have accepted that they must implement other forms of management that cannot be reduced to technical solutions – and thus transform themselves into managers so as to preserve their position in the field of water management.

As the forms of water management evolve, the goals to achieve and the nature of relevant accumulated resources required to be powerful also change: Wendy Espeland shows that the history of water policies can thus be understood as a competing evolution from technical to legal and finally political ways of management. A sociological approach to water management therefore needs to better understand and highlight these principles of structuring the field of water policy, which appear to be much more connected to the sphere of power than a technical approach might express.

REFERENCES

Abbey, E. (1968). *Desert Solitaire*. Tucson: University of Arizona Press.
Burawoy, M. (2009). *The Extended Case Method. Four Countries, Four Decades, Four Great Transformations, and One Theoretical Tradition*. Los Angeles: University of California Press.
Cronon, W. (1992). *Nature's Metropolis. Chicago and the Great West*. New York & London: Norton & Co.
Erie, S.P. (2006). *Beyond Chinatown. The Metropolitan Water District, Growth and the Environment in Southern California*. Stanford University Press.
Espeland, W. (1998). *The Struggle for Water. Politics, Rationality and Identity in the American Southwest*. Chicago/London: The University of Chicago Press.
Faulkner, W. (1965). On Privacy (The American Dream: What Happened to It). In W. Faulkner (Ed.), *Essays, Speeches and Public Letters*. New York: Random House.
Fowler, D. (2010). *A Laboratory for Anthropology. Science and Romanticism in the American Southwest* (1846–1930). Salt Lake City: The University of Utah Press.
Gagnon, B., Lewis, N., Ferraridu, S. (2008). Environnement et pauvreté: regards croisés entre l'éthique et la justice environnementales. *Ecologie & Politique*, 35: 79–90.
Gottlieb, R. and FitzSimmons, M. (1991). *Thirst for Growth. Water Agencies as Hidden Government in California*. Tucson: University of Arizona Press.
Gottlieb, R. (2007). *Reinventing Los Angeles. Nature and Community in the Global City*. Cambridge, Mass: MIT Press.
Hache, E. (Ed.). (2012). *Écologie politique: Cosmos, communautés, milieux*. Paris: Éd. Amsterdam.
Hiltzik, M.A. (2010). *Colossus: The Turbulent, Thrilling Saga of the Building of Hoover Dam*. New York: Free Press.
Jacoby, K. (2003). *Crimes against Nature. Squatters, Poachers, Thieves, and the Hidden History of American Conservation*. Berkley: University of California Press.
Kalaora, B. and Vassopoulos, C. (2014). *Pour une sociologie de l'environnement. Environnement, société et politique*. Seyssel: Champ Vallon.
Lascoumes, P. (1994). *L'éco-pouvoir. Environnements et politiques*. Paris: La Découverte.
Lascoumes, P. and Le Galès, P. (Eds.). (2004). *Gouverner par les instruments*. Paris: Presses de Sciences Po.
Leslie, J. (2005). *Deep Water. The Epic Struggle over Dams, Displaced People, and the Environment*. New York: Farrar Strauss & Giroux.
Lochet, F. and Quenet, G. (2009). L'histoire environnementale: origines, enjeux et perspectives d'un nouveau chantier. *Revue d'histoire moderne et contemporaine*, 56(4): 7–38.
Nash, R.F. (1967). *Wilderness and the American Mind* (4th ed.). New Haven & London: Yale University Press.
Nies, J. (2014). *Unreal City. Las Vegas, Black Mesa and the Fate of the West*. New York: Nation Books.
Ostrom, E. (2009). A general framework for analyzing sustainability of social-ecological systems. *Science*, 325(5939): 419–423.
Park, L. and Pillow, D. (2002). *The Silicon Valley of Dreams: Environmental Injustice, Immigrant Workers, and the High-Tech Global Economy*. New York: NYU Press.
Pincetl, S. (2011). Urban water conflicts in the western US. In Barraqué B. (Ed.), *Urban Water Conflicts*. Paris: UNESCO-IHP: 237–246.
Pisani, D.J. (1984). *From the Family Farm to Agribusiness: The Irrigation Crusade in California and the West, 1850–1931*. Berkeley: University of California Press.
Pisani, D.J. (2002). *Water and American Government: The Reclamation Bureau, National Water Policy, and the West, 1902–1935*. Berkeley/London: University of California Press.
Reisner, M. (1986). *Cadillac Desert. The American West and its Disappearing Water*. New York: Penguin Books.

Ross, A. (2011). *Bird on Fire – Lessons from the World's Least Sustainable City*. New York: Oxford University Press.
Schipper, J. (2008). *Disappearing Desert. The Growth of Phoenix and the Culture of Sprawl*. Norman: University of Oklahoma Press.
Stegner, W. (1953). *Beyond the Hundredth Meridian. John Wesley Powell and the Second Opening of the West*. Cambridge (MA): The Riverside Press.
Summit, A.R. (2013). *Contested Waters. An Environmental History of the Colorado River*. Boulder: University Press of Colorado.
Tarr, J. and Dupuy, G. (1988). *Technology and the Rise of the Networked City in Europe and America*. Philadelphia: Temple University Press.
Topalov, C. (2003). Ecrire l'histoire des sociologues de Chicago. *Genèses*, 51: 147–179.
Walton, J. (1993). *Western Times and Water Wars. State, Culture and Rebellion in California*. Berkeley/Los Angeles/Oxford: University of California Press.
Worster, D. (1985). *Rivers of Empire. Water, Aridity and the Growth of the American West*. New York/Oxford: Oxford University Press.
Worster, D. (1992). *Under Western Skies. Nature and History in the American West*. New York/Oxford: Oxford University Press.
Worster, D. (2001). *A River Running West. The Life of John Westley Powell*. New York/Oxford: Oxford University Press.

Narratives of urban growth

Chapter 8

The social logic of urban sprawl: Arizona cities under environmental pressure

Eliza Benites-Gambirazio
UMI iGLOBES, CNRS/University of Arizona, USA/Université Paris III, France

---

"Fortunes in post-war Arizona were not made of gold or silver mining but in real estate. Arizona became a vast Monopoly board of new land waiting to be developed."

—Thomas Sheridan (2012: 353)

INTRODUCTION: UNDERSTANDING URBAN SPRAWL IN A CONTEXT OF ENVIRONMENTAL PRESSURES

Urban researchers and policy makers view urban sprawl with increasing concern, as urban growth is considered to be one of the main factors contributing to the contemporary water crisis, especially in Western US. There have been countless attacks leveled against "urban sprawl", drawing attention to the negative effects it has on the environment, the economy, and on the social fabric of cities (*Langdon, 1997; Wiewel & Persky, 2002; Lindstrom & Bartling, 2003*). Most research to-date has been focused on the consequences of urban sprawl on environmental resources; these consequences include forcing people to drive extensively (thereby creating a high carbon footprint), placing pressure on water resources, increasing costs for urban networks and, ultimately, provoking overall ecological disaster (*Burchell et al., 1998; Johnson, 2001; Camagni et al., 2002; Squires 2002; Livingston et al., 2003*). Studies have argued that, when well implemented, dense forms of urbanization can help to lower the negative social, economic and environmental impacts of extensive development (*Camagni et al., 2002*).

Urban sprawl is often understood as the *result* of the societal demand to live in a suburban environment, in pursuit of the traditional post-World War II milestones of becoming a suburban homeowner. Traditional research does not uncover the *process* which creates sprawl. This chapter, rather than focusing on the consequences of urban sprawl, aims to address some of its mechanisms, with the primary assumption that population demand does not happen in a vacuum but lies at the conjunction between institutional and private actions both at the national and the local levels.

This chapter addresses the mechanisms by which urban sprawl is created within the growth machine (*Molotch & Logan, 1987*) of a "pro-growth culture." Traditional urban theory, developed by economists and political scientists, considers cities to be receptacles of population growth, organized by municipalities or driven by market forces. This framework has produced two sets of hypotheses; 1) that city officials demand growth in order to be financially stable, which is increasingly key in the context of scarce public money; and 2) that local elites and firms favor business growth, including employment, services and real estate residential and commercial projects. Both arguments support the idea of the existence of a "pro-growth" coalition composed of public and private actors, and the importance of the real estate industry, its financing and public advocates, in the overall economy. As claimed in a special issue on real estate history, growth is to be understood within the actions of the business elites:

> The process of city growth is not just a demographic process but involves crucial entrepreneurial choices and strategies of promotion and investment, and these processes in turn have profound consequences for the ethnic and social structuring of cities (*Tolliday, 1989: 1*).

This theoretical framework for the growth machine has produced interesting empirical cases but these are often limited to a particular historical period (pre 1990's). Further, no study has examined the processes and the coalitions through which urban sprawl is created, as well as the implications of these growth coalitions on urban forms. Despite the numerous historical studies regarding the American West, there are relatively few empirical sociological studies focused on the metropolitan areas of the American West – with the notable exception of Los Angeles (*Mike, 1990; Fulton, 2001; Gish, 2007*). In contrast, large cities such as Chicago (*Pierce 2007, Sampson, 2012, Spinney, 2000*) or New York (*Mele, 2000; Sze, 2007*), have been at the core of urban researchers attention. This is not surprising when one considers the relatively late development of the West. There are, however, some recent studies focused on the development of Phoenix, the 6th largest metropolitan area in the country, mostly critical of its urban development and impact on the environment (*Gammage, 1999; Heim, 2001; Ross, 2011; Shermer, 2009, 2013*). The hypothesis presented in this chapter is that urban development and its contribution to environmental pressures should be understood as a result of the suburban real estate industry.

1 SPRAWL SINCE THE 1950'S

1.1 Definition, process and implications

Urban sprawl refers to an extension of urban development beyond the city centers. It is characterized by *"a depopulation of city centers (downtowns) and a scattering of the suburbs, possible only through the extension of the network of highways and roads, the diversity of the tax system at the local level, the proliferation of municipalities, and the desire of people to own a house with land or a garden. Not only housing but also offices leave downtowns to go far in suburban areas"* (Paquot, 2006: 297). Three forms of urban sprawl can be distinguished: edge cities, planned communities and individual houses (*Nechyba & Walsh, 2004*).

The dynamics of urban sprawl are rooted within technical and socio-cultural evolution. Since World War II, two main factors are understood to be responsible for suburbanization and the decrease in urban densities in American metropolitan areas: income growth and the accessibility of means of transportation, supported by the development of the automobile industry and low gas prices (*Jackson, 1985; Bairoch, 1995; Newman, 2000*). In the West, the availability of air conditioning has further enabled urban and economic development.

Understanding the process of sprawl requires an investigation of the main producers of this shift, and how opportunities became available and used by agents working within the real estate industry (real estate developers, homebuilders, municipalities, planners, engineers, real estate agents and contractors). Private land was readily available and fairly cheap in the western United States, which provided the window for developing a profitable niche by promoting residential suburbs. The primary target was a single family with children who would potentially benefit from living away from ethnically and socially diverse city centers. Many of the marketing images of "ideal life" were conveyed via advertisements envisioning the concept of a happy life in the suburbs. This type of life became strongly connected with the new American lifestyle, promoted as part of the American dream. Happy families with children, owning a car and house with a garden, were key elements of the new American way of life and the suburban landscape of the postwar boom *(Hummon, 1990; Henderson, 1953; Jackson, 1985)*. "Suburbanism" was thus considered as a new emerging way of life (*Fava, 1956*), bringing to a next stage the classical 1938 essay of Wirth on "Urbanism as a way of life", where the sense of belonging embodies the socio-cultural function of the suburban environment.

Since the 1950's, the transformation of the city, through the increase of urbanization and the dynamics of suburbanization, prompted historians as well as urban researchers and political scientists to study the "city-building process" (*Weiss, 1989: 246*). A series of books focused on the relation between housing and urban development (*Wright, 1981; Edel, 1984; Fishman, 1987*). Urban historians attempted to give accounts of the genesis of American cities, and how the real estate industry impacted urban growth (*Goodkin, 1974; Abbott, 1981; Eichler, 1982; Stach, 1988; Weiss, 1988; Warner & Molotch, 1995*). Of course, real estate development should be understood in close relation with the modern history of business. The following section provides a historical overview of the suburban real estate development industry in the United States, so as to understand the process of urban sprawl with its attendant consequences of increased environmental pressure.

1.2 The suburban real estate development industry

The history of land acquisition and ownership in the Western United States is made up of purchases, treaties and cessions between the U.S. federal governments and foreign nations. These include the Louisiana Purchase in 1803 from the France of Napoleon, the treaties with Great Britain and Spain in 1817 and 1819, the cession of Mexican territory, the Mexican-American War in 1846–1848, and the treaty of Guadalupe Hidalgo in 1846, with the acquisition of Texas, New Mexico, Arizona and California being among the most important. Although the initial strategy of the federal government was to transfer of ownership to private actors and states, the federal government often used land ordinances to maintain ownership for specific purposes, such as beneficiating public schools. Thus, while the federal government continued selling and leasing its lands – private ownership increased from 12% in 1962 to 24% in 2003 – it was done with caution. The Public Land Review Commission of 1964 recommended that:

> The policy of large-scale disposal of public lands reflected by the majority of statutes in force today be revised and that future disposal should be only those lands that will achieve maximum benefit for the general public in non-Federal ownership, while retaining in Federal ownership those whose values must be preserved so that they may be used and enjoyed by all Americans.
> (Report by the Public Land Law Review Commission, 1970: 1)

According to 2003 Federal Real Estate Profile, the federal government still owns about 50% of the land in the Western States. From its beginning, the question of land pricing was central, and achieving the maximum possible revenue was encouraged through a series of conditions aimed at maximizing the success of land sales. Despite its status, land is acquired either through purchase or via a lease agreement for a period of time. Lands, or improved lands with houses or buildings, are valued and exchanged in a market operating on the basis of offer and demand. The time of mobility is thus the time where the land is being assessed.

> The day of free land is past. Naturally enough, the scarcity of land is most acute where men are crowded, in the great cities. The city bears one great point of resemblance to the frontier; it is a place of great mobility. Men, goods, and even buildings, circulate and push each other about. Mobility of men and scarcity of land have combined to make the real estate market. (*Hughes 1928: 13–14*)

Cities are associated with the process of urban population growth (*Burgess, 1928*). Cities and their metropolitan areas continuously receive people and increase their total areas. As this process happens, land use inevitably evolves and transforms. Since the late 18th century and the beginning of the 19th century, the city has been considered as a profitable place for business and market, in which people – mostly entrepreneurs and investors – invest and engage in speculation.

In the United States, places within cities have been considered as commodities that are exchangeable in a market for a price, depending on the nature of the commodity (raw vs. improved land), the value attributed to the real estate development, and the value attributed to the geographical area. The conditions for market exchange are both created and supported by governmental and business elites (*Bourdieu, 2000; Dobbin, 2004; Fligstein, 1996, 2001*). The conditions in which these alliances are made and thrive are what permit an understanding of the social construction of cities. Historians account for the rise of "metropolitan entrepreneurship" to build cities such as Los Angeles through businessmen (such as Henry Huntington – developer, planer and advocate for urban projects) and their use of financial power (*Friedricks, 1989*). These business elites are able to "promote their interests through municipal, locational, investment and regulatory decisions" and are helped by "a political entrepreneur" (*Fleischmann & Feagin, 1987: 209–210*).

A proper history of the real estate industry cannot be achieved without including the history of housing booms, government policy and reforms, and lending institutions. Banks, and legislators enacting banking regulation, have been a key support for real estate booms over the last 50 years:

> The financial institutions that have evolved to serve the needs of residential and commercial real estate have become some of the key financial institutions of the national economy with the widest ramifications in the field of insurance, brokerage, and institutional investment. *(Tolliday 1989: 1)*

The savings and loans industry provided significant support for commercial and residential projects until the late 1980's, and helped provide the financing for development. In the early 2000's, growth of the real estate market was influenced by a change in governmental legislation. After continuous lobbying from investors and bankers, the Glass-Steagall Act of 1933 (*Crawford, 2011*) was repealed in 1999, allowing investment banking activities to be performed by any bank. This provided an enormous avenue for banks and real estate investors and developers to invest in real estate.

Real estate development refers to the process of change in the use of land; from "raw land" (of lower value) to developed "improved land" (of higher value), or from developed land to its redevelopment at a higher use value. A developer (real estate development company or entity, which could range from private investors and financial companies to universities, hospitals or cities) will purchase raw land and convert it into developed or improved land such as a subdivision. To develop a subdivision, the developer needs to comply with requirements in place, including zoning, land use and physical aspect, build infrastructure (roads), and provide for utilities (water, sewage). At any part of the process of entitlement (including zoning), the developer can sell his land to a homebuilder or a company for the prior defined uses. Developers maintain control on how the land is used and occupied (architecture and built environment). Although risky, projects can provide a large return of investment and "large fortunes have been and continue to be made and lost in real estate development" (*Miles Berens & Weiss, 2000: 65*).

As seen in *Figure 1*, there are multiple private and public actors involved in the process of real estate development, as well as a myriad of professionals who support, and are part of, the real estate industry and the suburban growth machine. Developers sell the land to, or hire, a homebuilder or a commercial builder or contractor (the developer can also be the homebuilder or vice versa), who in turn is in charge of assembling a team of professionals (architects, engineers, consultants) that is paid to work on the project. The landowner will ultimately sell or lease the development. Lobbyists, by seeking to influence the decision-making process of the government, help developers negotiate with governmental institutions: "*The extensive public-review process for such projects is one of the reasons why real estate firms typically appear as the top spenders for lobbyists each year*" (Trio of developers paid $3 million to get projects going," *Crain's New York Business*, March 5, 2015). Lawyers help in mitigation processes and oversee the potential legal issues at all phases of the project: they can assist at the original purchase of the land, oversee environmental and zoning regulations, help with land use issues

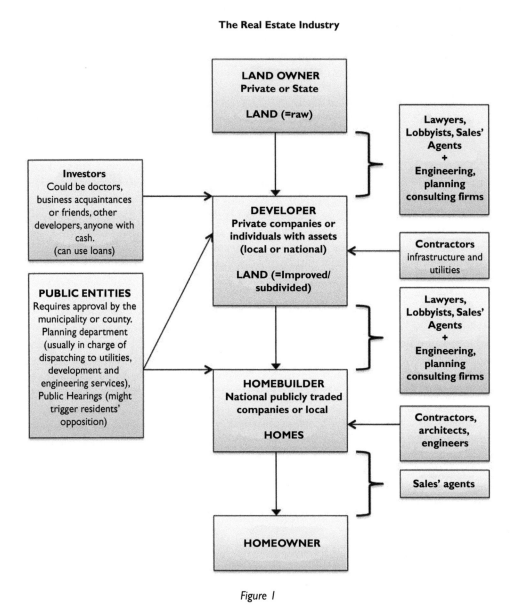

Figure 1

such as granting easements across the land for utilities, and act as negotiators for the contracts with the contractors, architects and banks. Developers can be compared to orchestra conductors: "*They don't play any instruments, they just direct the performance. But at the same time, developers assume 100% of the risk of the project. If the building fails (because you can't sell the condo units or lease out the space), that all falls on the developer (and his/her investors). All of the other team members are getting paid based on the services they provide. They're consultants*" (Brandon Donnelly, Architect This City Blog). Ultimately, the risk rests on the person who is financing the project, whether the developer, the bank, or a group of investors.

The government broadly affects the process of real estate development through its various federal, state or local agencies, divided into departments (Department of Real Estate, Department of Housing and Urban Development, Department of Health and Safety, Department of Water) and other institutions (cities, counties, etc.). Throughout the real estate development process, the developer must interact with these entities to obtain necessary permits and approvals for different aspects of the project. Although local institutions such as counties and cities must comply with regulations put in place, and must therefore, for example, ask developers to modify some of their original plans for landscape and water networks, they are anticipating through their zoning regulations which type of uses a land will be zoned for. Because they have stakes in increasing their network and the value of their land, they also have interests in having developers build on their "residential" zoning and for their projects to succeed. As their financial capacity diminished in the 1980's, in the era of neoliberal creed of transferring the burden of infrastructure and utilities needs to the private sector, public institutions transferred the costs of development to private actors.

Real estate development has been an important part of Arizona's economy for the past 50 years. The Arizona Sun Corridor (composed of Phoenix and Tucson and the area between the two cities) offers an interesting example of increasing environmental pressures and continued strong suburban growth. Local institutional systems of coalitions in Phoenix and Tucson sustain (or discourage at times) growth and development.

2 THE CASE OF ARIZONA: THE GROWTH MACHINE, SUBURBANIZATION AND THE ENVIRONMENT

Located in the Southwestern United States, Arizona is known for its dry climate and desert landscapes. Being the 48th to become a U.S. State – 6th in terms of area (295,234 km^2) and 15th in population (6,413,700 inhabitants as of 2010) – it is one of the Four Corners States and part of the Sun Belt. The cities of Tucson and Phoenix, and their suburbs, are the two largest metropolitan areas in Arizona, having approximately 982,018 and 3,823,000 inhabitants, respectively, in 2010. Until the end of the 19th century, the area that was to become the State of Arizona remained mostly rural and with very few inhabitants living mostly in Jerome, Flagstaff, Prescott, Tucson, Bisbee and (briefly) Tombstone. The onset of urban growth dates back to the early 20th century, but began to increase dramatically in the 1940's and the 1950's. Arizona's population increased by about 584% from the 1950's to the end of the 1970's, and by 89% since the 1980's. For comparison, the population of the United States grew by 86% and 24% in the corresponding periods. According to census bureau estimates (2000), Arizona is, the fastest growing state in the United States, with a population of over 6.41 million inhabitants in 2010.

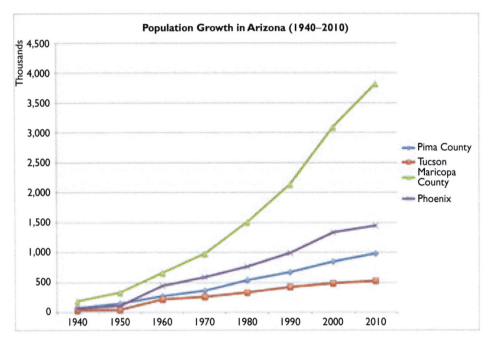

Chart 1

Phoenix and Tucson, as rapidly growing metropolitan areas, are believed to be shifting from metropolitan to "megapolitan" areas: "*The megapolitan concept is more than just two cities' suburbs spilling across each other's borders. It's a combination of land use, commerce, and transportation planning that looks at vast areas as single entities*" ("When Phoenix, Tucson Merge," *Arizona Republic*, September 4th, 2006). Although this was discussed actively before the 2008 recession, the encouragement of growth has not been stopped and there have been signs of recovery and growth since 2012.

Population growth has been well documented in the Sun Corridor. Growth has led to the development of the homebuilding industry with the in housing supply increasing to match anticipated demand. However the Morris Institute reveals that while Arizona's population increased by 25% from 2000 to 2010, its housing supply grew by 30% *(Gammage et al., 2011: p. 10)*. Interesting migration dynamics *(Forbes, 2014)* show that most of the migration to Tucson comes from Phoenix, which confirm the hypothesis of an urban corridor between the two metropolises and the necessity to address growth at the State level.

The impact of the housing crash and the financial crisis on the economy of Phoenix (and more broadly Arizona) is also well known: "*From 2005 to 2010, the prices of homes in Metro Phoenix, for example, fell by almost 50%. Arizona as a state went from creating 121,000 jobs between October 2005 and October 2006 to losing 183,000 jobs in 2009. The Sun Corridor's traditionally homebuilding-based economy saw housing construction plummet.*" *(Gammage et al., 2011: 10)*.

2.1 Suburbanization in Arizona

Suburbanization in Arizona dates back to the 1920's, but started to accelerate after World War II. Sun City was the first retirement community in the metropolitan area of Phoenix; it offered the particularity of being exclusively open to residents older than 65 years old. Constructed by Del Webb in the 1950's, it opened its door on January 1, 1960, offering the paramount of "resort" or "country club lifestyle" with a pool, a recreation center, a golf course, several shops and standardized homes with services provided by Maricopa County (maintenance of street and police) and private companies.

Not all newer developments were age-restricted. After the booms in Florida (1920's–30's), and the famous Levittown in Pennsylvania and New Jersey (1950's), master planned communities – planned residential housing subdivisions constructed by developers and homebuilders – became increasingly popular in California and Arizona (*Wattel, 1958; Gans, 1967; Kelly, 1993*). Since the 1970's, master planned communities have become the bulk of new residential developments. Created by private companies, and managed by investors and developers, these communities became increasingly popular during the real estate boom of the 1970's and 1980's. Examples of these communities in the Phoenix Metropolitan Area include Marley Park (2005, Market Park LLC, affiliate of DMB Associates, Inc.), Talking Rock Ranch (joint venture of The True Life Companies (TTLC) and Harvard Investments, Inc.), McCormick Ranch, Rio Verde (1973 by Rio Verde Development, Inc.), Tartesso (2007 by Stardust Companies) among others. There was also, to a lesser extent, similar urban development in the metropolitan area of Tucson and in the metropolitan area of Phoenix at the end of the 1980's and 1990's. Some of the most famous, in the Tucson area, include Saddlebrooke (by developer Ed Robson in 1986), Continental Ranch (1998), Dove Mountain (1997) and Gladden Farms in Marana or Rancho Vistoso in Oro Valley (by developer David Melh in 2003).

2.2 Land and planning

Local governments can use zoning and land-use planning to drive capital investments, creating (or not) local growth. Cities are responsible for coming up with a general plan to decide the uses of their lands, generally in agreement with their landowners. They in fact decide what their city is going to look like by expressing desires for future shape and organization. They can decide to annex portions of unincorporated areas from the County and generate more tax revenue. As Arizona does not have much land left undeveloped (only 24% is private), developers are looking towards alternative ways of finding land. One viable option is being provided by the State of Arizona, which owns about 13% of the total land and can maximize its revenue by leasing or selling lands. Leasing is the privileged strategy as there are expectations that real estate prices will keep rising, and that the value of the land can only increase.

The 1912 Arizona Constitution mandates that state public land should be *"managed and owned by the State for the benefit of public schools."* Trust land can therefore be leased to farmers, ranchers, miners, and businesses, with the benefits distributed to the school system. As the share of agriculture in Arizona's economy has decreased, and as population has grown, developers have argued that this land should be sold to private companies to generate revenue. In the 1980's, developers were able to trade private land for state trust land. In 1990, however, the Arizona Supreme Court determined that no constitutional authority exists for the Arizona legislature to authorize private exchanges and public auctions of school trust land. Several frauds were reported, as developers organized to agree on prices during the public auction, thereby preventing the State from receiving the highest revenue. Several proposals have been made to revise this ruling. The developers interviewed suggested that since most of the private land has already been bought and built upon, further development can happen only through the State progressively selling or leasing its land. Actions have been undertaken by developers to pressure the State, as demonstrated by the recent House Bill (2658) to create a committee to study the potential of public lands transfers.

The Sun Belt states (including Arizona) have, since the 1930's/1950's, been known for giving rise to a new generation of governmental leaders and politicians that created a "pro-business and growth climate" and fiercely opposing the New Deal politics. Business elites, with the help of politicians such as Barry Goldwater, promoted a pro – growth environment, by passing laws aimed at crippling labor unions and limiting industry regulation (*Shermer, 2013*). New flows of water were also brought to the region, through the Central Arizona Project (1968), and Congress-financed companies (such as Intel, McDonnell Douglas and Motorola) chose Phoenix to be one of their active business centers. A housing boom characterized the 1950's and the 1960's.

The differences in suburban development between Phoenix and Tucson provide a strong comparative case study to examine how the urban corridor has been evolving, and the different social logics of its construction. Phoenix, with exponential growth, serves as an example of the *"bird on fire"* (*Ross, 2011*), whereas Tucson provides a contrasting example where the government, with the support of its local residents, has curtailed urban development, more or less successfully.

2.3   Phoenix and the pro-growth coalitions

The metropolitan area of Phoenix can be considered as a representation of an "edge city", being that it is immediately adjacent to the surrounding towns of Tempe, Gilbert, Mesa, Scottsdale, Surprise, Goodyear and Glendale. Phoenix provides an excellent example to illustrate the issue of "shadow governments" that are characteristic of edge cities having multiple centers of power (*Garreau, 1991*). Shadow governments include the homeowners associations (HOAs) that are ubiquitous in newer developments. Scandals have arisen when real estate developers have tried to control their residents through their HOAs. Examples include the attempt by Charles Keating (associated with the Estrella project just outside of Phoenix,) to restrict the abortion rights of its residents and the case of John Lafferty (associated with Leisure World) who controlled a shadow government.

A shadow government was also involved in the origins of the Salt River Project (founded in 1903), which was developed to bring water and energy to Central Arizona. The project was started by business elites – real estate developers and bankers – who organized to promote the economic development of the downtown Phoenix area. However, because they weren't able to collect taxes, or have their own police force or cleaning services, they didn't have financial or enforcement capacity. So they created a public-private "shadow government" called the Management District. Its fees were collected through its public side, even though it operating as a non-profit. While the city managed the police department and cleaned the streets, the Management District paid the police department and was permitted to add more police officers or to ask them to perform specific tasks, and paid the city to clean the streets and to provide a transportation shuttle. *Garreau (1991)* commented that this endangered democracy because the State was thereby subordinated to the power of the private management group, and no distinction was established between the specific needs of the Management District and its execution by the State. Another example is the destruction of an entire Chicano neighborhood to expand the Phoenix airport and the subsequent dispute about residents *"trying to organize against the power of the State"* (here actually private corporations) which was overriding the "will" of the communities and neighborhoods to protect their health, care and culture.

2.4 Tucson, between growth and anti-growth dynamics

Historically, Tucson was the business center of Arizona. Being close to several mining sites, the Tucson Chamber of Commerce was (from the end of the 19th century on) led by several businessmen looking to encourage economic development and avenues for growth. These included real estate developers (due to the availability of open land) and manufacturers who could benefit from the low expenditures associated with a Hispanic labor force (*Logan, 2006*). When the post-World War II downturn hurt the manufacturing industry, the prospects for growth became centered on tourism. In cooperation with the Chamber of Commerce, organizations such as the Sunshine Climate Club actively promoted the image of an attractive "Western ranching lifestyle" (with its scenic views and abundant sunshine) by partnering with the railroads to offer tours of Tucson and its surroundings. In the 1940's and 1950's, Tucson became known for its ability to attracting both tourists and retirees (including veterans) and was often characterized as a of "retirement city" (*Abbott, 1981*). However, some public projects could not be undertaken in spite of the desire for growth and support from business and residents. For example, the 1936 Sabino Canyon Dam project was dismissed because Pima County refused to invest the amount required to build it, resulting in the resignation and departure of the manager of the Chamber of Commerce.

In the post-World War II period, resistance to urban growth began to arise among wealthy suburban citizens. Also, surprisingly, real estate developers opposed annexations proposed by the City because they did not want government regulation to apply to their lands and potential projects: "*Criticisms of the city's expansion became so vociferous that at one point Tucson's mayor seeking to deflect criticism with humor, wore a suit of armor to a public hearing on annexation*" (*Logan, 1995: 31*). Although this resistance may have slowed development, it was unable to prevent growth during the booming post-World War II decade of the 1950's during which the city population increased by 370% and the city limits expanded sevenfold from about 10 to 71 square miles. Instead, the resistance to growth caused the pro-growth coalitions (business elites, city and county agencies) to adapt by finding alternative ways to grow:

"The resistance never succeeded in stopping the city's expansion, but it did succeed in obstructing growth by forcing the city and county governments to maneuver around the resistance. The pro-growth forces in Tucson never lost their majority status during the 1950's, but resistance to growth remained part of the community ethos" (Logan, 1995: 31).

As reported in some newspaper accounts, real estate developers continued to powerfully organize the urban development of Tucson at the time: "*A look at the real estate business in Tucson not only reveals the growth of the city; it also mirrors a segment of the business community that is fairly typical of the courageous, pioneering and at times a bit unmanageable people who have settled this area. Real estate men traditionally have served on zoning committees, as members of school and church organizations, on civic promotion boards; in short, real estate people are where the action is*" ("Those Wide Open Spaces are Filling in" Tucson Citizen, October 19, 1970).

The construction of neighborhoods can be seen as the result of the real estate development, with strong encouragement and support from coalitions of business and governmental leaders. Concerns about the viability of development, and oppositions against some projects, emerged through the local voices of residents as they fought to save their homes and communities from destruction, and expressed environmental and lifestyle concerns to proposed development. Environmental organizations, such as the Citizens for Growth Management (CGM), acted to establish "urban growth boundaries," preserve "natural areas and neighborhoods" as well as "air and water" (Ruesga, 2000: 1064) and on November 7, 2000 pushed for adoption of the Citizen's Growth Management Initiative (CGMI). (Otero, 2010) describes the efforts to organize Chicano communities, who opposed speculators, developers and city officials that tried to profit from urban renewal by transforming neighborhoods into businesses or government offices. Logan (1995) reports that the earliest resistance to development actually occurred in South Tucson and was led by Mexican Americans. Also, in the 1970's, environmental concerns arose due to issues with CAP water and concerns for the health of the natural desert habitat.

While Phoenix did encourage the growth of residential and commercial development, Pima County and the City of Tucson have opposed development on numerous occasions. However, such anti-growth coalitions did not arise in cities such as Marana or Oro Valley, which have absorbed most of the suburban growth during the last decades. Although Pima County and the City of Tucson did not, initially, have strong planning policies in place, during the 1970's to the 1990's, the county and the municipality began to oppose residential and commercial developments, and companies that sought to become established within the city limits. One developer recalls the dynamics of "no growth" and its impact:

"In the 70's, 80's and 90's, they were people called "no growth" and people basically for good reasons realized that water is a huge problem, on one hand, but also realized that Tucson was really cool the way it was in a way, because it was so much space for so few people, some of them were from Massachusetts, with the densest cities in the United States. Tucson is the least dense place in the United States, so people who wanted to keep Tucson the same way it was in the 70's, very few people got elected in the office and turn the zoning code into what it is now, which is the thing that makes it very hard to actually build anything or fix anything and they did it on purpose to keep it so there were no density so the city talked all day long, they were in favor of density and in favor of public transportation but in reality, they are the biggest enemy to it, so that's part of what's going on" (…) "that's the main tension, so a lot of politicians are perfectly nice people, and a lot of them know that I'll be better if Tucson modernized and became a more modern place, but their constituent are people who don't want any growth and they want it to stay as it was, and they know they're not going to get reelected if they appear to be too modern". (*Interview, Developer Ross S., September 2014*).

For example, developers interviewed for this article pointed out that the failure of Motorola to become established in Tucson 30 years ago was due to political decisions and the "not in my back yard" (NIMBY) attitude of wealthy people in the north side of Tucson. Apparently, the location offered in the South of Tucson was rejected, implying that the company had a problem with "moving there" due to the population and the neighborhood being unsuitable for their activity (*Sonnichsen, 1987: 284*).

In terms of employment, todays Tucson's economy relies heavily on federal jobs including the Air Force Base. This amounts to 7.7% of GPD, compared to 1.9% of GDP in Phoenix and 3.8% of GDP as the nation whole, according to George Hammond, the director of the Economic and Business Research Center in the Eller College of Management at the University of Arizona. The air force also recently announced that there would be no more new missions, meaning that fewer people will move to Tucson. Developers complain about this lack of employment diversity, and the fact that young University of Arizona graduates are actually leaving town. With Tucson becoming famous for trying to slow development, other cities in the Tucson Metropolitan Area (such as Marana or Oro Valley) have, since the 1970's and 1980's, encouraged residential development via numerous master planned communities that sprawl across the desert. Pima County had also encouraged residential development to some extent. Again, the question of future planning for the metropolitan area has arisen – how much development do we want? At what point do we consider we have built enough? One developer accounts for how the real estate market depends heavily on politics.

"Much of real estate is involved with politics. Okay, let me give you some examples, so let's say that there are … 50 years ago, Pima County adopted what was called the Agua Caliente Area Plan, all of that property out there was zoned 3.3 acres or larger, well they adopted this plan which allowed for downzoning [to be] allowed if somebody wanted to…if they had a 3.3 parcel if they wanted to split it three ways and have three 1 acre lot the county will allow that, or if the developer wanted to put in a subdivision of 1 acre lot they say go ahead and do it, so that was politics because they wanted that area to build up, they wanted more population, and that's exactly what happened and there was a lot of subdividing and a lot of people buying one acre parcels … well now the population out in the North East part of town has built up quite a bit and the board of supervisors are not doing what they used to do, they are not just basically rubber stamping a new subdivision of one acre lots, they're trying to discourage that." (*Interview, Developer Donald R., April 2014*).

To some extent, the concerns for many politicians and Tucsonan residents (as can be heard expressed in many informal conversations) is the danger of becoming "like Phoenix"; i.e., a giant mega city, disrespectful of the environment and starving for growth and development. For example, according to the real estate agents interviewed, the story that often comes with a discussion of water conservation is that Phoenicians have yards with flowers and lawns, in comparison to Tucson which takes pride in developing desert gardens.

2.5 "There's no water crisis": increasing technology and securing water supply

Developers interviewed expressed a lack of concern regarding the future of water supplies. Although they certainly recognize that water as an important commodity – in that it's necessary to growth – they look at the progress made possible by technological advances, conservation and restoration. For developer Rick Denys, points to the great progress in water conservation that has been made: *"I look at it as a success story. In Tucson, the average water consumption right now is the same than in 1987."* For example, the fact that Tucson has 140,000 acre feet a year available to it, but only uses 100,000 acre feet a year is a good sign because that means that it can put 40,000 back into the ground. Another developer (Greg Mayo) also sees a positive trend: *"With the declining water consumption, even on a conservative trend, we have until 2040–2050 and then we can use the water stored."* The rationale for developers and homebuilders to believe that there is sufficient water seems to be largely based on the improvement of technology and the inevitable solutions that will be found to support future growth. Such technological success stories can be found in the implementations of systems to recycle water in Singapore or San Diego.

The "question of water" is thus resolved by a firm belief in technology and that without growth, communities and cities are doomed to die. One developer talks about the resistance of Pima County to the opening of a new mine. For him, it is a missed opportunity both for employment – it would have created a few thousand blue-collar jobs – but also the opportunity to bring more water to the Tucson Metropolitan Area via a company-proposed extension of the CAP canal towards the South of Tucson. From this point of view, undertaking such a costly infrastructure operation would be great for the community because it would make CAP water available. Developers argue they are not against the environment, as long as environmental action or inaction does not prevent the creation of jobs.

My interviews with developers and city officials clearly reveal that, for them, the concept of sustainability is to be understood within a "checklist approach", with cities making decisions or choosing programs that make them appear to be practicing "sustainability" without changing the growth politics. For example, supporting homeownership means promoting the new development of single-family housing units. Additionally, job growth is tied to the sprawl of areas, where housing is needed. The "growth" mindset does not seem to have been affected by the housing crisis of 2007–2008 (*Ross, 2011*).

There are also differences in the extents to which developers need to deal with water issues. Smaller developers in the central city areas or within the city limits do not have to justify their water needs because it is supplied by local water providers, whether through the municipal water system or through smaller water companies. Most of the new subdivision development in the Phoenix and Tucson metropolitan areas is subject to the Arizona Groundwater Management Act (GMA) that requires each developer to ensure 100 years of water supply. The GMA provides developers with several different methods to demonstrate an assured water supply. Assuming that the new subdivision is within a service area already serviced by an established water provider, the developer can demonstrate the necessary supply for a new subdivision by satisfying a multi-prong test. However, if seeking development of a subdivision that is not in an established water service area and does not have legal access to sufficient water, the developer may enroll the subdivision to be a member of a groundwater replenishment district and pay a set of fees in exchange for the groundwater replenishment district assuming the obligation to assure the necessary water supply. Rather than being constraints, environmental resources are perceived as adaptable to human needs. Suburban growth, although slowed during the last fifty years, is not being questioned. This is the consequence of the 19th and 20th century growth ideology in the Southwest:

> "Growth networks have made their larger political case and gained popular support by arguing that growth equals prosperity and that growth will create urban environment better suited to the western landscape and pace of life. Tucson billboards that asked: 'Does your job depend on growth?' and bumper stickers that declare « Construction Feeds My Family, » all expressed the basic argument that growth brought full employment. Growth, too, the argument went, would eventually decrease the share of the tax burden that any single taxpayer had to bear." (*White 2015*).

Growth is, in fact, perceived as "inevitable" and the question of planned growth is envisioned as the only path. This comes not only from developers, but also from city managers like Dave C:

> "If you aren't pro growth, you want to ask: ''Are you for death?'' because if you're not growing..., it's not that you need to keep adding more population in a sense you do, because if you have people leaving your community is gonna get smaller, there are countries, examples all over the world, there are declining population of cities and countries, and that has profound impact over generation, they are less people to pay into the systems, the infrastructure gets older and it's not able to be replaced, so, it's not that you have to keep growing out either, you don't have to keep growing big in terms of size but if you're not trying to attract new people that want to be a part of something."(*Interview, developer Dave C., May 2015*)

Thus, Dave underlines that growth has to be planned, and that keeping a city alive is synonymous with extended networks and increased population – accordingly, growth must be integrated with planning efforts and carried out by private projects led by developers and real estate agents. Developers and state administration, towards a mix between political will and negotiation with administrative rules, contribute to maintain urban sprawl.

3 CONCLUSION

Urban sprawl has been accused of provoking an ecological disaster, forcing people to drive extensively, causing high carbon footprint and increasing costs for urban networks. Studies have demonstrated the economic and environmental costs of sprawl, and argued that when well implemented, dense urban forms lower the negative social, economic and environmental impacts of development (*Camagni Gibelli & Rigamonti, 2002*).

Empirical studies can contribute to understand to which extent the sprawling is the result of the interrelations between public entities (government and local authorities) and private supported by a myriad of local service providers. The Arizona Sun Corridor offers an interesting example of increasing environmental concerns and continued strong suburban growth. Phoenix and Tucson have also experienced different types of growth and alliances to support or discourage growth. Growing at the different rates, they nonetheless experience similar models of growth.

Despite increasing concerns over sustainable allocation of resources, this chapter demonstrates that if the politico-legal framework of sprawling has shifted towards a more sustainable growth, its form has not changed and the pro growth coalitions, armed with better "conservation and sustainability apparatus," continue to promote an extensive urban development. Focusing on the *process* through which urban sprawl is created, this research reveals the institutional system in charge of urban development, and how developers and other professionals are able to design, plan and organize how cities grow in Arizona, and how environmental and water pressures are generated.

REFERENCES

Abbott, C. (1981). *The new urban America: growth and politics in Sunbelt cities*. Chapel Hill: The University of North Carolina Press.

Bairoch, P. (1995). *Economics and world history: myths and paradoxes*. Chicago: University of Chicago Press.

Bourdieu, P. (2000). *Les structures sociales de l'économie*. Paris: Seuil.

Burchell, R.W., Shad, N.A. and Listokin, D. (1998). *The Costs of Sprawl-Revisited*. Transit Cooperative Research Program Report 39. Washington, D.C.: National Academy Press.

Burgess, E.W. (1928). *The growth of the city: an introduction to a research project*. New York: Springer.

Camagni, R., Gibelli, M.C. and Rigamonti, P. (2002). Urban mobility and urban form: the social and environmental costs of different patterns of urban expansion. *Ecological Economics*, 40(2): 199–216.

Crawford, C. (2011). The repeal of the glass-steagall act and the current financial crisis. *Journal of Business & Economics Research (JBER)*, 9(1): 127–134.

Dobbin, F. (2004). Introduction: The sociology of the economy. In F. Dobbin (Ed.) *The Sociology of the Economy*. New York: Russell Sage Foundation, 1–26.

Edel, M. (1984). *Shaky Palaces: Homeownership and Social Mobility in Boston's Surburbanization*. New York: Columbia University Press.

Eichler, N. (1982). *The Merchant Builders*. Cambridge: MIT Press.

Fava, S.F. (1956). Suburbanism as a Way of Life. *American Sociological Review*, 21(1): 34–37.

Fishman, R. (1987). *Bourgeois utopias: The rise and fall of suburbia*. New York: Basic books.
Fleischmann, A. and Feagin, J.R. (1987). The Politics of Growth-Oriented Urban Alliances Comparing Old Industrial and New Sunbelt Cities. *Urban Affairs Review*, 23(2): 207–232.
Fligstein, N. (1996). Markets as politics: A political-cultural approach to market institutions. *American Sociological Review*, 61(4): 656–673.
Fligstein, N. (2001). *The architecture of markets: An economic sociology of twenty-first-century capitalist societies*. Princeton, NJ: Princeton University Press.
Friedricks, W.B. (1989). A Metropolitan Entrepreneur Par Excellence: Henry E. Huntington and the Growth of Southern California, 1898–1927. *Business History Review*, 63(02): 329–355.
Fulton, W. (2001). *The reluctant metropolis: The politics of urban growth in Los Angeles*. Baltimore: Johns Hopkins University Press.
Gammage, G. (1999). *Phoenix in perspective: Reflection on developing the desert*. Tempe, AZ: Herberger Center for Design Excellence, College of Architecture and Environmental Design, Arizona State University.
Gammage, G., Stigler, M., Clark-Johnson, S., Daugherty, D. and Hart, W. (2011). *Watering the sun corridor: managing choices in Arizona's megapolitan area*. Phoenix: Morrison Institute for Public Policy.
Gans, H.J. (1967). Levittown and America. In R. LeGates and F. Stout (Eds.) *The City Reader*. New York: Routledge, Vol. 2: 63–69.
Garreau, J. (1991). *Edge cities: Life on the new frontier*. New York: Doubleday.
Gish, T.D. (2007). *Building Los Angeles: Urban Housing in the Suburban Metropolis, 1900–1936*. Los Angeles: University of Southern California Press.
Goodkin, L.M. (1974). *When real estate and home building become big business: mergers, acquisitions, and joint ventures*. Boston: Cahners Books.
Heim, C.E. (2001). Leapfrogging, urban sprawl, and growth management: Phoenix, 1950–2000. *American Journal of Economics and Sociology*, 60(1): 245–283.
Henderson, H. (1953). The mass-produced suburbs. *Harpers*, 207: 25–32.
Hughes, E.C. (1928). *A study of a secular institution: The Chicago real estate board*. Dissertation submitted to the University of Chicago, Department of Sociology.
Hummon, D.M. (1990). *Commonplaces: Community ideology and identity in American culture*. New York: SUNY Press.
Jackson, K.T. (1985). *Crabgrass frontier: The suburbanization of the United States*. Oxford: Oxford University Press.
Johnson, M.P. (2001). Environmental Impacts of Urban Sprawl: a Survey of the Literature and Proposed Research Agenda. *Environment and Planning A*, 33(4): 717–735.
Kelly, B.M. (1993). *Expanding the American Dream: Building and Rebuilding Levittown*. New York: SUNY Press.
Langdon, P. (1997). *A Better Place to Live: Reshaping the American suburb*. Amherst, MA: University of Massachusetts Press.
Lindstrom, M.J. and Bartling, H. (2003). *Suburban sprawl: Culture, theory, and politics*. Lanham, MD: Rowman & Littlefield.
Livingston, A., Ridlington, E. and Baker, M. (2003). *The costs of sprawl, fiscal, environmental and quality of life impacts in low-density development in the Denver region*. Denver, CO: Environment Colorado Research and Policy Center.
Logan, M.F. (1995). *Fighting sprawl and city hall: Resistance to urban growth in the southwest*. Tucson: University of Arizona Press.
Mele, C. (2000). *Selling the lower east side: Culture, real estate, and resistance in New York City*. Minneapolis: University of Minnesota Press.
Mike, D. (1990). *City of Quartz: Excavating the Future in Los Angeles*. New York: Verso.

Miles, M.E., Berens, G. and Weiss, M.A. (2000). *Real estate development: principles and process* (3rd ed.). Washington, D.C.: Urban Land Institute.

Molotch, H. and Logan, J. (1987). *Urban fortunes: The political economy of place.* Berkeley: University of California Press.

Nechyba, T.J. and Walsh, R.P. (2004). Urban sprawl. *Journal of Economic Perspectives*, 18(4): 177–200.

Newman, P. and Kenworthy, J. (2000). The Ten Myths of Automobile Dependence. *World Transport Policy & Practice*, 6(1): 15–25.

Otero, L.R. (2010). *La Calle: Spatial Conflicts and Urban Renewal in a Southwest City.* Tucson: University of Arizona Press.

Paquot, T. (2006). *Terre urbaine: Cinq défis pour le devenir urbain de la planète.* Paris: La Découverte.

Pierce, B.L. (2007). *A History of Chicago, Volume III: The Rise of a Modern City, 1871–1893.* Chicago: University of Chicago Press.

Ross, A. (2011). *Bird on fire: Lessons from the world's least sustainable city.* Oxford: Oxford University Press.

Ruesga, C.M. (2000). Great Wall of Phoenix-Urban Growth Boundaries and Arizona's Affordable Housing Market. *Arizona State Law Journal*, 32:1063.

Sampson, R.J. (2012). *Great American city: Chicago and the enduring neighborhood effect.* Chicago: University of Chicago Press.

Sheridan, T.E. (2012). Arizona. A History. Tucson: University of Arizona Press.

Shermer, E.T. (2009). *Creating the Sunbelt: The Political and Economic Transformation of Phoenix, Arizona.* Santa Barbara: University of California.

Shermer, E.T. (2013). *Sunbelt Capitalism: Phoenix and the Transformation of American Politics.* Philadelphia: University of Pennsylvania Press.

Sonnichsen C.L. (1987). *Tucson: The Life and Times of an American City*, Tulsa: University of Oklahoma Press.

Spinney, R.G. (2000). *City of Big Shoulders: A History of Chicago.* DeKalb: Northern Illinois University Press.

Squires, G.D. (2002). *Urban sprawl: Causes, consequences, & policy responses.* Washington: Urban Institute Press.

Stach, P.B. (1988). Deed restrictions and subdivision development in Columbus, Ohio, 1900–1970. *Journal of Urban History*, 15(1): 42–68.

Sze, J. (2007). *Noxious New York: the racial politics of urban health and environmental justice.* Cambridge: MIT Press.

Tolliday, S.W. (1989). Entrepreneurs in Business History. *Business History Review*, 63: 1–4.

Warner, K. and Molotch, H. (1995). Power to Build How Development Persists Despite Local Controls. *Urban Affairs Review*, 30(3): 378–406.

Wattel, H.L. (1958). Levittown: a suburban community, In W. Dobriner (Ed.), *The Suburban Community*. New York: Putnam's Son, 287–313.

Weiss, M.A. (1988). Researching the History of Real Estate Development. *Journal of Architectural Education*, 41(3): 38–40.

Weiss, M.A. (1989). Real estate history: an overview and research agenda. *Business History Review*, 63(02): 241–282.

White, R. (2015). *It's your misfortune and none of my own: A new history of the American West.* Norman: University of Oklahoma Press.

Wiewel, W. and Persky, J. (2002). *Suburban sprawl: Private decisions and public policy.* Armonk, NY: ME Sharpe.

Wright, G. (1981). *Building the dream: A social history of housing in America.* New York: Pantheon.

OTHER SOURCES

Newspapers

Non-signed article. (October 19, 1970). Those wide open spaces are filling in. *Tucson Citizen*.
Anuta, J. (March 5, 2015). Trio of developers paid $3 million to get projects going. *Crain's New York Business*.
Reagor, C. (September 4, 2006). When Phoenix, Tucson Merge. *Arizona Republic*.

Reports

One Third of the Nation's Land: A Report to the President and to the Congress by the Public Land Law Review Commission, Washington, D.C.: U.S. GPO, June 1, 1970.

Interviews cited

Developer Ross S., September, 2014.
Developer Donald R., April, 2014.
Developer Dave C., May, 2015.

Chapter 9

Water and urban development challenges in the Tucson metropolitan area: An interdisciplinary perspective

Graciela Schneier-Madanes
Centre de Recherche et Documentation sur les Amériques CNRS, Université Paris 3 Sorbonne-Nouvelle, France

Juan B. Valdes
Department of Hydrology and Atmospheric Sciences, University of Arizona, USA

Edward F. Curley
Pima County Regional Wastewater Reclamation Department, Arizona, USA

Thomas Maddock III
Department of Hydrology and Atmospheric Sciences, University of Arizona, USA

Stuart E. Marsh
School of Natural Resources and the Environment, University of Arizona, USA

Kyle A. Hartfield
Arid Land Studies, University of Arizona, USA

INTRODUCTION

In today's urbanized world, there is a very significant relationship between growth and water/wastewater infrastructure, particularly in the rapidly growing but water scarce regions where the need to implement water uses/policies that are sustainable over the longer term is strong. In this chapter, we propose a new approach to analyzing the challenge of sustainable urban growth and water/wastewater development in the arid southwestern United States,[1] and especially in the Colorado Basin (*Figure 1*). In support of this, we assess the hydrological, historical and socio-economic interactions that have developed in this region, so as to explain today's problems while contributing to a comprehensive policy review.

Accordingly, we provide a general discussion of the nature of urban change and water resources and consider the extent to which different urban processes impact (or depend on) water. Urban development is analyzed in relation to the availability and supply of water resources in the context of varying urban settings and environmental conditions.[2]

---

1 The arid West is defined by rainfall, with Tucson averaging about 12 inches of precipitation annually between 1981 and 2010 (http://www.wrh.noaa.gov/twc/climate/tus.php).
2 For more details, see: https://swanproject.arizona.edu/work-packages/wp-4-urban-water-euusa-common-models.

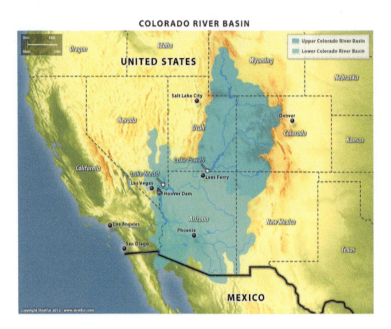

Figure 1 Colorado River Basin (*Image Source: Udall Center for Public Policy, University of Arizona*).

Indeed, to provide an integrated vision of the problem, the term "water" is used in this chapter to include water resources and potable water supply as well as wastewater.

As a specific case study, we address urban growth and water/wastewater supplies for the Tucson Metropolitan Region (population approximately 1 million people) in the state of Arizona. Arizona is especially illustrative of water/wastewater challenges in the context of urban sprawl. Urban population growth rates are among the highest in the nation, and the two largest cities, Phoenix and Tucson (located in Maricopa and Pima Counties, respectively), anchor a 200-mile urban corridor that is home to approximately 5.5 million people. Portions of this "Sun Corridor" (*Figures 2 and 3*) even extend south into Mexico.

In this chapter, we first examine urban development in the region by reviewing its urban and water history, with a focus on the end of World War II, which was an economic and demographic turning point for the area. Next, we analyze the availability and access to water and wastewater, and the landforms and the regulatory framework, which allow us to identify relevant illustrations. Our investigation includes both surface and groundwater regulations, as well as unregulated access to water. To add to this perspective, we discuss current and future water resource challenges such as safe yield, Colorado water shortages, droughts and climate change.

The chapter continues with an examination of urban growth and water and wastewater management interactions at the local scale. We use data from remote sensing imagery (1984–2010) to identify relevant illustrations from a land perspective, and discuss in detail the significant local case of the Continental Ranch in Marana, to

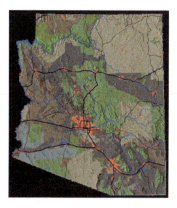

Figure 2 Land forms and population distribution in the State of Arizona (*Image Source: Maricopa Association of Governments*).

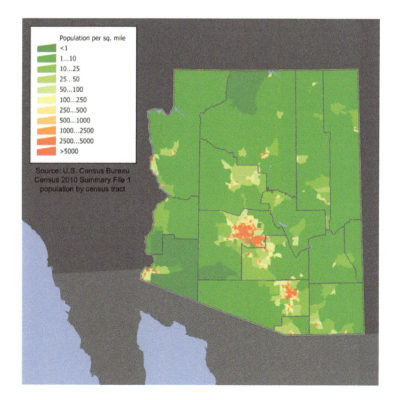

Figure 3 Arizona Sun Corridor is the high-density area in Central and Southern Arizona between Phoenix and Tucson (*Image Source: US Census Bureau*).

demonstrate the role of water and wastewater as drivers in the regulatory framework analysis. For this we provide illustrations through time-lapse and aerial photography, and point out some water/urban/regulation paradoxes.

Finally, we discuss future challenges, conclusions and recommended actions. The broad range of our backgrounds in urban planning, hydrology, and water/wastewater management enables us to provide a complex analysis of the interplay of urban growth and water/wastewater policies and practices. This includes causes and consequences, long-range implications and policy considerations for the future. Our multilevel approach highlights the complexity of urban growth and water interactions.

1 TUCSON METROPOLITAN AREA

Home to more than one million people, the Tucson metropolitan area covers more than 200 square miles in the Sonoran Desert, a semiarid region rich in biodiversity and culture. The metropolitan area includes the historical city of Tucson, the newly incorporated towns of Oro Valley, Marana, and Sahuarita, and the unincorporated urban areas of Pima County. (*Riebsame, W. E., & Robb, J. J. 1997*). Prior to becoming part of the United States in 1854, Tucson (also known as the Old Pueblo) was a classic Mexican frontier town. Gold and silver mining, the railroad, and warfare with Native American tribes changed the size and socio-economy of the initial frontier town (*Sheridan, 2012*). From the beginning of the 20th century, federal expenditures and economic activities associated with the five Cs: Copper, cotton, cattle, citrus, and climate fueled its growth.

World War II was a major turning point in Tucson's history. The population began to grow at an accelerated rate as migrants from all over the country swarmed to the "wide open spaces" of the Sunbelt. The postwar boom transformed not only the city, but also the whole region that would become the current Tucson Metropolitan Area. Federal military/industrial spending during and just after World War II fueled both economic development and population expansion. Attracted by climate and the natural landscape, retirees began to arrive in the area. Indeed, at the end of World War II, the metropolitan area included just two cities, the City of Tucson and the newly incorporated City of South Tucson (a square-mile community that ultimately would be enclosed within Tucson's boundaries while remaining independent). By the 1950s, 122,764 residents lived within the 25 square miles that formed the metropolitan area at the time (*Akros Inc., 2007*). Meanwhile, subdivisions quickly developed outside of the boundaries of both cities, expanding in all directions into unincorporated Pima County. About two thirds of the metropolitan area's population then lived outside the official city boundaries in new subdivisions, and in a few pockets of earlier settlements (*Akros Inc., 2007*). During this same post-war decade, Tucson's city limits evolved very little, with only two square miles annexed in 10 years.

Over the postwar decades, the Tucson Metropolitan Area's population continued to spread even further north and east across the basin floor toward the Catalina and Rincon Mountains, northwest toward the Tortolita Mountains, and west to the Tucson Mountains. In fact, between 1970 and 2010 the City of Tucson's population nearly doubled while Pima County's population tripled. As a consequence of this growth, the metropolitan region now includes (since the seventies) the adjacent cities of Tucson, South Tucson and three new towns: Oro Valley, Marana, and Sahuarita. It also includes the unincorporated urban areas in Pima County. (See *Figure 4*)

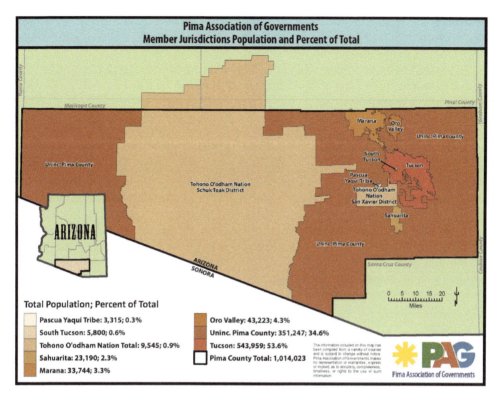

Figure 4 Pima County Jurisdictions (*Image Source: PAG 2015*).

2 WATER, WASTEWATER AND URBAN GROWTH

The relatively newly incorporated towns of Oro Valley, Marana, and Sahuarita, and the unincorporated urban areas of Pima County have experienced significant residential and commercial growth.

Oro Valley, incorporated in 1974, has 43,223 (*PAG 2015*) residents in 35 square miles and has become an affluent enclave to the north, emerging as a regional center for the biotech industry. Marana, incorporated in 1977, covers nearly 120 square miles, is home to 33,744 residents (*PAG 2015*), and is primarily an agricultural center for cotton and other crops that is rapidly developing into a suburban community.[3] Marana began to grow through an aggressive annexation policy (it currently has four times the surface area of Oro Valley and approximately the same population), which has important implications for water, wastewater, and reclaimed water management

---

[3] Indeed, the 2007 American Community Survey showed that, at that time, the median income for a household in the town of Oro Valley was $74,015, which was higher than the Tucson median ($36,752), although it is also higher than the U.S. median ($50,007). The estimated average household income in Marana is $64,332 per year (2007)—higher than Tucson median ($36,752).

in the metropolitan area. The latest of the newly incorporated areas is Sahuarita, located 15 miles south of Tucson and east of the Tohono O'odham Nation. Sahuarita was incorporated in 1994 and has 23,190 (*PAG 2015*) residents, but does not include nearby Green Valley, one of the first retirement communities in Arizona.

Even though the entire metropolitan area constitutes a functional unit, there is no centralized government for water and planning so each city and town has expanded following its own growth strategy based on its history, economy, and sociopolitical characteristics.

2.1 Water and wastewater: resources and access

The Tucson metropolitan area (See *Figure 5*) effectively illustrates water and wastewater issues in drylands, with its current water practices and policies stemming from complex resolutions and continuing debate about the questions of water availability, water quality, and water allocations.

Throughout its early history, Tucson relied on surface water but, beginning in the 20th century, and especially with the post-war urban development, it shifted to groundwater, located mostly between 100 and 500 feet deep. The groundwater is provided by two main aquifers: the Tucson Basin (or Upper Santa Cruz aquifer) and the Avra Valley aquifer west of the city. Since the 1950s, despite natural recharge of the aquifers, water demand became larger than replenishment and resulted in a general

Figure 5 LANDSAT image of Tucson metropolitan area (2010) / Tucson basin urbanization (*Image Source: University of Arizona School of Natural Resources, Remote Sensing Center*).

decrease in groundwater levels. The increases in population, along with copper mining and farming activities, also increased the demand for groundwater pumping, leading to subsidence and significant degradation of the riparian habitat that once existed. These conditions eventually led to the development of a large surface water transfer project from the Colorado River to Central and Southern Arizona.

Today, water management systems in Arizona are unique. They include large public works such as the Central Arizona Project (CAP)[4] that transports water from the Colorado River uphill via an open canal for 336 miles (541 km) to Phoenix and Tucson. (*Colby B. G., & Jacobs, K. L. 2007*). In addition to the CAP, Arizona and its governmental subcomponents – cities, towns, and counties as well as a special water and irrigation district—have adopted multiple strategies to establish diversified water supply sources: a) the use of reclaimed water on community parks and golf courses, which are important tourist and economic assets; b) reclaimed water for agricultural and industrial use; c) the CAP; and d) effluent aquifer recharge facilities.

The large number of different public and private providers in the Tucson metropolitan area forms a colorful mosaic of water production and distribution (*Figure 6*). The largest provider, Tucson Water, provides 75 percent of the water in the metropolitan area within and outside the city limits. Several other public water

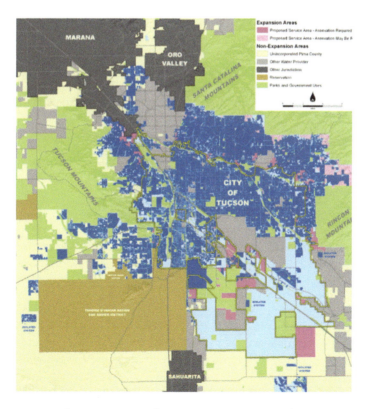

Figure 6 Tucson metropolitan area mosaic of water supply service areas (*Image Source: PAG 2015*).

4 For more details on CAP, see *supra* chapters 5 and 6.

providers, controlled by their own elected board of directors, and 20 very small private water companies serve the area. An early example of this is the Flowing Wells Irrigation District (founded in 1922 and today serving 16,000 customers). In 1992, the Metropolitan Water Improvement District (Metro Water District) was created to serve 45,000 people in a 26-square-mile unincorporated part of the metropolitan Tucson area. Both urban and rural homeowners may also pump groundwater in certain situations through domestic wells (limited to pumping no more than 35 gallons/minute) for their own personal use (*Valdes and Maddock, 2009*).

Wastewater services to about 250,000 households (comprising about 675,000 people) are provided by Pima County itself throughout a very large geographic area. This is accomplished by an integrated system of metropolitan and outlying water reclamation facilities with a capacity to treat 82 million gallons of wastewater per day. Marana and Sahuarita also provide wastewater services to a portion of their residents. The remaining population utilizes individual and collective septic disposal systems; most of these are in the rural and outlying areas, but a surprising number remain functional in the older areas of the urban core and the foothills of the Catalina Mountains.

The Tucson metropolitan area is also home to one of the first delivery systems in the U.S. for reclaimed water. Since 1984, treated wastewater has been utilized as a new water resource through an elaborate system known as "the purple (reclaimed) pipe network". Most golf courses, parks, and large turf areas in the metropolitan area use highly treated wastewater effluent for year-round irrigation, provided via a comprehensive conveyance system delivering up to 20 million gallons per day in the height of the summer. Reclaimed water has become the driver of urban growth through lifestyle villages built around golf courses.

The first CAP delivery in 1992 created a large social conflict because of issues regarding the "quality" of the water after it had flowed through the aging water infrastructure[5] and its direct use by Tucson Water in 1994. Tucson Water then developed a technology for groundwater recharge in the central Avra Valley, which enables it to recover a blend of water (groundwater and CAP), for delivery to customers after treatment. The purpose of the CAP utilization is to decrease groundwater mining and avoid land subsidence. Other water resources being explored more recently by local water utilities and public agencies in the area include storm water capture on a regional scale, and rainwater harvesting on an individual and neighborhood scale.

In terms of water demand, the main factor has been a shift since the 1980's from agricultural to municipal use. To control peak demand, Tucson initiated an intensive awareness campaign in the 1980s that made the city into a poster-child for water conservation in the West. In 2013, residential potable water use for Tucson Water customers was 88 gallons per capita per day (GPCD), which is less than for other major Southwest cities such as Las Vegas (212 GPCD) and Phoenix (106 GPCD).[6] Currently, new projects are being implemented in Tucson with several aims: increase of reclaimed water use, detention of storm water to control flooding and put the water to use and, at a smaller scale, development of water harvesting and gray-water use programs.

5 The pH level in the CAP water caused the minerals in the pipes to dissolve into the water so that by the time the water was delivered from the tap it was brown.
6 Source: Personal communications to the authors from the data management staff of the City of Phoenix Water Department, Tucson Water, and the Las Vegas Valley Water District (July 2015).

2.2 Patterns of urban growth: expansion, "leap-frog" and "wildcat developments"

Post-war growth in Tucson was largely a private endeavor. Developers simply acquired parcels of land and formed subdivisions with varying degrees of care and skill. By the beginning of the 1960s, real estate development and sales were already highly organized economic activities with large social and spatial implications. The new actors of urban development were local, regional and nationwide development companies, using subdivision trusts and integrated activities to attract buyers.[7] Such activities included land aggregation, subdivision platting and construction of streets and utilities, and incorporation of schools, parks, golf courses, community centers, and churches, etc.

Remote sensing maps over 26 years (1984–2010) demonstrate the significant impacts of this process and the built environment on the natural desert habitat. The remote sensing analysis (*Figure 7*) shows how growth clearly altered the natural environment (*Hartfield et al., 2013*). Housing developments replaced agriculture, and golf courses covered once desert landscapes. The population boom and urban sprawl impacted many aspects of the natural environment, particularly water resources and supply *(Benites-Gambiriaso E., Schneier-Madanes G., Cattan N. 2010)*. Some urban growth occurred on the edges of rivers and washes altering riparian areas. This process repeated itself throughout Pima County and beyond, and is evident in all the remote sensing study areas (*Hartfield et al., 2013; Schneier-Madanes et al., 2014*).[8] Three main patterns of urban growth can be seen to result from the combination of land development and water/wastewater access in the Tucson metropolitan area: urban expansion, "leap-frog" development and "wildcat" development.

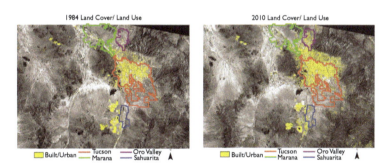

Figure 7 1984 and 2010 land cover classification map in Tucson metropolitan area (*Image Source: University of Arizona School of Natural Resources, Remote Sensing Center*).

---

7 In particular, the interest of California investors in the relatively inexpensive undeveloped land in Tucson and the advent of planned retirement communities helped sustain the growth. (Akros Inc., 2007).
8 This process was explored using historical Landsat 5 Thematic Mapper data from May and June of 1984, 1994, 2004, and 2010. The image data were radiometrically calibrated, atmospherically corrected, and ortho-rectified prior to performing a land use/ land cover classification. A CART algorithm was utilized to create a map with nine classes encompassing the natural and built environment within the study area. Maps were produced with overall classification accuracies above 91 percent. The built environment was then singled out to measure the growth and impact of humans on the natural environment within Tucson, Oro Valley, Marana, and Sahuarita.

2.2.1 Urban expansion development

Urban expansion development refers to those developments which either infill the urban core or extend out of the existing urban fringe. Developers and builders simply acquire the next piece of vacant land adjacent to existing development and build infrastructure that expands the urban core to a modest extent. Impact fees are assessed to all new subdivision developments as a means of paying for the public facilities and infrastructure needed to serve that development *(Morrison Institute 1998)*. With each new homeowner paying these fees to the applicable governmental entity. State law sets a legal framework regulating the determination of impact fees by municipalities' and counties', and all jurisdictions have guidance for assessment of impact and development fees.[9] This type of development has expanded the urban area by acknowledging the preference for single-family housing and the availability of cheap land, and by encouraging development *(Planning Department of the City of Tucson 2002)*. However, despite all the large-scale real estate activity, there has been only a minimal level of regional coordination with respect to land use planning.

Such development physically connects to, and utilizes, the integrated water and wastewater systems of the community, and meets the current planning and zoning regulations (with some special overlay zone criteria for large urban core infill projects). As the project plans are created and reviewed, development agreements are reached with the governing jurisdiction or utility agencies as to the requirements for: setbacks from the right-of-way; property heights; water and sewer connection points; and the need for any additional water and sewer construction to integrate these developments into the general community system. Although these projects have the potential for conflict regarding project density, character of the neighborhood, and noise and traffic issues for the adjoining neighborhoods, the water and wastewater infrastructure is typically integrated into the general community infrastructure in a seamless manner.

There are many examples, in the Tucson Metropolitan Area, of this very common type of urban growth pattern. We have selected one very graphic example, the Continental Ranch development in Marana, to analyze the interplay with water and wastewater challenges (see Section 3.4).

2.2.2 "Leap-frog" development[10]

Another major growth pattern is that of suburban "leap-frog" development, in which dense suburban developments skip over empty land to establish a new urban fringe *(Hayden, 2003)*. "Leap-frog" expands the boundaries of the involved urban area by jumping over established developments and initiating a new suburb as far as two or more miles away. Newly urbanized areas, along with water infrastructure, spring up

---

9 The city of Tucson assesses impact fees for water, roads, parks, police, fire, and public facilities. The county has impact fees for transportation only. Impact fees apply to all new developments and have specific benefit areas such as the Southwest Infrastructure Plan (SWIP) and the Houghton Area Master Plan (HAMP).
10 Leap frog is the game in which one player crouches down and another player vaults over the first.

beyond existing urban boundaries so that developers can avoid paying the higher costs of urban land while obtaining more flexibility in developing larger tracts of rural land.

One of the major factors driving "leap-frog" development in the Tucson Metropolitan Area was pressure from new residents *(Heim 2001, Kupel 2003)*. As noted previously, the real estate market began to boom after World War II, with the massive arrival of migrants from the eastern United States and Midwest. These customers came looking for either open-space single-family housing (the suburban and "ranchette" lifestyle) or a close-knit neighborhood atmosphere (the retirement community lifestyle). These preferences, together with the low costs of land and water, had important consequences.

Growth developed to the east and northeast, in non-urbanized area of Pima County where land was cheaper, and then to areas northwest of Tucson. It is estimated that, in 1992, two-thirds of all residential permits were issued north of the Rillito River, in the fast-growing Catalina Mountain foothills. Although these low-density developments are expensive to service, Tucson Water (along with Metro Water and Oro Valley Water) as well as Pima County provides utility services at no extra charge, as these developments become an integral part of the Tucson Metropolitan Area.[11]

2.2.3 "Wildcat" development

At the same time, "wildcat" subdivisions[12] became common on land outside of the urbanized metropolitan area, particularly to the south and west of Metropolitan Tucson, where little subdivision activity had occurred.

The creation of new residential developments, without the customary limitations of subdivision regulation, results in parcels devoid of any basic infrastructure or improvements—roads, water and wastewater, energy—which must typically be paid for by the developer. As noted in Section 3.2.1, impact or development fees are charged to each new development and new homeowners must pay these fees to the governing jurisdiction. The legal framework for how municipalities and counties determine impact fees is set by state law. In contrast, "wildcat" development, or subdivision lot splitting, is not directly banned by state law, which maintains that a parcel division of less than six portions is not considered a subdivision, and prohibits any governmental jurisdiction from denying approval or requiring a public hearing for these parcels.[13] As the value of land increases there is a growing trend for private

---

[11] In 1970 a slow-growth movement appeared in Tucson, with elected officials advocating infill and a limit to the expansion of the city. One facet of this movement was the effort to revise water utility practices, including raising water rates and charging residents' service fees that were related to delivery costs. The political resistance to growth lasted only a few years and ended with the electoral defeat of most of the slow-growth proponents.

[12] This use of the expression "wildcat" development draws analogy from the independent and solitary nature of the "wildcat", a medium-sized bobcat with a short tail and tufted ears.

[13] What is often not realized is that lot splitting can proliferate into many more "splits" of the same parcel. For example, if a property owner of 100 acres were to first lot split a parcel into five 20-acre parcels, the five subsequent owners would also have the right to lot split their 20 acre-parcels again five times, resulting in 25 property owners of four acres each. Depending on the minimum zoning, which could be as small as one acre per house, these four-acre parcels could be again split, perhaps resulting in a "wildcat" subdivision of as many as 80 to 100 parcels and perhaps 200 or more residents, all without basic improvements, particularly potable water.

ranchlands and rural holdings to be developed as "wildcat" subdivisions. In 1997, 41 percent of the new residential dwelling units in Pima County were not part of platted subdivisions and most of these were issued in rural areas outside the metropolitan area. However, "wildcat" development often devalues property, and can create significant hardships, and sometimes real hazards, for its residents.

3 WATER REGULATIONS

3.1 Effects of the Groundwater Management Act

During the 1980s, to cope with groundwater issues, the state of Arizona established the Groundwater Management Act (see *infra* Chapter 4) that divided central and southern Arizona into four Active Management Areas (AMA; see *Figure 8*) around the cities of Prescott, Casa Grande, Tucson, and Phoenix. Recently, a fifth AMA was established for Nogales. The goal of the Tucson Active Management Area (TAMA) is to provide approaches and mechanisms to help achieve water use and conservation targets. Meanwhile, land/water development practices, and climate conditions favorable to tourism and retirees (including so-called "snowbirds" who spend winters in southern Arizona), have resulted in southern Arizona becoming one of the fastest growing urban regions in the United States. The 1980s saw a transition from

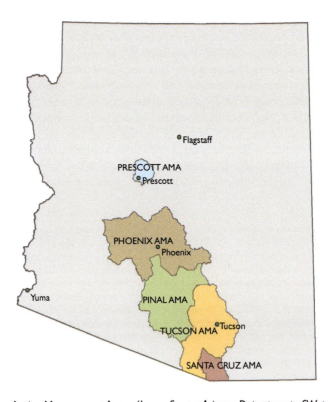

Figure 8 Arizona Active Management Areas (*Image Source:* Arizona Department of Water Resources).

agricultural to municipal use, with municipal demand representing 55 percent of the water supplied in the Active Management Areas (AMA) at that time.

The Groundwater Management Act, which aimed to prevent groundwater overdraft, had strong implications for developers in the AMAs. It established the requirement for an Assured Water Supply (AWS) Certificate, to be obtained by the developer from the Arizona Department of Water Resources after proving that water for a development will be physically, legally, and continuously available for the next 100 years. The aims of this requirement were twofold. One was to protect the buyers of new homes, by requiring the developers (prior to the sale of new homes) to demonstrate that sufficient water supplies are available for the new homes in these subdivisions. The second was to control water demand so as to reach "safe-yield" within the AMA limits; i.e., a balance between withdrawal (including metropolitan, mining, and agricultural activities) and replenishment into the aquifer *(Arizona Department of Water Resources (ADWR), 1999)*.

The Arizona Department of Water Resources (ADWR) created the Assured and Adequate Water Supply Programs to address the problem of limited groundwater supplies. The Assured Water Supply Program functions to protect and preserve limited groundwater supplies within Arizona's five Active Management Areas (AMAs). Outside the AMAs, the Adequate Water Supply Program (not as protective as the Assured Water Supply Program) acts as a consumer advisory program, ensuring that potential real estate buyers are informed about any water supply limitations (two counties outside of AMAs which made satisfying Assured Water Supply requirements a condition of building).

As many neighborhoods were built beyond the reach of existing water and sewer services, builders created private water companies to serve the development's homeowners or negotiated agreements to privately finance the extension of wastewater infrastructure *(Logan 2006)*. The Rita Ranch development, in far southeast Tucson, is a good example of this; the area that now contains this housing development had no water connections in 1984 and by 2010 had nearly 10,000 connections, due to the construction of a privately funded five million dollar (M$5) wastewater interceptor extension.

In fact, in the Tucson metropolitan region, two kinds of water exist: wet water and paper water. The latter includes water credits, assured water supply designations, and water rights. That separation can lead to hydrological contradictions, such as the practice of recharging an aquifer in one place to obtain the right to withdraw water at another location.

3.2 An example of land/water and wastewater interactions: Continental Ranch in Marana

The Continental Ranch development, located in old Cortaro Farms (in Marana) on the west side of Interstate 10 (along the Sun Corridor north/south axis), is a relevant example of water/wastewater and urban growth interactions. The 3000-home development was made possible by construction of a very significant and expensive wastewater pumping system by the developer *(Clavreul et al. 2010)*. This system allows for Continental Ranch wastewater flows to be delivered against gravity several miles to the Pima County Tres Rios Water Reclamation Facility. This pattern of constructing

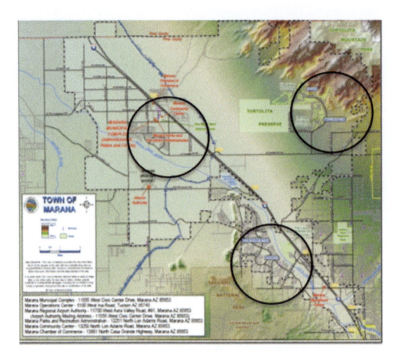

Figure 9 Marana urban development: Three poles of growth (downtown Marana in the northwest, Continental Ranch in the south, and Dove Ranch in the northeast. (Clavreul et al., 2010).

water and wastewater facilities to facilitate development repeats itself throughout Pima County and is evident in all four of the study areas: Tucson, Oro Valley, Marana and Sahuarita.[14]

The Continental Ranch Development illustrates the pattern of conversion of farm and ranch lands into urban developments (*Figure 9*). It represents the classic configuration of slow growth from the interior, out towards new urban boundaries having immediate access to water and wastewater services in the developed areas of the urban core. The Ranch receives water services from Tucson Water and the other urban water companies, and is also provided complete wastewater services by Pima County Regional Wastewater Reclamation. *Figure 10* demonstrates the dramatic increase in development (which will be considered infill development) based on the construction of the pumping station.

Most real estate developers in the Tucson area follow the regulated zoning and subdivision platting process to ensure the value of their product, and to give the public and the buyers of these homes important assurances that their developments have met a variety of environmental and regulatory standards. People pay to be in the "urban core."

14 The increase in the amount of water connections was very high in all four jurisdictions. The number of water/sewer connections in Oro Valley, for example, increased from about 2,000 in 1984 to approximately 11,000 in 2010. Wastewater infrastructure (new treatment plants and conveyance lines) expanded, as did the reclaimed infrastructure that allowed golf courses and resorts to multiply (Clavreul et al. 2011).

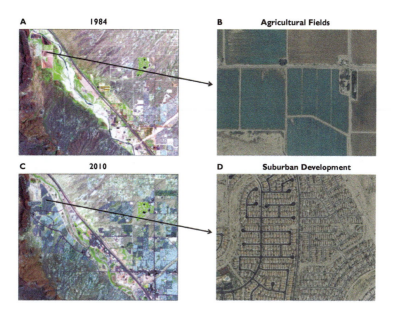

Figure 10 A) LANDSAT image of Continental Ranch area in 1984; B) Aerial view from 1984 showing a portion of that property which is agricultural fields; C) A LANDSAT image showing the same area 16 years later; D) An aerial view of the same section in 2010. (*Image Source: University of Arizona School of Natural Resources, Remote Sensing Center*).

Urban expansion development has been done best via a master plan development approach with community input and acceptance (*Tucson (Arizona) and Pima County (Arizona) 2008*). Many urban expansions in the Tucson Metropolitan area have been small scale, piece-meal, developments that left many parcels undeveloped in the first push of development to the east and then to the northwest. Subsequent attempts to "infill" in these areas have been markedly unpopular with established property owners due to the scale, size of units, number of stories, and anticipated noise and traffic. However, the Continental Ranch Development provides an example where developers worked closely with government agencies and utilities to produce residential and commercial development that complemented the existing areas and led to an increase in desirability and property values for all concerned.[15]

4 CONCLUSION

In arid and semiarid regions, the availability of water and wastewater in rapidly growing urban areas in the Arid Southwest, and other dry lands around the world, raises significant challenges to urban sustainability and water resources. This chapter discusses the different urban growth forms and their implications in the study

---

15 The infill development in the University of Arizona and downtown areas are totally different from these development types and deserve their own study and analysis.

of water and land issues. In this context, we have simultaneously addressed both surface/groundwater water supplies and wastewater access and urban planning models.

Our investigation illustrates the fact that a major challenge for water policy and urban planning is the diversity of land and water/wastewater situations. The complexity of these very large issues will demand multi-disciplinary skills and a sophisticated data acquisition and evaluation approach. It is clear that data collection and resources must be dramatically improved to allow appropriate management and evaluation of these issues. This chapter illustrates how a detailed local approach to urban development is needed to effectively meet water supply and wastewater needs. Such an approach will face specific methodological challenges such as the lack of data on "wildcat" development and domestic wells; e.g., there is no easily available systematic data on the usage and consumption of water from domestic wells across the Southwest.

As a whole, our investigation highlights certain challenges at the different levels of governance (local, state, regional and federal). At a local level, existing water resources management systems will need to be changed to better integrate patterns of urban growth and water and wastewater access. At higher levels of governance, more sophisticated choices will need to be made between obtaining new resources through the construction of new public works and the preservation of scarce local water resources through demand management.

Finally, even though water resources governance has been extensively studied, interest in governance in the field of urban water supply is only now beginning to happen.[16] Our investigation suggests that this governance challenge will increase in the coming years, due to the diversity of the land, water and wastewater situations that calls for more integrated local jurisdictional approach rather than a solely technical, sectorial approach. From this perspective, our investigation contributes to the identification of themes that generalize to drylands around the world, and to the development of lessons for best management practices that may be applied and adapted to other arid regions and situations. It also points to the need for a more specific discussion on water resources availability and wastewater collection/treatment, land and the nature and dynamics of urban growth under conditions of scarcity.

REFERENCES

Akros, Inc. Wilson Preservation Coffman Studios, LLC HDR (2007). *Tucson Post World War II Residential Subdivision Development 1945–1973*, prepared for the City of Tucson, Urban Planning and Design Department Historic and Cultural Resources. http://cms3.tucsonaz.gov/sites/default/files/imported/resources/publications/wwii_102207.pdf.

Arizona Department of Water Resources (ADWR) (1999). *Third Management Plan: 2000–2010 for the Tucson Active Management Area.*

---

16 Paris, for example, is only after the municipalization of the water company, Eau de Paris, that an « upstream approach » has been developed. Taking into consideration that half of the water distributed in Paris is groundwater, the company has developed resource protection programs to avoid important treatments. However, downstream, for wastewater, the company does not have competences (*Schneier-Madanes G., Radioscopie de l'expérience parisienne de l'eau, City of Paris-CREDA, 2014*).

Benites-Gambiriaso, E., Schneier-Madanes, G. and Cattan, N. (2010). *Water and Urban Sprawl: A State of the Art*. Research Report. Tucson, UMI3157-Partners University Fund— PIRVE CNRS.

Clavreul, D., Duczinski, D. and Schneier-Madanes, G. (2010). *Water and Urban Sprawl: Marana Case Study*. Research Report. Partners University Fund & CNRS/UA Joint International Unit UMI 3157.

Colby, B.G. and Jacobs, K.L. (Eds.) (2007). *Arizona Water Policy: Management Innovations in an Urbanizing Arid Region*. Washington: RFF Press.

Hartfield, K., Marsh S., Schneier-Madanes, G. and Curley, E. (2013). *Remote sensing of urban change: water in the Tucson basin of Arizona*. In M.-F. Courel, T. Tashpolat and M. Taleghani (Eds.), WATARID 3: Usages et politiques de l'eau en zones arides et semi-arides. Paris, Hermann.

Hayden, D. (2003). *Building Suburbia: Green Fields and Urban Growth, 1820–2000*. New York, Pantheon Books.

Heim, Carol E. (2001). Leapfrogging, Urban Sprawl, and Growth Management: Phoenix, 1950–2000. *American Journal of Economics and Sociology*, 60(1): 245–283.

Kupel, D. (2003). *Fuel for Growth: Water and Arizona's Urban Environment*. Tucson, University of Arizona Press.

Logan, M. (2006). *Desert cities: the environmental history of Phoenix and Tucson*. Pittsburgh, University of Pittsburgh Press.

Maddock III, T. and P.W. Barroll (2012). Domestic Wells in New Mexico. In D. Brookshire, H. Gupta and O.P. Matthews (Eds.), *Water Policy in New Mexico: Addressing the Challenges of an Uncertain Future*, RFF Press.

Morrison Institute for Public Policy (1998). *Growth in Arizona: The machine in the garden*, edited by J.S. Hall, N.J. Cayer and N. Welch. Tempe, Arizona State University.

Planning department of the City of Tucson (2002). *Census 2000: The Basics – Focus on Tucson*. Available at: http://www.tucsonaz.gov/planning/data/census/census2000basic/basicpubs/funfacts2000.pdf.

Riebsame, W.E. and Robb, J.J. (1997). *Atlas of the new West: Portrait of a changing region*, New York, W.W. Norton & Co.

Schneier-Madanes, G. (Ed.) (2014). *Globalized Water: a question of governance*. Dordrecht: Springer.

Sheridan, T. (2012). *Arizona: a history*. Tucson, University of Arizona Press.

Tucson and Pima County, Arizona (2008). *Water & Wastewater Infrastructure, Supply & Planning Study: a City of Tucson and Pima County Cooperative Project*. http://www.tucsonpimawaterstudy.com/Reports/Phase2/FinalReports_Ph2.html.

Valdes, J. and Maddock III, T. (2009). Conjunctive Water Management in the U.S. Southwest. In G. Schneier-Madanes and M.-F. Courel (Eds.), *Water and Sustainability in Arid Regions*. Dordrecht, Springer: 221–224.

Chapter 10

Comprehensive urban planning: Implications for water management in Pima County (Arizona)

Sergio Segura Calero
Department of Human Geography, University of Seville, Spain

INTRODUCTION

Comprehensive urban planning is an important method for analyzing future scenarios and the contextual historical, institutional and territorial conditions necessary for sustainable development. This chapter examines the evolution of comprehensive planning related to water resources in Pima County, Arizona, and in the City of Tucson, in order to evaluate its implications for water management. The investigation aims to answer the following key research questions: How has the content of comprehensive planning evolved in Pima? What are its implications for water management? And finally, based on an analysis of the latest comprehensive plans, has substantial progress been made in this area?

With more than three generations of development plans in Pima County, there is a relatively solid tradition of, and experience, in the field of comprehensive planning. Therefore, as might be expected, the planning procedures are rather well developed. Our study consists of an analysis of current comprehensive planning legislation, the relevant academic literature and the comprehensive planning documents of Pima County in Arizona. This regulatory information is supplemented with practical insights provided by staff interviews with the departments of Pima: Services Department, Planning Division-Comprehensive Plan (Pima County) and Housing & Community Development Department (City of Tucson).

After a brief introduction to the concepts of comprehensive spatial and water planning, we provide an overview of the current situation of comprehensive planning in the United States, including its origins, definitions and links with water management. Next we characterize the nature of comprehensive planning in Arizona. Finally, we consider the contents of the most recent comprehensive plans for Pima County and the City of Tucson, with a view to answering the research questions and developing our conclusions.

1 COMPREHENSIVE SPATIAL AND WATER PLANNING

Spatial planning, which includes a comprehensive territorial model and associated strategic visions and goals, is an important way of analyzing future scenarios and the contextual historical, institutional and territorial conditions necessary for sustainable development. In the United States, other names for comprehensive spatial planning include 'master plans', 'general plans', 'regional area plans' or 'local government plans'. According to the American Planning Association, the term "comprehensive planning" (local\regional) refers to *"the adopted official statement of a legislative body of a local government that sets forth (in words, maps, illustrations, and/or tables) goals, policies, and guidelines intended to direct the present and future physical, social, and economic development that occurs within its planning jurisdiction and that includes a unified physical design for the public and private development of land and water"* (Meck, 2002). In most of the United States, it is the only planning document that both considers multiple programs and takes into account the various activities occurring on all of the land located within the planning area (whether that property is public or private) (*Kelly, 2010*).

Across the Atlantic, the European Water Framework Directive (WFD) (*EC, 2000*) has highlighted a growing need to explore the connections between spatial planning, water use, conservation and hydrological risk management (*White & Howe, 2003; Hernández-Mora & Del Moral, 2015*). A core concept is that of *integration*, understood as a key tool for the protection of water and associated ecosystems, where the integration is of scientific disciplines, approaches and experiences within a coordinated interdisciplinary dialogue between the social, natural and engineering sciences. The term *Integration* embraces the need for cooperation and coordination between administrative bodies, along with coordination of different decision levels (local, regional, national) and countries. It also refers to users, stakeholders and civil society involved in decision-making processes, promoting a process of social learning aimed at identifying and implementing the most appropriate alternative measures. The end goal is an integration of water management strategies in related sector specific and territorial policies (*Del Moral, 2009*).

In recent years, attention has been focused on the spatial mismatch between different planning scales, highlighting difficulties that frequently arise in relation to the adjustment of the different hydrographic, socio-economic and jurisdictional aspects involved (*Cohen, 2012; Del Moral & Do Ó, 2014*). Examples range from the experiences of the polder system in the Netherlands to the management of desert groundwater by the Arizona Department of Water Resources. In countries having a spatial planning tradition that is based on a comprehensive integrated approach or on land use regulation, the connection with water planning is more easily made (*Bouma & Slob, 2014*). The existing literature on flood hazard mitigation in Arizona shows that community resilience is best achieved when mitigation strategies are integrated with land use and comprehensive planning. Further, comprehensive plans can promote public information and education campaigns so as to increase citizen awareness of potential hazards and the best use of natural resources (e.g. water), and to educate homeowners and builders about the best location and building practices (*Rahul, 2006*). Comparing spatial planning and river basin management plans, *Carter (2007)* and the *EEA (2012)* highlight other general potential synergies such as a long-term strategic focus and the ability to influence a broad range of economic sectors that affect water consumption, pollution and water body status.

2 TOWARDS COMPREHENSIVE PLANNING IN THE UNITED STATES

To understand the origins of comprehensive planning in the United States, it is necessary to appreciate the importance of water in the context of its planning tradition and to be familiar with the current definition and its regulations.

2.1 From original planning to comprehensive planning

One of the basic elements underlying territorial planning in the United States dates from just after the American Revolution, when there was strong concern regarding rights relating to personal private property. In the western states, the same applies in relation to water according to the maxim *"first in time, first in right"*, under which the first person who puts the water to a beneficial use has the higher priority right to its future use than do later users. That principle arose because local authorities had few options to manage early territorial and urban developments.

There was no effective intervention in organic urban development until the mid-nineteenth century, when the Sanitary Reform Movement of 1840 occurred (*Peterson, 1979*). The Sanitary Reform began in order to influence urban and environmental development. For example, the location of the city and the need for large open spaces and parks were considered, based on the ability to achieve hygiene and leisure goals. A classic example of this is the renowned Central Park in New York, designed by Frederick Law Olmsted (*Campbell & Fainstein, 2012*). In the late 19th and early 20th century, the increase in industrial development in large American cities led to the emergence of the City Beautiful Movement. Later, the Garden City urban movement propounded by Ebenezer Howard from the UK also arose in the United States, creating the need for a new comprehensive territorial planning approach.

The federal and state constitutions and laws that provide for the protection of health and welfare constitute the legal basis for the first American comprehensive spatial planning. However, in most states comprehensive plans are not required by law. Two documents, drafted by the United States Department of Commerce in the early decades of the twentieth century, provided a framework for comprehensive planning at the state level. While the first of these was the Standard State Zoning Enabling Act of 1926, which was never approved by Congress, many states adopted it as a framework to institutionalize comprehensive planning. Subsequently, the Department of Commerce drafted the Standard City Planning Enabling Act of 1928. These acts demonstrate the considerable interest shown by the United States government regarding this area, and reflect the lack of planning that existed in the country at the beginning of the 20th century. The federal government also supported local authorities that wanted a comprehensive plan in many states where comprehensive plans were not required (*Cullingworth & Caves, 2013*).

Along with these two enabling acts, the 701 Planning Program of the Department of Housing and Urban Development[1] (1954), and the 208 Areawide Wastewater

---

1 Funds for comprehensive planning under section 701 of the Housing Act of 1954.

Treatment Management Program of the Environmental Protection Agency[2] (1948) reinforced the commitment to the kind of comprehensive planning that focused on long-range planning, defined area, and emphasis on physical environment. Together, they established the terms of comprehensive planning relating to land use, transportation, housing, the environment, parks and open spaces (*green infrastructure*), educational infrastructure, social equality and preservation of historic neighborhoods.

During the sixties and early seventies, spatial planning in the United States was enriched by federal legislative developments such as the National Historic Preservation Act (1966), the Coastal Zone Management Act (1972) and the requirement to assess the environmental impact of government action under the National Environmental Policy Act (1970). Moreover, there were also professional advances as a result of the "(Louis B.) Wetmore Amendment" (1967), whereby the American Institute of Planners added comprehensive planning and regulation of land use to its declaration of purpose of 1938, and also included the concept of the physical planner and the social planner (*Kelly & Becker, 2000*). Meanwhile, the publication of *Design with Nature* (*McHarg, 1969*) at the end of the sixties constituted a major international milestone, introducing a new approach that merged planning processes with consideration of natural environmental characteristics. A few years earlier (1961), Lewis Mumford had explored the development of urban civilizations in his influential book *The City in History*. Harshly critical of urban sprawl, the author argued that the structure of modern cities was partially responsible for many social problems seen in western society and that urban planning should emphasize an organic relationship between people and their living spaces.

2.2 Comprehensive planning: definition and regulation

Early planners in the United States, who were mainly engineers, architects, urban planners and lawyers, agreed on the need for a long-range plan for urban development as public policy. Comprehensive planning is a good way to strengthen relationships between functional elements of a plan over time and space. The three main factors included in a comprehensive plan are (*Kelly, 2010*) a) consideration of the whole jurisdictional territory, b) inclusion of all matters related to physical-territorial community development, and c) a long-term horizon for the plan. Early examples of comprehensive planning are the works of celebrated planners such as the Burnham Plan of Chicago in 1909 (*Smith, 2006*). They provided a vehicle by which to involve society in the development and land use planning process, although early plans typically only involved certain members of the elite in the decision-making processes.

These plans were supported by the federal and state administrative levels, as they were the only ones that considered the breadth of different programs and activities in the area regardless of whether they involved public or private property. This fact is quite significant in the context of the evolution of private property in the United States. In addition, comprehensive planning could be seen as a democratic tool at

---

2 Clean Water Act 1974 SEC. 208 [33 U.S.C. 1288] Areawide Waste Treatment Management. The basis of the Clean Water Act was enacted in 1948 and was called the Federal Water Pollution Control Act.

the local level, since participation is both desirable and necessary for comprehensive planning.

Moreover, there are two main levels of government that dominate planning and control of land use in the United States: municipalities (cities and towns) and counties (similar to European regions). However, another administrative unit exists, namely townships. This is a reference plot of 6 square miles, which in turn is divided into 36 sections. It is used as a reference map for development of detailed local tasks but is also sometimes used for planning land use and zoning (*Kelly & Becker, 2000*).

While comprehensive planning is typically a local government matter, states can also formulate comprehensive plans encompassing the whole state. For example, the State of Oregon has a Statewide Plan, consisting of a comprehensive plan that provides a framework for regional and local comprehensive planning. Furthermore, although there is no federal jurisdiction in this matter, comprehensive planning is affected by sector specific legislation. An example of sector specific laws affecting comprehensive planning in the United States is the mandatory review of environmental conservation established by the Federal Congress (US Code 16 Conservation) for Comprehensive General Management Plans of American National Parks.

The approved Standard City Planning Enabling Act of 1928 focuses on the creation of a Master Plan and a planning commission for local governments. This new legislation developed important concepts, content and procedures to be performed by planning commissions. One such procedure is the collection of information through comprehensive studies and surveys to meet present and future conditions relating to urban growth. It also includes elements to enhance democratization and transparency, such as requirement to hold public hearings before adopting or readopting the plan or amendments to the same.

These first laws and regulations, such as the Planning Program 701 of the Department of Housing and Urban Development, include examples of comprehensive planning based on land use. The subsequent development of state laws and regulations affecting comprehensive planning in the 1980s led to the strong emergence of ideas and laws relating to Smart Growth cities, as will be seen in the following section.

3 COMPREHENSIVE PLANNING IN ARIZONA

In Arizona, the Arizona Revised Statutes (ARS) have (since 1973) required most cities, towns and counties to develop comprehensive plans to regulate their land use, traffic, housing, public services and facilities and the conservation, rehabilitation and reconstruction of their territories, among other structural elements. This boost for comprehensive planning was a result of the Smart Growth urban theory. According to the Arizona Office of Smart Growth,[3] Arizona has been working over past two decades to manage the smart growth of its cities and to preserve the State's open spaces. In the next subsection, we discuss the origins of '*Smart Growth*' and how it has influenced comprehensive plans, and present brief summary of State of Arizona's comprehensive

---

3 http://old.azcommerce.com/SmartGrowth/ (Arizona Office of Smart Growth, Arizona Commerce Authority, March 2015).

plans. The context of the State of Arizona is a good framework to assess the comprehensive plans developed by Pima County, and to examine their implications for water management.

3.1 The influence of smart growth regulations

Population growth rates in many American cities increased significantly during the 1990s and a general socio-political movement emerged to offer solutions to manage this rapid growth. Arizona is one of the fastest-growing states in the nation, and it took a rather conservative approach to growth management and open space preservation policies until the late 1990s (*Zinn, 2004*). Through coordination among concerned citizens, members of the Arizona State Legislature, and the Office of the Arizona Governor, a comprehensive effort was made to address issues relating to urban growth, leading to passing of the "*Growing Smarter Act*" of 1998. This law strengthened aspects of comprehensive planning and added four new elements: open spaces, new growth areas, modern environmental planning, and cost of development. The subsequent Growing Smarter Plus Act of 2000 further developed these aspects and others related to land use planning in Arizona.

Together, these laws comprise the Growing Smarter Legislation of Arizona. They support counties, cities, towns and citizens with a variety of tools and restrictions to manage their growth, such as the right to approve municipal comprehensive plans by referendum. These principles have been enshrined in the Arizona Revised Statutes, which establish a legal basis for declaring and preserving open spaces that goes beyond the requirements laid down by the State Land Department. Their main objective is to ensure that future territorial development in the State is carried out in a rational, efficient and environmentally friendly manner, and to foster public interest by promoting natural heritage protection in balance with economic competitiveness.[4]

In addition to the strong urban orientation of Smart Growth laws, the Growing Smarter Planning Grant program was created to provide assistance through subsidies for rural communities and counties to draft their own comprehensive plans meeting the new legal requirements. Only rural communities or counties engaged in the development of a comprehensive plan are eligible for these grants. The Growing Smarter Legislation also led to the development of Conceptual State Land Use plans by the State of Arizona, although these are not yet in force. These conceptual, spatially comprehensive, plans are for open spaces designated by the State to be Urban State Trust Lands. Basically, these are open spaces located within a mile of cities and towns having less than 250,000 inhabitants (within three miles of larger cities). As per the Arizona Revised Statutes, these spaces are defined and regulated via general plans. The State is required (among other functions) to constitute a commission to prepare and review the plans regularly (at least once every ten years). For example, the following requirements are established in relation to water:

---

[4] http://old.azcommerce.com/SmartGrowth/ (Arizona Office of Smart Growth, Arizona Commerce Authority, March 2015).

- The department commission must cooperate with the Arizona Department of Water Resources to determine that urban lands have the quality and quantity of water needed for urban development.
- There must be consideration of proximity to and impact on public facilities, including streets and highways, water supply systems, wastewater collection and treatment systems and other public facilities and services necessary to support development.

The Growing Smarter Legislation grants powers to the State to determine practices and common interests in areas declared to be Urban State Trust Lands. Theoretically, if there was no comprehensive plan, the conceptual land use plans developed by the State Commission could designate these areas as suitable for urban development or to conserve resources if necessary. The legislation establishes criteria consistent with the Growing Smarter Legislation to prevent urban sprawl and to promote common interests and long-term planning. These State plans must be reviewed by local and regional authorities and planning agencies. The State plans must also be consistent with other local and regional public policies and plans.

3.2 Characterization of comprehensive plans

The benchmark for comprehensive plans in the United States was established by the American Planning Association in 2002. The following table (*Figure 1*) shows the content and basic characteristics of the comprehensive plan for the State of Arizona.

Although the Pima County plans have been or are in the process of being renewed, they still incorporate many elements of the characteristics presented above. For example, the county comprehensive plans are closely related to the 1920s statutes and have a weak State presence. Also they identify elements in the fields of agriculture, forests and preservation of open spaces. In this regard, it should be remembered that Pima County has a large region of agricultural production closely linked to water management.

The relationship between comprehensive plans and models to regulate land use planning is evident. The principles expressed by the Arizona Growing Smarter Legislation have not yet been fully incorporated into the comprehensive plans approved before 2002. As a clear example, features such as institutional coordination and cooperation and policy implementation are not significantly developed.

Finally, according to the Arizona Revised Statutes, the comprehensive plan may include studies and recommendations relative to water quality and floodplain zoning. The county comprehensive plan is required to have a water resources element that addresses:

- The known legally and physically available surface water, groundwater and effluent supplies.
- The demand for water that will result from future growth projected in the comprehensive plan, added to existing uses.
- An analysis of how the demand for water that will result from future growth projected in the comprehensive plan will be served by the water supplies identified in subdivision (a) of this paragraph or a plan to obtain additional necessary water supplies.

Similarity to 1920s Statutes	Moderately updated (many significant changes but still resembles the 1920s model planning laws in some way)
Local Plan Mandated By Municipality	It's mandated if a precondition is met such as when a planning commission is created (Cities & Towns); Mandatory for Counties.
Local Plan Mandated	It's mandated if a precondition is met such as when a planning commission is created.
State Land Use Policy Basis	NO (an exception with Conceptual State Land Use plans)
Model Dev.Code Influenced	YES
Strength of State Role	WEAK
General Citation	A.R.S. Cities & Towns 9-461; A.R.S. Counties 11-801
Municipality Covered in caps	Cities &Towns, and Counties;
Internal Consistency Required	No

ELEMENTS in Comprehensive Plans:
Land Use (2)
Housing (2)
Agriculture, Forest Land, Open Space Preservation (3)
Natural Hazards (1)
Redevelopment (1)
Recreation (2)
Energy (1)
Air Quality (1)
Transportation (2)
Community Facilities (1)
Policy (2)
Visioning or Public Participation (2)
Other Plans: Specific (Optional_3)

Economic Development **NO**
Urban Growth Limits **NO**
Critical & Sensitive Areas **NO**
Human Services **NO**
Community Design **NO**
Historic Preservation **NO**
Implementation **NO**
Local Coordination **NO**
Other Elements **NO**

Note: Detail 1 = little; Detail 2 = Moderate; Detail 3 = substantial.

Figure 1 Summary of State of Arizona Statutory Requirements for Comprehensive Plans.
Source: Compilation based on Meck (2002).

For counties having more than 200,000 inhabitants, the environmental planning element must contain analyses, policies and strategies to address anticipated effects (if any) of plan elements on air quality, water quality and natural resources associated with the proposed development under the comprehensive plan.

According to the Arizona Revised Statutes, cities and towns with a population between 2,500 and 10,000 persons having growth rates exceeding an average of two per cent per year for the ten year period before the most recent United States decennial census, and cities and towns having a population greater than 10,000 person, their plans must satisfy the same water resource element and environmental element requirements.

For cities with a population in excess of 50,000 people, a conservation element must be included that addresses the conservation, development and utilization of natural resources, including forests, soils, rivers and other waters, harbors, fisheries, wildlife, minerals and other natural resources. Further, a safety element for the protection of the community from natural and artificial hazards must be included.

4 PIMA COUNTY COMPREHENSIVE PLANS

In this section, we briefly discuss the comprehensive plans developed for Pima County and the City of Tucson, present their basic terms, and compare and evaluate the changes and treatments of water management.

4.1 Pima County Comprehensive Plan (2001)

The current Pima County comprehensive plan was approved in 2001, and updated through amendments until 2012. The Plan covers the entire county and almost one million inhabitants. The initial plan consists of four documents:[5]

- Sonoran Desert Conservation Plan (SDCP)
- The Conservation Lands System (CLS): Regional Land Policy map
- The Land Use Intensity Legend (revised in 2012): residences per acre restrictions
- Rezoning (RP) and Special Area (S) Policies Plan (revised in 2003): local actions

Within these documents, there is little emphasis on water resources. However, the following restrictions exist in regards to riparian areas under the Natural Resources element of the Conservation Lands System (CLS):

- Important riparian areas are essential elements in the CLS, and are valued for their higher water availability, vegetation density, and biological productivity. In addition to their inherent high biological value, these water-related communities (including their associated upland areas) provide a framework for linkages and landscape connections. They are characterized hydro-riparian areas (where vegetation is supported by perennial watercourses or springs), meso-riparian areas (where vegetation is supported by perennial or intermittent watercourses or shallow groundwater) and xero-riparian areas (where vegetation is supported by an ephemeral watercourse).
- At least 95 percent of the total acreage of lands within this designation shall be conserved in a natural or undisturbed condition. Every effort should be made to protect, restore and enhance the structure and functions of *'Important Riparian Areas'*, including their hydrological, geomorphological and biological functions. Areas within an Important Riparian Area that have been previously degraded or otherwise compromised may be restored and/or enhanced. Such restored and/or enhanced areas may contribute to achieving the 95% conservation guideline for Important Riparian Areas.

These four documents, updated in 2012, comprise the Regional Plan Policies. It is an interesting classic example of comprehensive planning integrating the topic of land use with environment and natural resources, circulation, growing areas and open spaces management, among other issues. The section devoted to water resources has

---

5 http://webcms.pima.gov/government/development_services/land_planning_and_regulation/ (March, 2015).

five main objectives: information, efficiency, reliance upon renewable water supplies, groundwater-dependent ecosystems protection and minimize impacts.

Additionally, the county staff that handle planning are required to conduct a Water Supply Impact Review (which may recommend comprehensive plan amendments), and the Pima County Regional Flood Control District staff are required to conduct a Water Resource Impact Assessment in relation to any re-zoning that requires a site analysis. Approval of the proposed development must be considered in terms of potential mitigation of increased water demand projected between existing zoning and proposed re-zonings.

Management tools may also be used by Pima County in moving towards a more sustainable water future based on a set of 20 policies that address water use requirements, coordination with other counties, reduced per capita consumption, limited turf water, pumping and human groundwater use, revised design and construction standards, and utilization of effluent and surface water for riparian restoration. Apart from these, there are another 18 policies concerning conservation water and groundwater quality measures. According to the land use element of the plan, 12 policies are established for wastewater treatment and flood control.

In addition, there is growing awareness of the need for policy measures to address the institutional dimensions of the water-energy nexus (*Scott et al., 2011, 2015*). The relationship between energy and water is clearly highlighted in the document, particularly in relation to solar energy. Power plants require fresh water for cooling, and energy is required to pump, move and treat water. Solar energy systems conserve water that would have been lost during power generation, and conserving water helps reduce energy demands. More specifically, the combined use of passive solar energy systems with water conservation methods (e.g. planting shade-producing landscape that is irrigated with harvested rain water) creates more sustainable development.

Finally, the document highlights the need to establish a Concurrency Management System with the aim of integrating all specific plans for water, wastewater and other urban elements.

4.2 Pima Prospers (Draft 2015)

The updated comprehensive plan for Pima County, called '*Pima Prospers*', was recommended to the Board of Supervisors of Pima County by the Planning and Zoning Commission. The Commission recommended several amendments in relation to land use. According to the Pima County Government, the Pima Prospers plan is the result of an 18-month process, including extensive community involvement with participation at all levels, coordinated efforts with other departments, and the opinion and support of community leaders, residents, businesses and other stakeholders, including other regional and State agencies.[6] The Pima Prospers 2015 proposal, which is yet to be approved by the Pima board of supervisors, has structural elements that are consistent with the requirements of the Arizona Revised Statutes such as land use, circulation, physical infrastructures, economic development, and cost of development.

---

6 See: http://webcms.pima.gov/government/pima_prospers/ (March 2015).

Water resources, wastewater management and flood risk are specifically discussed in the infrastructure chapter of the proposal. The State-mandated water resources element requires counties to perform a basic comparative analysis of known water supply and demand, to establish whether proposed new developments have an impact on the overall water supply.

The plan has five water resource objectives, and briefly develops a series of policies to meet them. The key issues discussed include future water security, efficient water use and groundwater quality protection.

Wastewater management is effectively addressed by the plan, resulting in an efficient and operational wastewater treatment system in Pima County. Furthermore, Pima Prospers 2015 maintains important relationships between water, clean energy and reduction of energy consumption. Fossil fuel energy production requires water and it is therefore necessary to promote energy efficiency. Renewable energy (especially solar energy) also has major sustainability potential. In addition, the Pima County Regional Flood Control District strives to use forward-looking floodplain management practices to minimize flood and erosion damage for all county residents, property and infrastructure. To achieve the Pima Prospers goals for flood control, Pima County is involved in a variety of flood monitoring, flood control and natural resource management activities.

According to the Pima County Planning Department, physical planning and land use planning were traditionally the most important elements of the plans. In contrast, the new plan is much more comprehensive in nature, in part as a result of the legal requirements that govern such plans. For example, while land use planning is still a fundamental element, the plan also includes comprehensive planning such as public health, an area over which Pima County has significant power.

Pima Prospers further develops goals and action policies, and establishes implementation measures, which constitute one of the main difficulties when carrying out the policies. The main mechanism for implementation of comprehensive planning is *re-zoning,* as it constitutes a municipal ordinance (cities and towns) to manage land use. Although a much localized ordinance, re-zoning has to be consistent with the county comprehensive plan.

The new plan further reinforces the idea that coordination and cooperation is a fundamental condition, and helps to guarantee the success of these plans. Pima County Development Services Department staff (Planning Division) recognized a significant lack in coordination at different administrative levels. All of the authorities now work together and give opinions about growth developments in Pima County cities and towns. Moreover, the coordination, cooperation and sector specific planning between Pima County and the State go beyond the legal requirements, including collaboration on various comprehensive planning works.

The Pima County Development Services Department staff appreciate that, in the current situation, there are fewer economic resources. An economist has been hired to work on the efficiency of the Pima Prospers 2015 plan. Even so, according to people interviewed for this chapter, the new county comprehensive plan is the largest effort to produce and manage a comprehensive plan in Pima to date. The officials that were interviewed highlighted the fact that there has been a noteworthy development in terms of participation projects. Pima County has improved public participation and decision-making relations in the area of comprehensive and water management far beyond the minimum legal requirements. They have conducted numerous public hearings for large new developments and substantial amendments of the previous plan, although referendums are not required in county cases as is the case in municipalities (cities and towns). It is also significant that the new plan format is much more user-friendly and understandable for the general public.

4.3 City of Tucson: from General Plan (2001) to Plan Tucson (2013)

The City of Tucson is the largest city in the region and, consequently, Tucson is the main case study to follow in regards to the development of a municipal comprehensive plan in Pima County (*Table 1*).

In 2015, the City of Tucson approved a new comprehensive plan. The previous one was known as the City of Tucson General Plan 2001, the general features of which were based on the Smart Growth Legislation of Arizona in the late 1990s. The plan proposes guidelines and policies for the city to move forward as a whole, and comprises the framework for Tucson's local regulations and sector specific plans. However, the end goal of this project is not simply adoption of a plan, given that the policies need to be implemented through legal action by public authorities. The plan document itself identifies the need to develop a series of indicators and yardsticks in order to control the complete process.

Accordingly Chapter 7 of the Plan briefly discusses flood risks, and Chapter 8 establishes efficiency targets relating to resource consumption and water quality conservation. Fifteen primary action policies are outlined that include, among others, administrative inter-coordination, public participation in water processes, water quality and quantity conservation guarantees and the need for water reuse programs. Other water quality and conservation policies are mentioned very briefly in the final chapter of the plan, in a document relating to Environmental Planning and Conservation.

In 2013, the City of Tucson comprehensive plan, known as Plan Tucson, was subject to a referendum vote and approved. By law, the Plan must be updated every ten years. Plan Tucson is, of course, a new document in comparison to the General Plan of 2001. According to City of Tucson Housing and Community Development Department staff, a decision was made to not update the previous document. Consequently, a new plan was created to adapt to the new situation in Tucson that has emerged during the interim period that spans more than ten years. The new compre-

Table 1 Pima County municipal comprehensive plans.

Cities & Towns	Population (est. 2013)	Comprehensive plan	Date
City of Tucson	526.116	Plan Tucson	2013
Oro Valley	41.627	General Plan	2005
Marana	38.290	Strategic Plan II	2012
Sahuarita	26.870	General Plan	2003
South Tucson	5.696	Zoning map 2011	2011

Source: Own elaboration based on United States Census Bureau 2013 (http://www.census.gov/).

Arizona State Statute Required Elements (ARS 9-461.05)

Plan Tucson Elements	Land use	Circulation	Open Space	Growth Area	Environmental Planning	Cost of Development	Water Resources	Conservation	Recreation	Public Services & Facilities	Public Buildings	Housing	Conservation, Rehabilitation, & Redevelopment	Safety	Bicycling	Energy	Neighborhood Preservation & Revitalization
Housing												✓					✓
Public Safety	✓													✓			✓
Parks & Recreation	✓	✓							✓	✓	✓						
Arts & Culture	✓		✓		✓				✓	✓	✓		✓				✓
Public Health	✓	✓												✓			✓
Urban Agriculture	✓																
Education	✓								✓	✓	✓		✓				✓
Governance & Participation																	
Jobs & Workforce Development	✓			✓													
Business Climate	✓	✓	✓	✓		✓			✓	✓	✓	✓		✓			✓
Regional & Global Positioning	✓	✓		✓						✓							
Tourism & Quality of Life	✓	✓							✓	✓	✓				✓		
Energy & Climate Readiness					✓											✓	
Water Resources					✓		✓										
Green Infrastructure			✓		✓			✓									
Environmental Quality					✓												
Historic Preservation													✓				✓
Public Infrastructure, Facilities, & Cost of Development					✓	✓				✓	✓						
Redevelopment & Revitalization													✓				✓
Land Use, Transportation, & Urban Design	✓	✓	✓	✓								✓			✓	✓	✓

Figure 2 Plan Tucson 2013 Mandated Elements Matrix.
Source: Plan Tucson 2013.

hensive plan document consists of a strategic general plan and a sustainability plan. It is divided into three sections or areas that contain different elements, objectives and action policies. There is also another important chapter devoted to implementation measures. These different elements under Plan Tucson correspond to the requirements established by State law in Arizona (*Figure 2*).

The plan identifies the need to ensure water supply of sufficiently high quality for human activities and the natural environment over a long-term period. The document also promotes renewable water supplies to avoid over-exploitation of local aquifers. It includes Colorado River resources from the Central Arizona Project (CAP) (*Figure 3*).

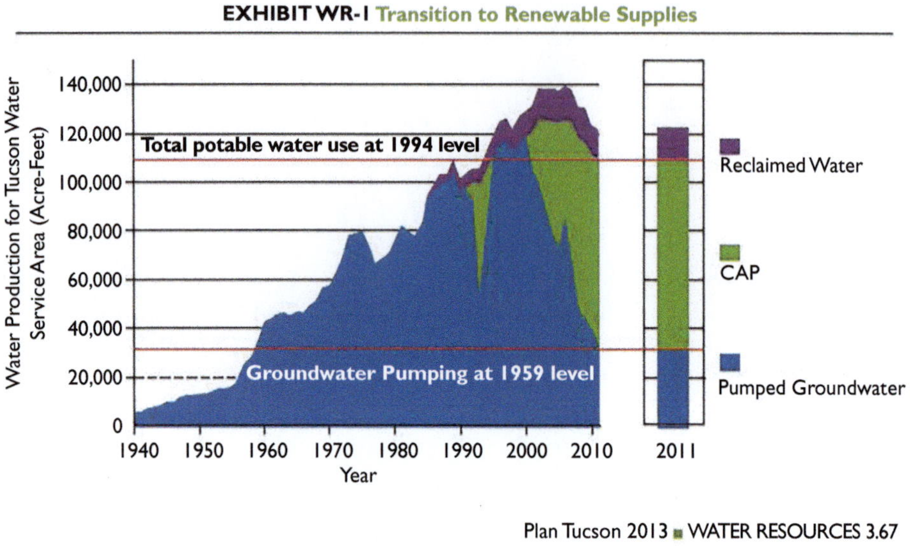

EXHIBIT WR-1 Transition to Renewable Supplies

Figure 3 Transition to Renewable supplies of water.
Source: City of Tucson Water Department. Plan Tucson 2013.

The general goal of water management (in terms of water resources) under Plan Tucson is the habitual *safe yield* established in the Groundwater Management Act of 1980, meaning that no more water is withdrawn from the groundwater aquifer than is replenished.

Actually, according to other authors, there is no sector specific or territorial plan that aims to achieve the goal of *safe yield* in the Tucson basin. However, three main management strategies have been combined to pursue *safe yield*: conservation, growth control and substitution of groundwater by additional CAP supply. The main management strategy to pursue *safe yield* is to substitute other resources to diminish overdraft of fossil groundwater. These authors also identify disconnect between recharge and recovery, which has important local implications to condition of the water table. Currently, these growth control strategies have been implemented only in the agricultural sector, and conservation measures have only been effective for residential water consumption of larger urban areas. Moreover, the technical achievement of *safe yield* at the basin level is uneven and there are significant areas in which overdraft continues, especially in new development locations (*Cabello et al., 2015*).

In addition, the responsibility for water management is somewhat relaxed under the Pima County comprehensive plan (Pima Prospers 2015 draft) because the County is not a water resource provider. Although at this regional scale, water management and planning is a matter of interest, full powers are not granted in this area. The city of Tucson does, however, have these powers and it is also a major water provider. In particular, the Tucson Water Department is the largest municipal water provider in

WATER RESOURCES

WR1	Continue to plan and manage the City's water supplies, quality, and infrastructure for long-term reliability and efficiency.
WR2	Expand the use of alternative sources of water for potable and non-potable uses, including rainwater, gray water, reclaimed water, effluent, and stormwater.
WR3	Expand effective water efficiency and conservation programs for City operations and for the residential, commercial, and industrial sectors.
WR4	Ensure an adequate amount of water to meet the needs of riparian ecosystems.
WR5	Protect groundwater, surface water, and stormwater from contamination.
WR6	Integrate land use and water resources planning.
WR7	Collaborate on multi-jurisdictional and regional water planning and conservation efforts.
WR8	Integrate the use of green infrastructure and low impact development for stormwater management in public and private development and redevelopment projects.
WR9	Provide opportunities to supply alternative water sources for sewer system flush.
WR10	Continue to manage the City's Water Service Area, considering service area expansion only when it furthers the long-term social, economic, and environmental interest of City residents.
WR11	Conduct ongoing drought and climate variability planning.

Figure 4 Plan Tucson Water Resources Policies.
Source: Plan Tucson 2013.

the region and plays an important role in assuring long-term, high quality, dependable water supply. Plan Tucson 2013 proposes policies to manage these objectives and infrastructures for long-term reliability and efficiency (*Figure 4*):

Another noteworthy aspect of the new plan is the emphasis on administrative cooperation and coordination. The new Plan Tucson explains the competences of some water institutions and departments and also integrates the content of the Groundwater Management Code with the results of the first regional joint project study to plan the area's water future between Pima County and City of Tucson: Water & Wastewater Infrastructure, Supply & Planning Study 2008–2010. This joint project study defines a new paradigm for water resource management, by:

1 Recognizing scarcity and uncertainty.
2 Including the natural environment as a recipient of water.
3 Balancing water supply and demand.
4 Building upon the link between urban form and water use.
5 Elevating public discussion of water resource planning to a central position in the future.

The study also identifies the 3 pillars of the joint project to be:

1 Aggressive demand management.
2 Development of new water supplies.
3 Guiding growth in terms of urban form, density, and location.

This joint study highlights a historic disconnection between land use planning and water planning. This problem has had negative impact on the region (declines in the groundwater level, growth in wrong places, habitat loss or degradation, groundwater contamination, and increased flooding). The cost of this growth has been born by local governments, other service providers, and taxpayers.[7] The action policies of Plan Tucson 2013 focus on these problems.

In fact, Plan Tucson has many elements in common with the previous plan, mainly due to state regulatory requirements. The main objectives relating to supply guarantee and quality of water are also common. According to Housing & Community Development Department of Tucson staff, other issues such as inter-coordination are further reinforced and assured. Moreover, excellent cooperative efforts relating to water have been carried out between the Tucson Water Department and the Pima County authorities under Plan Tucson. In addition, the relationship between sustainability, energy and the decrease of water consumption is again a key element, requiring significant coordination among several public departments and the private sector.

Further, efforts have also been made to improve the implementation phase of the Plan. Its recommendations go beyond the legal requirements and powers contemplated, because it is able to establish zoning for new land use. It details measures for implementation of policies. This makes the plan highly useful for Tucson, because the implementation phase typically constitutes the main problem.

The staff of the Housing & Community Development Department of Tucson mentioned the significant progress that has been achieved in the area of public participation. They confirmed that the plan takes a bottom-up approach, and was developed following a long and intense process with public participation as its central axis, whether at a neighborhood level, with NGOs, or the Chamber of Commerce. The Department also highlighted the fruitful dialogue among all the stakeholders regarding the different policies, which constituted an extraordinary interchange of ideas and experiences and an opportunity to learn more. In addition, the Plan is very user-friendly in nature and is easily understood by the general public.

5 CONCLUSION

These two scales of comprehensive planning (*Figure 5*) – City and County – are comparable to the kind of spatial planning carried out in European countries, exploiting the benefits of a statewide comprehensive planning as a framework for regional and sub-regional planning. In Europe, it is recognized that even in cases where these state plans are not supported by state constitutions they can provide coherence and coordination, and form the basis of strategic guidelines for spatial plans. In the case of Pima County, although there is no statewide comprehensive plan, there are certain commonalities between the plans analyzed. The Pima County planning experience spans three generations of plans. There is a clear change in the comprehensive planning model towards more integrated, sustainable, strategic and flexible comprehensive planning. These new plans are clearly also more user-friendly, and exceed the public participation require-

---

7 City of Tucson and Pima County. "Integrating Land Use Planning with water resources and infrastructure". Technical Paper, July 2009. Source on-line: http://webcms.pima.gov/UserFiles/Servers/Server_6/File/Government/Wastewater%20Reclamation/Water%20Resources/WISP/071609-Integrating.pdf.

	Pima County Comprehensive Plan		City of Tucson Comprehensive Plan	
	Pima County Comprehensive Plan 2001	Pima Prospers 2015	City of Tucson General Plan 2001	Plan Tucson 2013
Style	Land Use Planning (specific topics)	Comprehensive Planning	Smart Growth Planning	Strategic & Sustainable Planning
Administrative Coordination & Cooperation	Concurrency Management System	Good relations, beyond law requirements	Support for efforts to improve regional cooperation and communication	Good relations, beyond legal requirements in the absence of conflict
Water Resource Element	Water infrastructure, natural hazards, reduce demand and conserve quality	Detailed policies, comprehensive management	Water infrastructure, natural hazards, reduce demand and conserve quality	Detailed policies, comprehensive management against scarcity and uncertainties
Energy & Water	Solar energy systems coordination with water conservation strategies	Renewable energy alternatives to conserve water	Energy and water conservation measures	Sustainability, energy and reduce water demand
Implementation Measures	Not incorporated	Incorporated (Interagency Monitoring Team)	Not incorporated	Incorporated (The STAR Community Rating Systems[8])
Public Participation	Without a specific project	Important public participation project	Extensive public participation program (Grow.Smart.Leg. requirement) +referendum	Important public participation project +referendum
Document Design	Technical Reports	User-Friendly Booklet	Report	User-Friendly Booklet

Figure 5 New comprehensive planning challenges in Pima County and City of Tucson.
Source: Own elaboration.

ments of 1990s Smart Growth Legislation. Moreover, commitments have been made to pursue cooperation and coordination in an effort to resolve this key problem.

The comprehensive planning experience of Pima County and the City of Tucson over the years has also positively influenced water management under the plans. Administrative coordination, cooperation and public participation have been improved and there are more detailed contents and action policies. However, as is frequently the case in Europe, the comprehensive planning aims are limited to coordination of pre-existing sector specific policies and plans rather than their integration. This is the case of the water policy aspects under the Pima comprehensive plans relating to the *safe yield* objective. At the same time, important progress remains to be made in the area of comprehensive planning to achieve this goal, and to conduct planning in a truly comprehensive manner. Safe yield measures require a truly integrated approach because water and territorial relations involve more complex measures than simply reducing residential consumption or other specific water policies. In conclusion, there is a lack of integration in comprehensive planning that needs more adequate resolution. Consequently, comprehensive planning requires overarching measures to resolve spatial inequalities relating to water management, for example disconnections between aquifer recharge and recovery that have a local impact on water tables or the continuous aquifer overdraft as a result of the water demands of extensive urban areas or new developments.

On the one hand, while the flexible nature of comprehensive planning allows for the development of guidelines and policy goals, obligations are not effectively established. Meanwhile, comprehensive planning cannot be considered effective without effective implementation. This contradiction usually causes a real conflict when policymakers assume that legal matters and responsibilities have to be defined. There have been important efforts to incorporate implementation measures for proposed policies in recent Pima County plans. The Pima County and City of Tucson comprehensive plans have mechanisms to monitor their progress and to verify the achievement of goals, but not to assign specific responsibilities to guarantee their accomplishment.

Greater effort still must be made to effectively engage society in comprehensive planning decision-making processes, and public participation stands out as one of the biggest achievements of the latest comprehensive planning process for Pima County, especially in relation to water in this arid region. It goes far beyond passive public participation (such as municipal referendums for plans) and supports active public participation. The departments responsible for this confirm that they have received valuable contributions as a result of these experiences. Although there are technical matters that cannot be resolved through participation, a comprehensive planning approach with a user-friendly format for citizens (that allows them to understand the reality of these problems) is essential to ensuring effective public involvement in a successful common territorial project.

REFERENCES

Bouma, G. and Slob, A. (2014). How Spatial Planning Can Connect to River Basin Management. In J. Brils et al. (Eds.) *Risk-Informed Management of European River Basins. The Handbook of Environmental Chemistry*, 29: 321–345.

Cabello, V., Hernández-Mora, N., Serrat-Capdevila, A., Del Moral, L. and Curley, E. (2015). Water use and sustainability in the Tucson basin: Implications of a spatially neutral groundwater management (cf. *infra* Chapter 15).

Carter, J.G. (2007). Spatial planning, water and the Water Framework Directive: insights from theory and practice. *Geographical Journal*, 173(4): 330–342.

Campbell, S. and Fainstein, S.S. (Eds.). (2012). *Readings in planning theory* (3rd Ed.). Oxford: Wiley-Blackwell.

Cohen, A. (2012). Rescaling environmental governance: Watersheds as boundary objects at the intersection of science, neoliberalism, and participation. *Environment and Planning A*, 44: 2207–2224.

Cullingworth, B.J. and Caves, R. (2013). *Planning in the USA: Policies, Issues, and Processes*. London: Routledge.

Del Moral, L. (2009). Nuevas tendencias en gestión del agua, ordenación del territorio e integración de políticas sectoriales. *Scripta Nova, XIII* (285). Retrieved from http://www.ub.edu/geocrit/sn/sn-285.htm.

Del Moral, L. and Do O, A. (2014). Water governance and scalar politics across multiple-boundary river basins: states, catchments and territorial powers in the Iberian Peninsula. *Water International,* 39(3): 333–347.

EC (2000). Directive 2000/60/EC of the European Parliament and of the Council establishing a framework for Community action in the field of water policy. OJ L327, 22.12.2000.

European Environment Agency [EEA] (2012). *Territorial cohesion and water management in Europe: the spatial perspective.* Luxemburg: Publications Office of the European Union.

Hernández-Mora, N. and Del Moral, L. (2015). Evaluation of the Water Framework Directive implementation process in Europe. *Sustainable Water Action Network Project.* Deliverable 3.2. Retrieved from http://swanproject.arizona.edu/sites/default/files/Deliverable_3_2.pdf.

Kelly, E. (2010). *Community Planning: an introduction to the comprehensive plan* (2nd edition). Washington, D.C.: Island Press.

Kelly, E. and Becker, B. (2000). *Community Planning: an introduction to the comprehensive plan.* Washington, D.C.: Island Press.

Meck, S. (2002). *Growing Smart Legislative Guidebook: Model Statutes for Planning and the Management of Change.* Chicago: American Planning Association Publications Office.

McHarg, I.L. (1969). *Design with Nature.* New York: American Museum of Natural History [by] the Natural History Press.

Mumford, L. (1961). *The city in history: its origins, its transformations, and its prospects.* New York: Harcourt, Brace & World.

Peterson, J.A. (1979). The Impact of Sanitary Reform upon American Urban Planning, 1840–1890. *Journal of Social History,* 13(1): 83–103.

Rahul, L.L. (2006). Natural hazard mitigation in local comprehensive plans. *Disaster Prevention and Management: An International Journal*, 15(3): 461–483.

Scott, C.A., Kurian, M. and Wescoat Jr., J.L. (2015). The Water-Energy-Food Nexus: Enhancing Adaptive Capacity to Complex Global Challenges. In M. Kurian and R. Ardakanian (Eds.) *Governing the Nexus. Water, Soil and Waste Resources Considering Global Change.* New York: Springer, 15–38.

Scott, C.A., Pierce, S.A., Pasqualetti, M.J., Jones, A.L., Montz, B.E. and Hoover, J.H. (2011). Policy and institutional dimensions of the water–energy nexus. *Energy Policy*, 39(10): 6622–6630.

Smith, C. (2006). *The Plan of Chicago: Daniel Burnham and the Remaking of the American City.* Chicago: University of Chicago Press.

White, I. and Howe, J. (2003). Policy and Practice: Planning and the European Union Water Framework Directive. *Journal of Environmental Planning and Management*, 46(4): 621–631.

Zinn, J.A. (2004). *State Policies to Manage Growth and Protect Open Spaces.* New York: Novinka Books.

LAWS

Advisory Committee on Zoning (1926). Standard State Zoning Enabling Act of 1926 (SZEA), U.S. Department of Commerce, Washington, D.C.

Advisory Committee on Planning and Zoning (1928). Standard City Planning Enabling Act of 1928(SCPEA), U.S. Department of Commerce, Washington, D.C.

United States Congress (1954). The Housing Act of 1954, Public Law 83–560; 68 Stat. 590, August 2, 1954.

United States Congress (1966). The National Historic Preservation Act of 1966, Public Law 89–665; 16 U.S.C. ch. 1 A, subch. II § 470 et seq., October 15, 1966.

United States Congress (1970). The National Environmental Policy Act of 1970, Public Law 91–190; 42 U.S.C. § 4321 et seq., January 1, 1970.

United States Congress (1972). The Coastal Zone Management Act of 1972 (CZMA), Public Law 92–583; 16 U.S.C. 33 § 1451 et seq., October 27, 1972.

Legislature of the State of Arizona (1998). The Growing Smarter Act of 1998, May 22, 1998.

Legislature of the State of Arizona (2000). The Growing Smarter Plus Act of 2000, May 18, 2000.

Chapter 11

Potential impacts of the continuing urbanization on regional climate: The developing Phoenix-Tucson "Sun Corridor"

Zhao Yang
Department of Hydrology and Atmospheric Sciences, University of Arizona, USA

Francina Dominguez
Department of Atmospheric Science, University of Illinois, USA

Hoshin Gupta
Department of Hydrology and Atmospheric Sciences, University of Arizona, USA

Xubin Zeng
Department of Hydrology and Atmospheric Sciences, University of Arizona, USA

Laura Norman
U.S. Geological Survey, Western Geographic Science Center, USA

INTRODUCTION

The metropolitan region of Phoenix Arizona has, since the 1950's, been one of the fastest-growing urban areas in the United States (*Chow et al., 2012*). It has undergone substantial land use and land cover change (hereafter LULCC) since the Second World War, by shifting economic priorities from a mostly agrarian lifestyle to an urbanized one. As the baby-boom generation reached adulthood in the 1970's, an era of very rapid development began, with a large number of job opportunities becoming available in the metropolitan area. By 2010, Phoenix (the largest populated city in Arizona) reached a population of 1.4 million, and had an urban extension of 1,338.26 km² (*US Census, 2010*). Meanwhile, Tucson has become the second-largest city, with an area of 588 km² and a population of about 520,116 (*US Census, 2010*). With continuing development, both cities are projected to grow towards each other along what has been called the "Arizona Sun Corridor". By the year 2050, under a high intensity development scenario, the "Sun Corridor" is projected to develop as shown in *Figure 1*, with urban landscapes replacing agricultural and native semi-desert landscapes. Recent research by *Georgescu et al. (2012)* has suggested that projected urbanization could impact summer-season local to regional temperatures to a level that is *as significant* as those induced by large scale-climate change.

Records indicate that since 1970, under the synergistic effects of global warming and the urban heat island effect, summers in U.S. urban areas have been becoming

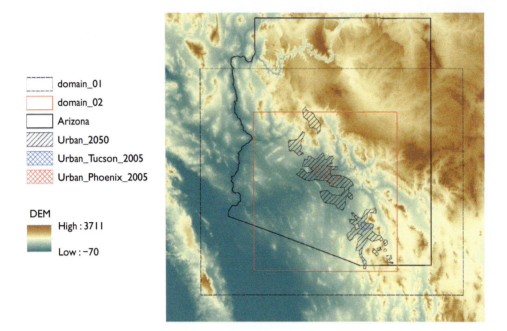

Figure 1 In 2005, Phoenix and Tucson are represented as the red and blue crossed region. The Phoenix-Tucson Corridor in 2050 is represented as the black slash area. Domain 1 and 2 are represented as the black dash and red box. Elevations are shown in meter.

progressively warmer (*NCDC*). Of the 60 largest U.S. cities, as many as 57 have experienced measurable increases in urban heat island effects from 2004 to 2013 (*Kenward et al., 2014*). For Phoenix and Tucson, the mean temperatures are now 1.8°C and 0.22°C warmer, respectively, than in their surrounding rural areas (*Kenward et al., 2014*), while temperatures in the Phoenix urban core have been reported to be ~2.2°C higher than the surrounding areas (*Brazel et al., 2007*). The most significant effect of urbanization on temperature (due to the urban heat island effect) has been found to occur at night, with minimum temperatures in Phoenix and Tucson being (on average) 3.8°C and 1.3°C respectively warmer than in the surrounding rural area (*Brazel et al., 2007*).

In addition to the heat island effect, urbanization can also affect regional climate by changing local circulation and precipitation patterns. Several observational studies have suggested that urbanization can modify rainfall patterns both over and downwind of cities (*Changnon et al., 1977; Burian and Shepherd 2005; Shepherd 2005, 2006*). The Metropolitan Meteorological Experiment (METROMEX) showed that increased precipitation during the summer season can be linked to urbanization (*Changnon et al., 1977; Huff and Vogel, 1978*), with 5–25% increases in observed precipitation over and within 50–75 km downwind of the urban area (*Huff and Vogel, 1978; Changnon, 1979; Changnon et al., 1981*). Examining a 108-year precipitation historical data record for Arizona, *Shepherd (2006)* reported statistically significant 11–14% increases in mean precipitation in the Lower Verde basin (northeast of Phoenix), from a pre-urban (1895–1949) to post urban (1950–2003) period associated with the expansion of the Phoenix metro area. Studies also suggest that urbanization can lead to modifications in the diurnal distribution of precipitation. For instance, *Balling and Brazel (1987)* reported a higher frequency in the occurrence of late-afternoon storms in Phoenix; however, they did not find evidence for significant changes in the mean precipitation amounts during the summer monsoon season.

Over our region of interest, *Georgescu et al. (2008)* used the Regional Atmospheric Modeling System (RAMS) to simulate 3 different dry and 3 different wet years with land surface data circa 1973, 1992, and 2001 over the Phoenix metro region. They concluded that the signal of increasing precipitation due to LULCC is present only during dry years. Further investigating the mechanism of precipitation enhancement, *Georgescu et al. (2009a, 2009b)* concluded that precipitation recycling, rather than the direct or indirect effect of the urbanization, may be the actual cause for the precipitation increase.

While numerous studies (such as those mentioned above) suggest an enhanced signature of precipitation over and downwind of metropolitan areas, there remain reasons to be skeptical. Precipitation anomalies at the La Porte station, Indiana (studied extensively in the METROMEX program), began to shift locale in the 1950s and then disappeared in the 1960s (*Changnon et al., 1980*). Despite decades of work the reasons for why the UHI can enhance precipitation in some regions, while seemingly having no effect (or even leading to decreases) in other regions, still remains unclear.

This chapter investigates temperature and precipitation variations over the state of Arizona that may arise due to projected urban expansion of the Phoenix-Tucson "Sun Corridor". To understand the causes for potential temperature and precipitation changes over the urban and downwind regions, we employ a numerical modeling framework to examine the changes that may arise due to projected future expansion. We use the non-hydrostatic, compressible Weather Research and Forecasting model (WRF), and account for urban characteristics by incorporating an Urban Canopy Model (UCM). Our simulations have a relatively high resolution (2 km), thereby enabling the model to accurately simulate convective events during the monsoon season.

In the next section we introduce the numerical model and experimental design. Section 3 presents and discusses results from two land use scenarios, and reports additional analysis regarding the potential impacts of LULCC on future water and energy demand variation. A summary and our conclusions appear in section 4.

1 METHODOLOGY

The study region extends over the state of Arizona, covering the area between latitude 30.7°N to 35.7°N and longitude 115.2°W to 108.2°W. We use the WRF model to simulate regional climate under two different land use scenarios—one with historical observed land cover corresponding to 2005, and the other for a projected future land cover representing 2050. The simulation covers a 10-year period during the peak summer monsoon season (July and August) for each year from 1991 to 2000.

1.1 WRF Model and configuration

The WRF model version 3.4 is coupled to a land surface and urban modeling system designed to investigate emerging issues in urban areas (*Skamarock et al., 2008*). In our experiment, we used the Noah land surface model (LSM) to model the land surface (*Chen & Dudhia, 2001*), thereby providing surface energy fluxes and surface skin temperatures that serve as the boundary conditions for the atmospheric model. While the original version of Noah LSM has a bulk parameterization for urban land use, our experiment uses a single layer urban canopy model (UCM) to better represent the energy fluxes and temperature within the urban region (*Kusaka et al., 2001; Kusaka & Kimura, 2004*). The lateral boundary conditions were obtained from the North American Regional Reanalysis (NARR) data (*Mesinger et al., 2006*). We simulated each of the years 1991–2000, beginning on June 15 1200Z and running through the end of August (August 31 1200Z) to cover the peak of the Arizona monsoon season. We used two nested domains with an outer grid spacing of 10 km and an inner grid spacing of 2 km (see *Figure 1*). The inner domain, chosen to correspond to the urban area and regions that may be affected by the urban corridor (i.e., including areas at least 75 km far from the urban border), has 175 grid cells in the zonal direction and 190 grid cells in the meridional direction.

Land use characteristics used as lower boundary conditions for 2005 (hereafter LULC_2005) were obtained directly from default MODIS land use data available in the WRF model. The projected land use characteristics for 2050 (hereafter LULC_2050) were derived by combining three different datasets, under a high intensity urbanization scenario: 1) a future current-trends scenario generated by the SLEUTH model assuming unmanaged exponential growth of land cover change into the year 2050 in the Santa Cruz Watershed (Tucson area); 2) MAG, a dataset developed by the Maricopa Association of Governments (*MAG, 2005*) using a "red-dot" algorithm based on land ownership and census information to establish areas that are available for urbanization; and 3) the 2005 North American Land Cover (NALC) data, classified using MODIS data (250 m), describing current land use/land cover (*Commission for Environmental Cooperation, 2013*).

1.2 Observational temperature data

To correlate urban temperatures with electricity load within Tucson and Phoenix, so as to estimate the effect of future temperatures on electricity load under urban expansion, we used near surface air temperatures collected by the Arizona Meteorological Network (AZMET). The observation station for Tucson (32°16'N and 110°56'W) is located very close to the city center, and the station for Phoenix (33°28'N and 112°05'W) is located at the Phoenix Encanto site. Both sites are located within the urban region in the LULC_2005, and therefore provide a good representation of actual urban temperatures. The sites provide hourly air temperatures, relative humidities, wind speeds and other measurements (data available online at *http://ag.arizona.edu/azmet/*).

Electric load data for Tucson was provided by the Tucson Electric Power (TEP) public utility company, and for Phoenix by the Arizona Public Service (APS) and the Salt River Project (SRP). These data were aggregated from one-minute timestep to hourly timestep to correspond to the observed temperature data.

2 RESULTS

2.1 Urban impacts on temperature

The simulated effects of urban expansion on temperature are shown in *Figure 2*. Daily mean, maximum and minimum temperatures were obtained from hourly temperature data, and then averaged over the simulation period to compute the seasonal

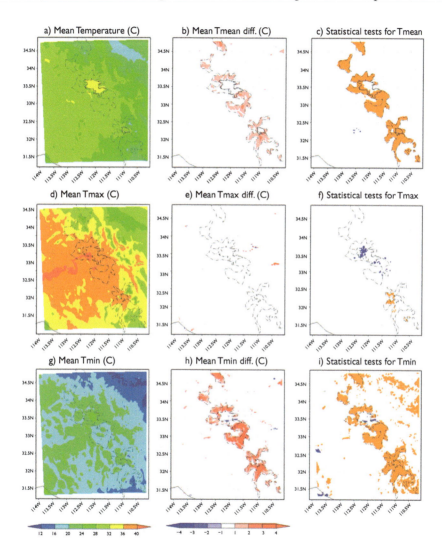

Figure 2 July and August simulated temperature difference between LULC_2050 and LULC_2005, a) mean temperature in LULC_2005, b) mean temperature difference between LULC_2050 and LULC_2005, c) student t test and bootstrap test for mean temperature difference, orange (blue) indicates significant increase (decrease) in both statistical tests, d), e) and f) are similar to a), b) and c) except for maximum temperature, g), h) and I) are similar to a), b) and c) except for minimum temperature.

mean, maximum and minimum temperatures. On average, changes in the mean, maximum and minimum temperature over the urbanized area are +1.27°C, −0.07°C and +3.09°C respectively. The difference in mean and minimum daily temperatures between LULC_2050 and LULC_2005, averaged over 10 years of simulation, show statistically significant increases over the urbanized area, while the change in daytime daily maximum temperature is not significant. Changes in the urban mean and minimum temperatures can be explained by changes in surface thermal properties such as heat capacity and thermal conductivity. This explanation is supported by *Figure 3* which shows the diurnal energy cycle averaged over the regions transformed from native vegetation in 2005 to urban in 2050 (newly urbanized area). Urbanization results in more energy being stored as ground heat flux during the daytime from 7am to 4pm local time. This results in a more positive ground heat flux during the night

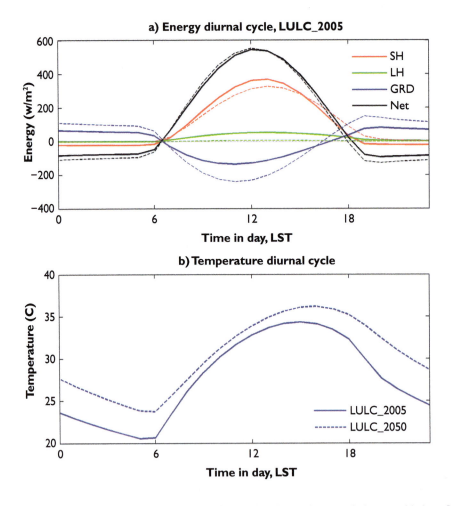

Figure 3 On the upper, energy diurnal cycle over the newly urbanized area, including sensible heat flux (SH), latent heat flux (LH), ground heat flux (GRD) and net radiation (Net) for LULC_2005 (solid line) and LULC_2050 (dash line). On the bottom, temperature diurnal cycle over newly urbanized area for LULC_2005 (solid line) and LULC_2050 (dash line).

so that the energy is released (warming the atmosphere) and leading to warmer air temperatures. At the same time, urbanization results in decreased daytime latent heat fluxes (reduced evaporation), which reduces the atmospheric column water available for convection (similar to the finding by *Georgescu et al., 2012*). Correspondingly, and as expected, the sensible heat flux difference is increased during the night, causing urban nighttime air temperatures to be larger. *Figure 3* shows that, while the magnitude of difference in maximum and minimum temperature is consistent with our analysis, the times of maximum and minimum temperatures are delayed by about 1 hour and 30 minutes respectively. This longer duration of higher temperatures during the night can be a potential source of stress for the population in the urban areas.

The spatial patterns of sensible and latent heat fluxes are shown in *Figure 4*. Sensible heat fluxes are larger in the western part of the domain (the lower desert), and smaller in the northeast (higher elevations). In contrast, latent heat fluxes are greatest in the northeastern mountains and gradually decrease toward the western desert. The differences in the mean sensible heat flux due to urbanization indicate

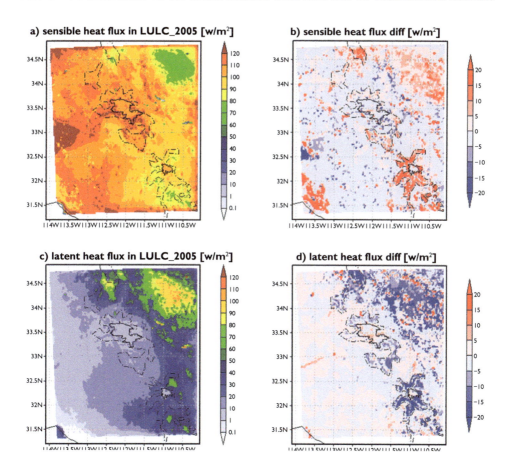

Figure 4 a) Average sensible heat flux in July and August from 1991 to 2000 in LULC_2005. b) Average sensible heat flux difference between LULC_2050 and LULC_2005. c) Similar to a) but for latent heat flux. d) Similar to b) but for latent heat flux difference.

an increase in sensible heat flux (and decrease in latent heat flux) over Tucson and surroundings and the northeast high mountains, indicating a change in the partitioning of available energy. While increased sensible heat flux can indicate less stable conditions that could lead to increased precipitation, the reduced evapotranspiration results in lower availability of water for convection; i.e., the two processes act in opposite directions. As shown below, the overall effect of decreases in moisture availability is reduced precipitation.

2.2 Urban impacts on precipitation

Figure 5 shows the 10-year average summertime precipitation differences between LULC_2005 and LULC_2050. A pattern of decreasing precipitation dominates the majority of the domain, including the northeastern mountains and some portions

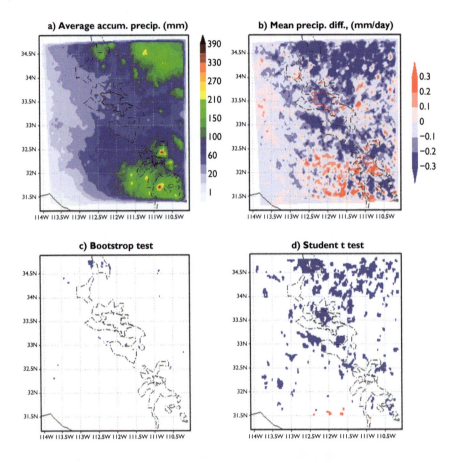

Figure 5 a) Average July and August accumulated precipitation from 1991 to 2000 in LULC_2005. b) July and August precipitation difference between LULC_2050 and LULC_2005 from 1991 to 2000, normalized by number of days. c) Bootstrap test of difference in daily precipitation between LULC_2005 and LULC_2050, blue (orange) indicates statistically significant decrease (increase). d) Similar to c) except for student t test, blue indicates statistically significant decrease.

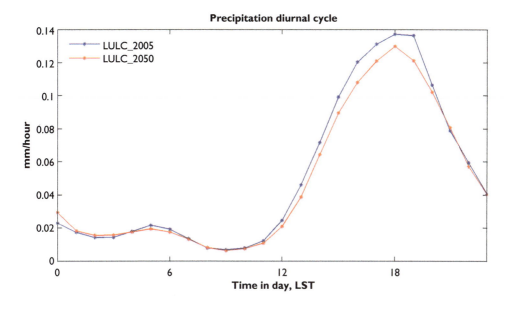

Figure 6 Diurnal cycle of precipitation over the newly urbanized area in LULC_2050 (red) and LULC_2005 (blue).

of the urban corridor. A standard statistical test of the difference in daily mean precipitation at each grid cell suggests that this change is significant, while an alternative bootstrap test suggests the opposite; accordingly the results must be deemed inconclusive.

Focusing on the diurnal cycle of precipitation, the precipitation differences for newly urbanized regions (*Figure 6*) indicate that while the timing of the peak remains unchanged, there is a marked decrease precipitation (of about 6%) during the late afternoon and early evening (1400–2100 LST). This suggests that precipitation in the newly urbanized areas will be reduced, primarily due to reduced levels of evapotranspired water.

2.3 Urban impacts on water and energy demand

Overall, our results (as simulated by WRF) indicate that projected expansion of the urban corridor in the Phoenix-Tucson area is likely to affect both temperature and precipitation, thereby affecting the demand for both water and energy in the region. The Arizona Department of Water Resources (ADWR) provides water demand information and assessments for each active management area (i.e., area that heavily relies on the groundwater supply), with the goal of ensuring that (by 2025) groundwater withdrawal rates will equal those of recharge due to natural processes. Tucson and Phoenix are located in the Tucson Active Management Area (i.e. TAMA) and Phoenix Active Management Area (PAMA), respectively. Historically, Tucson municipal water demand has increased by 68% from 1985 to 2006 (from 1.39×10^8 to 2.33×10^8 m^3 yr^{-1}). During the same period, Phoenix municipal water demand grew by 76% (from 7.82×10^8 to 1.38×10^8 m^3 yr^{-1}). The increase in total water demand for both cities was roughly 75%. At the same time, the urbanized extent of Tucson grew by 124% while that of Phoenix grew by 50% – so that the combined area of the two cities grew by 68%. Taken together, we see that (during this period) the combined water demand for the two cities grew approximately linearly with area (although Tucson grew more, it also conserved more water per area than Phoenix). Based on these estimates, together with the projected 7-fold increase in urbanization, and an assumption that water usage intensity will remains the same in 2050 as in 2005, water demand can be expected to grow (linearly) to around 1.12×10^{10} m^3 for the entire corridor. Of course, this projection is based on an assumption of high rates of urban expansion, and it is quite unlikely that this level of water demand can be easily satisfied given the water-limited nature of Arizona.

In this regard, Arizona relies heavily upon groundwater as an important water source; in 2006 groundwater accounted for 64% of water supply in TAMA and 31% in PAMA. ADWR projections of future water demand are based on population growth and water use rates, and assume that groundwater can be utilized when all other water sources are unable to meet demand. Their highest water demand scenario estimates municipal water demands to be 3.80×10^8 m^3 and 2.59×10^9 m^3 for TAMA and PAMA respectively, requiring groundwater overdrafts in 2025. In contrast, our estimates indicate much larger demands by 2050, suggesting there will simply not be enough water to sustain the projected levels of urbanization. Accordingly, urban growth to the extent portrayed by LULC_2050 is likely to be unsustainable, with the availability of water being an important limitation to future urbanization unless alternative sources are found.

Meanwhile, temperature increases within the urban area (along with greenhouse gas-induced global climate change) are likely to increase electrical energy demands for cooling (*Georgescu et al., 2013*). Since cooling demands are known to account for more than 50% of the total electricity demand in urban areas, with this ratio climbing to as high as 65% during the hot season evening hours in semi-arid urban environments (*Salamanca et al., 2013*), this can place heavy demands on urban infrastructure. *Figure 7* shows that the historical diurnal pattern of air conditioning (AC) consumption follows a very diurnal pattern that is similar to the diurnal temperatures observed within the urban areas. The figure is based on electric load data for Tucson and Phoenix (provided at the intra-daily timescale by Tucson Electric Power for Tucson and Arizona Public Service (APS) and Salt River Project (SRP) for Phoenix), with the assumption that the ratio of AC consumption to total electric load follows the diurnal pattern as shown in *Salamanca et al. (2013)*. The diurnal AC consumption in each city is obtained by multiplying the diurnal total electric load to the corrected ratio.

To estimate AC consumption under projected warmer temperatures, temperature and electric load data were fitted to a polynomial function (with linear correlation coefficient of 0.85 for Tucson and 0.85 for Phoenix); the results clearly suggest that temperature plays a significant role in influencing the AC consumption (see *Figure 8*).

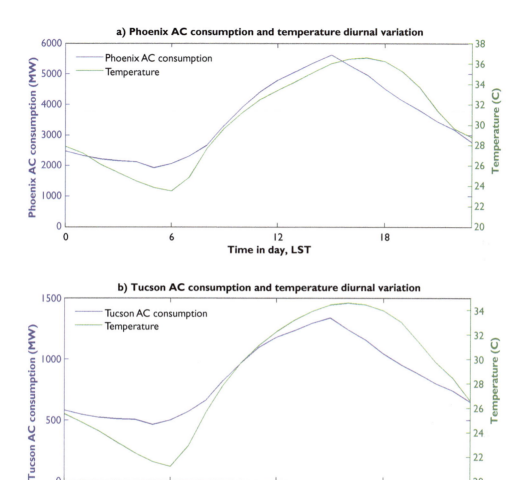

Figure 7 Diurnal cycle of Air Conditioning (AC) consumption and temperature at Phoenix metropolitan and Tucson metropolitan.

For each hour of the day, the increased energy load due to UHI-induced temperature increase can be estimated using the fitted function. The overall projected future total AC consumption therefore accounts for both increases in temperature and growth due to urban expansion. Even though many other factors may affect energy consumption, it is safe to assume that areal enlargement and temperature are likely to be among the most important. The results are shown in *Figure 9*. Overall, the dominant factor affecting energy consumption in the future appears to be the areal enlargement due to urban expansion.

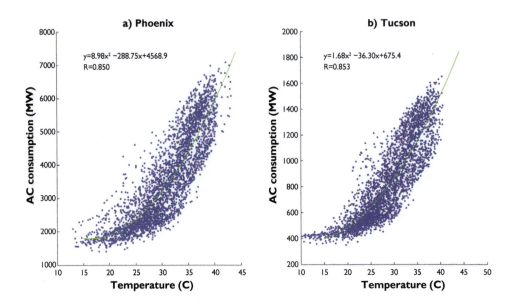

Figure 8 Scatter plot of the AC consumption and temperature data, the green line indicates the fitted polynomial function.

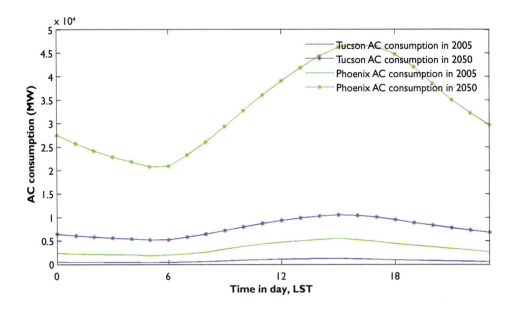

Figure 9 AC consumption of Tucson and Phoenix in 2005 and 2050, considering both areal enlargement and warmer temperature.

3 CONCLUSIONS

This study has examined the climatological impacts on summer monsoonal (July and August) climate in Arizona that can be associated with anticipated expansion of the Phoenix-Tucson urban corridor. To be clear, we have only investigated the impacts of urbanization, and have not considered the expected additional effects of global warming. Given that the Phoenix metropolitan area has been one of the most rapidly developing areas in the United States during the past 30 years, and that the Phoenix-Tucson "Sun Corridor" is expected to add another 5 to 6 million inhabitants between 2000 and 2030 (*US Census, 2005*), it is useful for planners to understand the ways in which urbanization is likely to affect the hydroclimate of the region.

Our results suggest that while urbanization will probably not significantly alter the daily maximum temperatures, it is very likely to result in significant increases in daily minimum (night time) temperatures over the urban corridor. Further, the daily maximum and minimum temperature are expected to occur about an hour and 30 minutes later in the day, respectively. This will result in longer periods of significantly hotter temperatures, increase the risks of heat related health issues and even death (*McGeehin & Mirabelli, 2001*). Based on this, planners may wish to explore the possibilities of implementing "cool-roof" and other technologies to help reduce the strength of the UHI phenomenon (*Georgescu et al., 2013*). Meanwhile latent heat fluxes are likely to decrease dramatically throughout the day, resulting in reduced evaporation over urban regions and downwind mountainous areas. The overall effect of urbanization is likely to be less moisture available for convection. Accordingly, significant decreases in precipitation can be expected over the mountains at higher elevations in the northern parts of the domain and parts of the newly urbanized region.

Overall, we estimate that about 7 times the current water supply would be needed to sustain such a scale of urban expansion, which would require extensive (and unsustainable) access to groundwater storage. Meanwhile, energy supplies would have to be expanded to meet future air conditioning needs to deal with the longer durations of high temperatures. Clearly, urban expansion to the extent projected here may be very difficult to achieve without access to new sources of water and energy.

There are, of course, limitations to our study that must be considered in any analysis involving urban planning. In particular, we did not consider the impacts of aerosol, even though the role of aerosols in urban environments is likely to be an important factor. Other limitations include the lack of irrigation effects in our land surface representation, which can be expected to influence the water and energy budgets over the region by increasing latent heat fluxes while decreasing sensible heat fluxes, leading to the so-called "oasis" effect (*Georgescu et al., 2011*). However, a preliminary investigation (not reported here) conducted by us indicated that urban irrigation can cause precipitation to increase only slightly over the region, and that the effects are negligible compared to the much larger effects of urbanization growth. Finally, an important factor affecting local precipitation may be regional and continental scale precipitation recycling (*Georgescu et al., 2009b*).

ACKNOWLEDGEMENTS

We thank APS, TEP and SRP for providing electric load data.

REFERENCES

Brazel, A., Gober, P., Lee, S.J., Grossman-Clarke, S., Zehnder, J., Hedquist, B. and Comparri, E. (2007). Determinants of changes in the regional urban heat island in metropolitan Phoenix (Arizona, USA) between 1990 and 2004. *Climate Research*, 33(2): 171–182.

Cao, M. and Lin, Z. (2014). Impact of urban surface roughness length parameterization scheme on urban atmospheric environment simulation. *Journal of Applied Mathematics*, 14: 1–4.

Changnon, S.A. (1979). Rainfall changes in summer caused by St. Louis. *Science*, 205: 402–404.

Changnon, S.A. (1980). More on the La Porte Anomaly: a Review. *Bulletin of the American Meteorological Society*, 61: 702–711.

Changnon, S.A., Huff, F.A., Schickedanz, P.T. and Vogel, J.L. (1977). Summary of METROMEX, 1: Weather anomalies and impacts. *Bulletin of the Illinois state water survey*, 62, Urbana: State water survey division.

Changnon, S.A., Semonin, R.G., Auer, A.H., Braham, R.R. and Hales, J. (1981). METROMEX: A review and summary. *Journal of the Amer. Meteor. Soc.*, 40: 181.

Chow, W.T.L., Brennan, D. and Brazel, A.J. (2012). Urban heat island research in phoenix, Arizona theoretical contributions and policy applications. *Bulletin of the American Meteorological Society*, 93: 517–530.

Kenward, A., Yawitz, D., Sanford, T. and Wang, R. (2014). Summer in the city: Hot and getting hotter. climate central. Retrieved from http://assets.climatecentral.org/pdfs/UrbanHeatIsland.pdf.

Georgescu, M., Mahalov, A. and Moustaoui, M. (2012). Seasonal hydroclimatic impacts of Sun corridor expansion. *Environmental Research Letters*, 7: 034026.

Georgescu, M., Miguez-Macho, G., Steyaert, L.T. and Weaver, C.P. (2008). Sensitivity of summer climate to anthropogenic land-cover change over the Greater Phoenix, AZ, region. *Journal of Arid Environments*, 72(7): 1358–1373.

Georgescu, M., Miguez-Macho, G., Steyaert, L.T. and Weaver, C.P. (2009a) Climatic effects of 30 years of landscape change over the Greater Phoenix, Arizona, region: 1. Surface energy budget changes. *Journal of Geophysical Research Atmospheres*, 114: D05110.

Georgescu, M., Miguez-Macho, G., Steyaert, L.T. and Weaver, C.P. (2009b) Climatic effects of 30 years of landscape change over the Greater Phoenix, Arizona, region: 2. Dynamical and thermodynamical response. *Journal of Geophysical Research Atmospheres*, 114: D05111.

Georgescu, M., Moustaoui, M., Mahalov, A. and Dudhia, J. (2011). An alternative explanation of the semiarid urban area "oasis effect". *Journal of Geophysical Research: Atmospheres*, 116: D24113.

Georgescu, M., Moustaoui, M., Mahalov, A. and Dudhia, J. (2013). Summer-time climate impacts of projected megapolitan expansion in Arizona. *Nature Climate Change*, 3: 37–41.

Huff, F.A. and Vogel, J.L. (1978). Urban, topographic and diurnal effects on rainfall in the St. Louis region. *Journal of Applied Meteorology*, 17: 565–577.

Kusaka, H. and Kimura, F. (2004). Coupling a single-layer urban canopy model with a simple atmospheric model: Impact on urban heat island simulation for an idealized case. *Journal of the Meteorological Society of Japan*, 82: 67–80.

Kusaka, H., Kondo, H., Kikegawa, Y. and Kimura, F. (2001). A simple single-layer urban canopy model for atmospheric models: Comparison with multi-layer and slab models. *Boundary-Layer Meteorology*, 101(3): 329–358.

McGeehin, M.A. and Mirabelli, M. (2001). The potential impacts of climate variability and change on temperature-related morbidity and mortality in the United States. *Environmental Health Perspectives*, 109: 185–189.

Mellor, G.L. and Yamada, T. (1982). Development of a turbulence closure model for geophysical fluid problems. *Reviews of Geophysics*, 20: 851–875.

Mesinger, F., DiMego, G., Kalnay, E., Mitchell, K., Shafran, P., Ebisuzaki, W., Jovic, D., Woollen, J., Rogers, E., Berbery, E., Ek, M., Fan, Y., Grumbine, R., Higgins, W., Li, H., Lin, Y., Manikin, G., Parrish, D. and Shi, W. (2006). North American regional reanalysis. *Bulletin of the American Meteorological Society*, 87: 343–360.

Salamanca, F., Georgescu, M., Mahalov, A., Moustaoui, M., Wang, M. and Svoma, B.M. (2013). Assessing summertime urban air conditioning consumption in a semiarid environment. *Environmenal Research Letters*, 8: 034022.

Shepherd, J.M. (2005). A review of current investigations of urban-induced rainfall and recommendations for the future, *Earth Interactions*, 9: 1–27.

Skamarock, W., Klemp, J., Dudhia, J., Gill, D., Barker, D., Duda, M., Huang, X.-Y., Wang, W. and Powers, J. (2008). A description of the advanced research WRF version 3. *NCAR technical note*. Boulder: Mesoscale and Microscale Meteorology Division.

Swaid, H. (1993). The role of radiative-convective interaction in creating the microclimate of urban street canyons. an canopy model for atmospheric models: Comparison with multilayer and slab models. *Boundary-Layer Meteorology*, 64(3): 231–259.

OTHER SOURCES

Reports

Maricopa Association of Governments. (2005.) Arizona Growth Maps and Algorithm. Unpublished report.

Websites

NCDC. Climate at a Glance (Climate at a Glance). http://www.ncdc.noaa.gov/cag/

AZMET: The Arizona Meteorological Network – The University of Arizona (AZMET: The Arizona Meteorological Network – The University of Arizona). http://ag.arizona.edu/azmet/

COMMISSION FOR ENVIRONMENTAL COOPERATION (Commission for Environmental Cooperation). http://www.cec.org/Page.asp?PageID=924&SiteNodeID=495&AA_SiteLanguageID=1

US Census Bureau, Population Projections (2005, 2010 Interim State). http://www.census.gov/population/projections/data/state/projectionsagesex.html

Ecosystem services
and biodiversity

Chapter 12

Quantification of water-related ecosystem services in the Upper Santa Cruz watershed

Kremena Boyanova
National Institute of Geophysics, Geodesy and Geography,
Bulgarian Academy of Sciences, Bulgaria

Rewati Niraula
Department of Hydrology and Water Resources, University of Arizona, USA

Francina Dominguez
Department of Atmospheric Sciences, University of Arizona, USA

Hoshin Gupta
Department of Hydrology and Atmospheric Sciences, University of Arizona, USA

Stoyan Nedkov
National Institute of Geophysics, Geodesy and Geography,
Bulgarian Academy of Sciences, Bulgaria

INTRODUCTION

The ongoing drought in the Southwestern United States places pressure on both scientists and practitioners to find new solutions to water-related issues. In the state of Arizona, this situation requires that the present state of the ecosystems and natural resources be re-evaluated to assess their capacity to sustain the future flow of ecosystem services to society. Ecosystem Services (ES) are the contributions of ecosystem structures and functions – in combination with other inputs – to human well-being (*Burkhard et al., 2012a*). The availability of water as a benefit provided to people by nature is dependent on multiple human and non-human factors. Human activities change the environment in ways that alter its structure and functioning. By using hydrological models of the system, we can develop quantitative simulations of the ways in which existing environmental conditions influence the hydrological cycle. Different elements of the hydrological cycle influence the supply of Water-Related Ecosystem Services (WRES) to society. It is, therefore, important for decision-makers to quantitatively understand how various human activities can influence the functioning of those natural processes.

In this chapter, we focus on the Upper Santa Cruz watershed located mainly in southern Arizona in order to propose a methodology for spatially explicit quantification and evaluation of the WRES within the watershed. Using the Soil and Water

Assessment Tool (SWAT) hydrological model to derive a set of hydrological indicators from model simulation for the period 1987–2006, we analyze how the different land use types within the watershed influence the hydrological cycle and, thereby, the supply of WRES. We therefore assess and map impacts by analyzing the average annual values of the hydrological variables for each land use type. To support planning, we assess the hypothetical influence of three urban growth scenarios on the supply of WRES via an area weighting of the contribution of the different land use types to the supply of WRES. A decreasing trend in the supply of almost all services is observed under all three scenarios. Our methodology and results can provide support for water and land management and decision-making in areas experiencing water scarcity.

1 MATERIALS AND METHODS

1.1 Case study – Upper Santa Cruz watershed

One methodologically challenging aspect of any transdisciplinary research effort is the differences in the concept of "space" adopted by different disciplines. Most often the case study boundaries are derived based on the functions and structures of interest. For example, hydrologic science most commonly uses the "watershed" as the system boundary, since it encompasses a natural area over which the hydrological cycle can be studied. In this study, we focus on the Upper Santa Cruz watershed located mainly in semi-arid southern Arizona, USA (85%), with a small portion extending into northern Sonora, Mexico (15%) (*Figure 1*). The river originates in southern Arizona, flows south into Mexico and bends northwards to re-enter the USA, where it drains into the Gila River. About half of the watershed coincides with the Tucson Active Management Area (AMA), overlapping with its east aquifer. Elevations within the watershed vary between 610 m and 2884 m. Much of the river is ephemeral and only a few sections have perennial flows fed by discharge from waste water treatment plants. The average annual precipitation varies between 11 inches (280 mm) and 25 inches (637 mm). The average annual daily maximum and minimum temperatures are 82° F (28° C) and 50° F (10° C) respectively. Soils are dominated by loam and sandy loam. The biggest urban area within the watershed is the city of Tucson.

The land use map used for our model simulations relates to 1999 (*Figure 2*) and is referred to as '*current*' land use in the present paper. It indicates that approximately 59% of the watershed is covered by shrubland, 15% by evergreen forest, 12% by urban areas and 9% by grassland. Other land use classes have less than 3% representation, with open water areas being the smallest (0.04%), followed by forested wetland (0.29%) and agricultural land (0.9%).

The Upper Santa Cruz watershed is part of the Colorado River watershed, which is presently threatened by an ongoing drought and a severe potential for future water shortage (*Eden et al., 2015; Cayan et al., 2010; Dominguez et al., 2010*). A significant share of the water supply within the Tucson AMA comes from the Colorado River, delivered via the Central Arizona Project (CAP). According to the 2010 report of the Water Resource Development Commission (WRDC), the Tucson AMA groundwater basins may require development of additional water supplies. In view of the

Quantification of water-related ecosystem services in the Upper Santa Cruz watershed 199

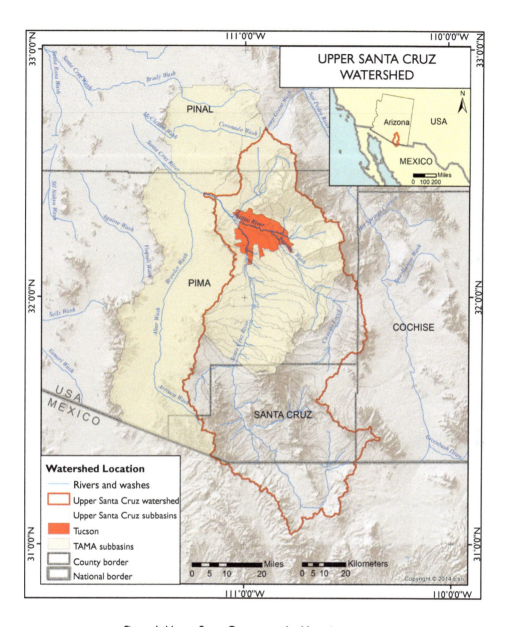

Figure 1 Upper Santa Cruz watershed location map.

continuing drought, the CAP water delivered to south Arizona becomes an extremely vulnerable and insecure resource. The report by the Arizona Department of Water Resources (ADWR) entitled *"Arizona's Next Century: A Strategic Vision for Water Supply Sustainability (Strategic Vision)"*, projects a supply-demand gap in Arizona sometime in the next 20 to 100 years (ADWR, 2015).

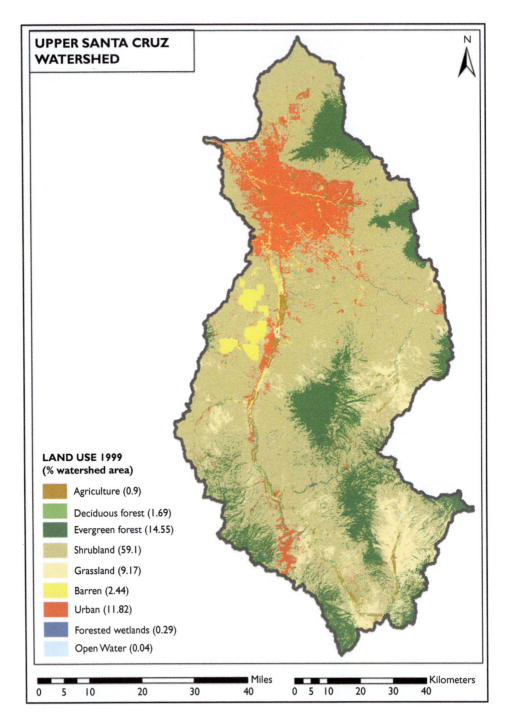

Figure 2 Upper Santa Cruz watershed land use map for the year 1999 and percentage representation of the different land use types.

1.2 Ecosystem services and their quantification and mapping

The concept of ES has gained popularity since the end of the 20th century, and has been institutionalized by the Millennium Ecosystem Assessment (*MA, 2005*). ES are defined as the contributions of ecosystem structures and functions – in combination with other inputs – to human well-being (*Burkhard et al., 2012a*). The "*combination with other inputs*" refers to both human alterations of the environment and anthropogenic inputs, the latter being largely inseparable from nature-based contributions to human well-being (*Burkhard et al., 2014*).

ES can be classified and categorized in different ways. Popular classification have been provided by the The Economics of Ecosystems and Biodiversity (*TEEB, 2010*), the Millennium Ecosystem Assessment (*MA, 2005*) and the Common International Classification of Ecosystem Services *(CICES)*[1] research initiatives, while other classifications developed by independent research groups also exist (*Burkhard et al., 2009, 2014; de Groot et al., 2010; Kandziora et al., 2012*). Here, we apply the most common categorization of ES as belonging to three main themes: regulating, provisioning and cultural. According to *Haines-Young & Potschin (2013)* these can be defined as follows:

- Regulating services include the ways in which ecosystems control or modify biotic or abiotic parameters that define the environment of people; these are ecosystem outputs that are not consumed but affect the performance of individuals, communities and populations and their activities;
- Provisioning services include all material and biota-dependent energy outputs from ecosystems; they are tangible things that can be exchanged or traded, as well as consumed or used directly by people in manufacture;
- Cultural services are all non-material ecosystem outputs that have symbolic, cultural or intellectual significance.

The ES framework provides a multi-perspective approach to the human-environmental system. A simplified representation of this system, and the role of the ES within it, is shown in *Figure 3*. It represents the strong direct interconnections between the society and the environment, and depicts their cause-effect relationships – any changes in one component of the system can lead to changes in all of the other components. The system boundary is defined by the boundary of the structures and processes that underpin the supply of ES. It can be natural (watershed) or anthropogenic (county, municipality), depending on the management practice and system of interest. The potential of the ecosystems to supply ES depends on their structures and functions, which are often influenced by the present land use practices. The ES used by the population and economy in certain area predetermine the human benefits and well-being. The demand for benefits creates a feedback loop to the supply of ES and respectively the existing policies and management practices. In the present text, only the components shown within the red dashed line of *Figure 3* are analyzed. The underlying condition for the supply of ES is the ecological integrity within an area, which is deter-

---

[1] http://cices.eu/.

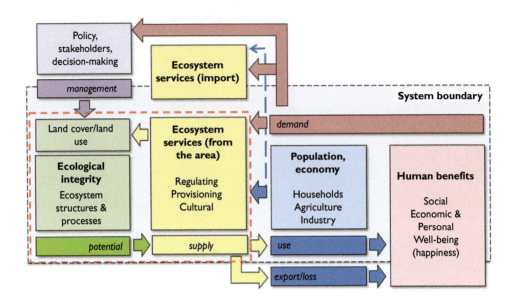

Figure 3 Theoretical framework (adapted after Burkhard et al., 2012b, 2014).

mined by its ecosystems structures and processes that are strongly influenced by the land cover and land use practices. The quantification of the supply of services from the different land use types is important for the understanding of the ways changes in land management influence the flow of services to the society.

Proper quantification and mapping of ES is recognized as being very important for policy and decision-making support *(de Groot at al., 2010; Burkhard et al., 2012a; Brauman et al., 2007; Daily & Matson, 2008; Daily et al., 2009)* and many studies have focused on the development of applicable methods and models *(Crossman et al., 2013; Martínez-Harms & Balvanera, 2012; Bagstad et al., 2013)*. In Europe, the concept has been strongly integrated within the Biodiversity Strategy 2020 and as result several major initiatives for assessment and mapping of ES have arisen (Mapping and Assessment of Ecosystems and their Services (MAES), Enhancing ecoSysteM sERvices mApping for poLicy and Decision mAking (ESMERALDA), OperationalisatioN of Natural Capital and Ecosystem Services (OpenNESS). In the USA, the most significant initiative for mapping of ecosystem services is the EnviroAtlas, and the concept has been incorporated into activities of the US Environmental Protection Agency (EPA), United States Geological Survey (USGS) and United States Department of Agriculture (USDA). Important global initiatives that incorporate the concept are United Nations Statistics Division (UNSD) System of Environmental-Economic Accounting (SEEA) and Intergovernmental Platform on Biodiversity and Ecosystem Services (IPBES).

While multiple tools for assessment, modeling and monetary valuation of ES exist, most have limited scope, functionality and applicability and many are still under development (Bagstad et al., 2013). As with other fields of the natural science, the field of Hydrology has been developing improved computer-based models that seek to simulate the natural world so as to better understand its elements and processes. While many previous studies have applied hydrological models for spatial quantification and mapping of WRES (*Boyanova et al., 2014; Nedkov et al., 2015; Nedkov & Burkhard, 2012; Stürck et al., 2014*), most have focused on water yield (discharge) and water quality (*Vigersol & Aukema, 2011; Crossman et al., 2013; Logsdon & Chaubey, 2013; Maes et al., 2009*) while lacking spatially explicit representation of WRES supply.

1.3 Application of hydrological modeling for quantification and mapping of water-related ecosystem services

Hydrologic (water-related) ES are the benefits to people that are produced by the effects of terrestrial ecosystems on freshwater (*Brauman et al., 2007*). Hydrological and ecosystem service models both find broad application for WRES assessment and visualization but, depending on the data availability and research needs, the application of hydrological models is preferable in cases where sufficient data and expertise are available (*Vigerstol & Aukema, 2011*). Furthermore, as it passes through the environment, water is affected in one way or another by all ecosystems, and hence all ecosystems can be viewed as suppliers of WRES.

Often the ES source areas do not coincide with the area where the beneficiaries are located and services are used (*Villamagna et al., 2013; Brauman et al., 2007*) and spatial ES assessments should refer to the areas affected by the related processes instead of to the administrative units which usually mark artificial system boundaries (*Burkhard et al., 2014*). Therefore, the spatially explicit simulation of water movement through the ecosystems provided by a watershed hydrological modeling provides a solid foundation for the quantification of WRES.

Some studies differentiate the ES potential and the actual ES flow to society. ES "potential" refers to the hypothetical maximum yield of a selected service while the ES flow is the used set (bundle) of ecosystem services and other outputs from the natural system in a particular area within a given time period. Such a distinction can be made relatively easily for most provisioning services, but for many regulating and cultural services this distinctions tends to be more difficult to establish (*Burkhard et al., 2014*). As water is relevant in a substantial way for all ecosystems, and because each ecosystem directly or indirectly contributes to human well-being through the supply of ES, the distinction between potential and actual flow of WRES can be difficult to establish. In this chapter, we quantify the sum of both and referred to it as the WRES supply.

In the following discussion we focus on two regulating WRES (water flow regulation, water purification) and one provisioning (freshwater) WRES. Our discussion will not take cultural services into consideration. According to the definitions provided by *Burkhard et al. (2014)*:

- Freshwater provision is the fresh and processed water available for drinking, domestic use, industrial use, irrigation;
- Water flow regulation is the water cycle feature maintenance (e.g. natural drainage, water storage and buffer, irrigation and drought prevention);

- Water purification is the ecosystem ability to purify water, e.g. from sediments, pollutants, nutrients, pesticides, disease-causing microbes and pathogens. In the present study the water purification supply is considered as the ability to dilute (dissolve) contaminants.

In order to quantify the respective services, we must select the appropriate indicators. Indicators are variables that provide aggregated information regarding certain phenomena (*Wiggering & Müller, 2004*). In most cases a single indicator is not sufficiently representative and the choice of multiple indicators is necessary (*Niemeijer & de Groot, 2008; van Oudenhoven et al., 2012*). As the literature provides no ultimate reference set of indicators for assessment of ES (there are some suggested sets – see *Kandziora et al., 2012; Burkhard et al., 2014; Maes et al., 2013, Staub et al., 2011*), indicators are typically chosen based on the research questions, data availability and case study specifics.

Each WRES is represented by multiple indicators and includes a diversity of processes. Here, we analyze the services and map them per indicator as well as per service. This set of indicators can be adapted to conform with the specific interest of a potential stakeholder in relation to the supply of a particular service. A more thorough understanding of the provision of the service can be provided by the indicator based analysis and mapping, because the aggregation of the values from different indicators used for the assessment of the services leads to loss of information, as discussed below.

To quantify the supply of WRES, we employ the Soil and Water Assessment Tool (SWAT) hydrological model. *Table 1* provides a list of the WRES investigated here, along with their respective indicators selected from the SWAT output hydrological variables and a clarification of their contribution to the supply of WRES.

Conceptual scheme of the applied methodology is provided in *Figure 4*. The model has been calibrated to our case study (the Upper Santa Cruz watershed), and its uncertainties have previously been statistically analyzed. The discharge from waste water treatment plants was taken into account during the model calibration (*Niraula*

Table 1 SWAT output hydrological variables as indicators of WRES supply.

Water-related ES	Indicator	Contribution to WRES supply
Freshwater provision	Evapotranspiration	Water used by plants
	Soil moisture	Water available for plants (without irrigation water)
	Water yield	Streamflow available for riparian vegetation
	Percolation	Recharge to the aquifer
Water flow regulation	Percolation	Decrease of surface runoff; recharge to the aquifer
	Evapotranspiration	Water used by plants and going back to the atmosphere
	Surface runoff	Water input to streamflow (without effluent water)
	Groundwater flow	
	Lateral flow	
Water purification	Surface runoff	Water input to dissolve contaminants in the soil and surface water, if contamination is observed
	Groundwater flow	
	Lateral flow	

Quantification of water-related ecosystem services in the Upper Santa Cruz watershed

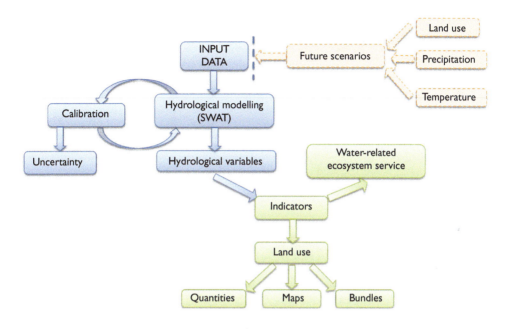

Figure 4 Conceptual scheme of the research methodology.

et al., 2012a, 2015). While the SWAT model enables different future land use, precipitation and temperature scenarios to be used as input data (*Niraula et al., 2015*) so as to analyze their influence on the supply of WRES within the watershed, such an investigation is not pursued here.

1.4 Hydrological modeling – Soil and Water Assessment Tool (SWAT)

SWAT is a physically based, deterministic, continuous-time, watershed scale simulation model developed by the USDA Agricultural Research Service to assist water managers in the assessment of management practices and climate on the water resources (*Arnold et al., 1998; Arnold & Fohrer, 2005; Neitsch et al., 2011*). It was developed to assess the impacts of land management practices on water, sediment and agricultural chemical yields in large complex watersheds.

The model operates at a daily time step. The watershed is divided into subbasins, which are further divided into Hydrological Response Units (HRUs) composed of unique combinations of land use, soil type and slope. The major inputs required by the model include a Digital Elevation Model (DEM), land use and land cover information, soil properties and climate. Description of the applied input data can be found in *Niraula et al. (2012a, 2012b, 2013, 2015)*. The simulation of the hydrological cycle in SWAT is based on the following water balance equation:

$$SW_t = SW_0 + \sum_{i=0}^{t}\left(R_{day} - Q_{sr} - E_a - W_s - Q_{gw}\right)$$

where SW_t is the final soil water content on day i, SW_0 is the initial soil water content on day i, t is the time in days, R_{day} is the amount of rainfall on day i, Q_{sr} is the amount of surface runoff on day i, E_a is the amount of evapotranspiration on day i, W_s is the amount of water entering the vadose zone from the soil profile on day i, and Q_{gw} is the amount of baseflow or return flow on day I; see *Neitsch et al. (2011)* for a detailed description.

Here, we use a 20-year (1987–2006) hydrologic simulation made with the SWAT to help us understand how different land use types affect the various elements of the hydrological cycle and the consequent supply of WRES. For purposes of this investigation, we use only the estimates of hydrological variables simulated by the model (sediment and chemical movements within the watershed *(Niraula et al. 2012b, 2013)* are not examined). The variables relevant to our investigation, and their definitions, are shown in *Table 2*. Estimates of hydrological variables provided by the model are used as indicators for the quantification of WRES supplied by different land use types within the watershed. Maps of the WRES supply quantities per indicator are then created, and the bundles of WRES supplied by different land use types are analyzed.

We provide a spatially explicit analysis of the SWAT model output variables, thereby going beyond the usual application of SWAT to predict changes in streamflow, soil moisture and evapotranspiration *(Schilling et al., 2008; Stone et al., 2001)* and its less frequent use for the analysis of Ecosystem Services (ES) *(Bekele et al., 2014; Notter et al., 2012)*. Although hydrological modeling can improve the understanding of hydrological processes, its integration within the ES framework creates multiple opportunities and challenges (Guswa et al., 2014). Norman et al. (2013) applied some of the SWAT

Table 2 Definitions of SWAT output hydrological variables used in the present research.

Name	Definition
Evapotranspiration	Actual evapotranspiration from the subbasin during the time step (mm).
Soil moisture	Soil water content (mm). Amount of water in the soil profile at the end of the time period.
Percolation	Water that percolates past the root zone during the time step (mm). There is potentially a lag between the time the water leaves the bottom of the root zone and reaches the shallow aquifer. Over a long period of time, this variable should equal groundwater percolation.
Water yield	Water yield (mm). The net amount of water that leaves the subbasin and contributes to streamflow in the reach during the time step.
Surface runoff	Surface runoff contribution to streamflow during time step (mm).
Groundwater flow	Groundwater contribution to streamflow (mm). Water from the shallow aquifer that returns to the reach during the time step.
Lateral flow	Lateral flow contribution to streamflow (mm). Water flowing laterally within the soil profile that enters the main channel during time step.

model outputs that are also used in this study to the Santa Cruz Watershed Ecosystem Portfolio Model (SCWEPM), which uses the concept of the three "pillars" of sustainability as sub-models: the Ecological-Value Submodel (EVM), the Human Well-Being Submodel (HWB) and the Market Land-Price Submodel (MLP). They applied SWAT within the EVM to quantify the visible water (discharge and aquifer recharge) that is available for riparian vegetation and its replacement costs under different scenarios. We employ a spatially explicit method for quantification and visualization of the WRES supplied by the different land use types within the Upper Santa Cruz watersheds, based on the analysis of all SWAT output hydrological variables, thereby enhancing the analysis of results provided by the SCWEPM. Accordingly, we extend the possibilities provided by the SWAT model for conducting an ES based analysis.

1.5 Land use change scenarios and influence on the supply of WRES

The land use change data for the Upper Santa Cruz Watershed is based on three plausible future scenarios for the year 2050, developed by *Norman et al. (2012)* with the SLEUTH urban growth model (*Clarke et al., 1997; Clarke and Gaydos, 1998*). Details about the SLEUTH model and its application in the Upper Santa Cruz Watershed can be found in *Norman et al. (2012)*. The simulated scenarios correspond to "conservation", "current trend" and "megalopolis", and all suggest significant growth of about three times their current area by 2050 in the urban area (*Figure 5*). It is interesting to follow the pattern of change for the different scenarios for the three most represented land use types within the watershed – shrubland, evergreen forest and urban area. In the conservation scenario, the growth of urban areas is accompanied by a decrease in scrublands but evergreen forests keep relatively high representation. In the current trend and megalopolis scenarios the area of shrubland stays similar to the current period but evergreen forests decrease significantly.

Of the main factors on which the hydrological processes depends (land use, soil, slope and climate), only land use can be directly altered by human decisions and activities, while all of the other factors can be indirectly affected. It is, therefore, important to have a better understanding of the consequences of land use change decisions on the supply of WRES. The projected changes in land use will clearly influence the supply of WRES in a significant manner. To assess this impact, the contribution of indicators' supply was reweighted to correspond to the changed areas of the different land use types. The area weighted supply index is calculated through the multiplication of the WRES indicators supply classes and the area of the respective land use type. Note that these calculations of area weighted supply index provide only a hypothesis regarding the impacts of land use change to the supply of WRES. The values of the simulated hydrological variables outputs by the hydrological model are strongly influenced by neighboring land use types within the model subbasins. Therefore, a significant land use change will influence the entire water cycle within the watershed, which will lead to changes in the values of the model output variables. For a more precise calculation of the influence of land use change scenarios on the hydrological variables and the supply of WRES, the model should be actually run with the respective scenarios.

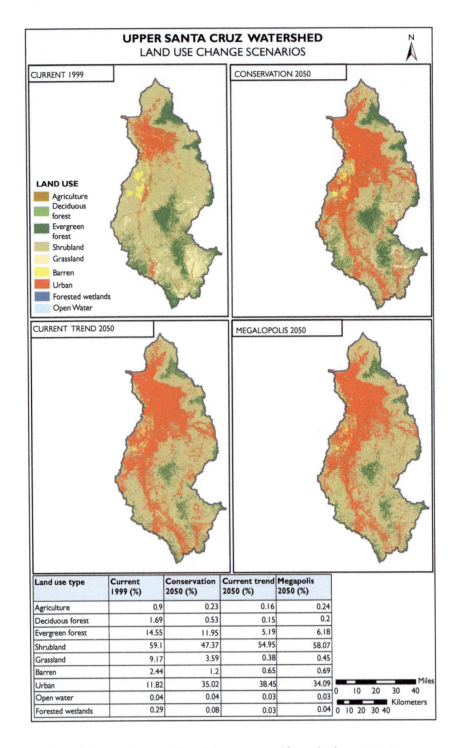

Figure 5 Maps and composition of the current and future land use scenarios.

2 RESULTS AND DISCUSSION

2.1 WRES supply quantities per indicator

The SWAT model was calibrated to reproduce monthly observed behaviors of the system; for more details see *Niraula et al., (2012a, 2015)*. The average annual values of selected indicators for the period 1987–2006 were calculated from the SWAT output table and extracted per land use type, instead of at the subbasin level as originally provided by the model. The average annual values of hydrological variables aggregated over the whole watershed (*Figure 6-a*) indicate that 87% of the annual average precipitation (16 inch/year; 402 mm/year) leaves the watershed in the form of evapotranspiration (14 inch/year; 349 mm/year). The other hydrological variables have values below 10% of the average annual precipitation. The water that percolates into the soil to recharge the aquifer over long periods of time (0.9 inch/year; 22 mm/year) constitutes only 5% of the precipitation. The water yield of the river (1.5 inch/year; 37 mm/year) is just 9% of the average annual precipitation and is composed of 74% surface runoff (1.1 inch/year; 27 mm/year), 17% groundwater flow (0.3 inch/year; 6 mm/year) and 9% lateral flow (0.1 inch/year; 3 mm/year). Soil

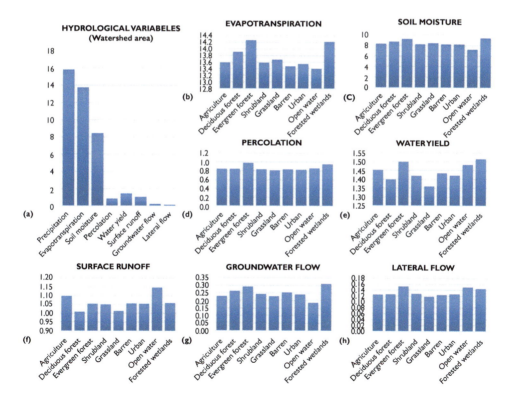

Figure 6 Average annual values of the simulated SWAT hydrological variables within the Upper Santa Cruz watershed (a) for the period (1987–2006) for the different land use types (b, c, d, e, f, g, h). All values are in (inch/year).

moisture, stored in the soil as green water, represents an important part of the water resources within this water-scarce area; it constitutes 8.5 inch/year (215 mm/year) for the investigated period.

We analyze the supply of WRES by land use types based on the average annual depth of each hydrological variable for the respective land use type, resulting from the SWAT model. Almost all hydrological variables show significantly different levels of supply depending on land use type (*Figure 6-b, c, d, e, f, g, h*), while percolation shows the least dependence on land use (*Figure 6-d*) having similar values for almost all land use classes.

Within the Upper Santa Cruz watershed the highest amounts of evapotranspiration are in the evergreen forests and forested wetlands, where the vegetation is most dense, followed by deciduous forests, grasslands and agricultural areas (*Figure 6-b*). It is important to note that, through the process of evapotranspiration, the water does not simply return to the atmosphere – in doing so it contributes to the growth and productivity of terrestrial plants. Evapotranspiration is often used as an approximation of the water used by plants (*Hoekstra et al., 2011*); it is the quantity of water transpired by the plants during their growth or retained in plant tissue, plus the moisture evaporated from the surface of the soil and the vegetation (*Michael, 1978*).

The soil moisture, also referred to as green water (*Falkenmark, 2003; Falkenmark & Rockström, 2006.*), is the water that is available for use by plants. The highest supply is found in forested wetlands and evergreen forests, followed by deciduous forests, agriculture and grasslands (*Figure 6-c*). For both evapotranspiration and soil moisture, the supply provided by shrub land is slightly below that of agricultural areas. Nevertheless, this is the most represented land use type (59% of the watershed) and its contribution to the total supply of WRES within the watershed is of very high importance.

Percolation shows the highest supply within evergreen forests and forested wetlands (*Figure 6-d*), which makes these land use types very important for the recharge of natural groundwater within the watershed. The percolation supply of all the other land use types has very similar values, headed by open water.

Water yield, formed by the sum of surface runoff, groundwater flow and lateral flow contributions to streamflow, is a variable with very high importance for riparian vegetation, which is a supplier of multiple ecosystem services (*Norman et al., 2013; Jenkins et al., 2010*). Forested wetlands provide the highest supply of water yield, followed by evergreen forests and open water (*Figure 6-e*). The forested wetlands and open water have significantly smaller areal representation within the watershed (*Figure 2*), and so their contribution to the overall supply of water yield is relatively low. Nevertheless, they are of very high importance for the ecological integrity of the watershed and supply multiple significant ES that are not necessarily water-related. The highest supply of surface runoff is from open water areas, followed by agriculture (*Figure 6-f*). Forested wetlands, evergreen forests, shrublands, barren and urban areas have similar supplies of surface runoff, while the lowest supply is from deciduous forests and grasslands. The highest groundwater flow supply is generated by the forested wetlands, followed by evergreen and deciduous forests (*Figure 6-g*). The lateral flow, even though having the lowest contribution to the water yield, has highest supply within the evergreen forests, followed by open water areas and forested wetlands (*Figure 6-h*).

2.2 WRES supply classes per indicator

The average annual values of the WRES indicators per land use type are categorized into classes labelled from 0 to 5, following the matrix method for mapping ES, develop by *Burkhard et al. (2009, 2012b, 2014)*. The classification is conducted within a Geographical Information System (GIS) using the equal interval method. The range is determined by the minimum and maximum value of the respective indicator, where 0 indicates very low supply, 1 = low supply, 2 = relevant supply, 3 = medium supply, 4 = high supply and 5 = very high supply. Each class has a range of values with class 0 being the minimum value and class 5 being the maximum value for the area. For a certain indicator, the supply class of a land use type is determined by the range within which the indicator's value lies. It is important to note that the supply classes represent the respective potential of the land use types to supply WRES based on the indicators' magnitude (inch/year; mm/year) simulated by the SWAT model, without reflecting their areal representation within the watershed.

Many land use types show very low (0) and low (1) supply of WRES (*Table 3*). Grasslands show very low and low supply for almost all indicators of WRES with the exception of soil moisture, for which they show medium (3) supply. Similarly, the urban areas and shrub lands show very low to relevant (2) supply for all indicators, but no medium (3) or higher supply. Barren areas show medium supply of groundwater flow, which contributes to the relevant supply of water yield, and for all the other indicators the supply is relevant or very low. Open water and deciduous forests show very different supplies of WRES depending on the indicator. Open water has very high (5) supply of surface runoff and lateral flow, but very low supply of groundwater flow, all of which contributes to the high (4) supply of water yield. However, its supply of evapotranspiration and soil moisture is very low, and supply of percolation is low. These results are probably due to the specifics of the land use type open water, which is essentially a water body. The deciduous forests show high supply of soil moisture and medium evapotranspiration, while percolation and water yield are low. The low supply of water yield is due to the very low supply of surface runoff and low supply of lateral flow, despite the medium supply of groundwater flow. Evergreen forests and forested wetlands show the highest supply of WRES with very high supply of almost all indicators besides surface runoff.

Table 3 WRES supply classes of the different land use types per indicator. The classes indicate the supply as follows: 0 – very low (minimum); 1 – low; 2 – relevant; 3 – medium; 4 – high; 5 – very high (maximum) (after Burkhard, 2009, 2012b, 2014).

Land use	Evapotranspiration	Soil moisture	Percolation	Water yield	Surface runoff	Ground water flow	Lateral flow
Agriculture	1	2	1	3	3	2	1
Deciduous forest	3	4	1	1	0	3	1
Evergreen forest	5	5	5	5	1	5	5
Shrubland	1	2	0	2	1	2	1
Grassland	1	3	0	0	0	1	0
Barren	0	2	0	2	2	3	0
Urban	1	2	0	2	1	2	1
Open water	0	0	1	4	5	0	5
Forested wetlands	5	5	5	5	2	5	4

212 Water bankruptcy in the land of plenty

The WRES supply classes per indicator and their respective ranges are represented spatially via the maps in *Figure 7*. From these, the areas of high and low supply of all indicators and the spatial distribution of the classes and the respective quantities are clearly visible. For example, the map of percolation indicates

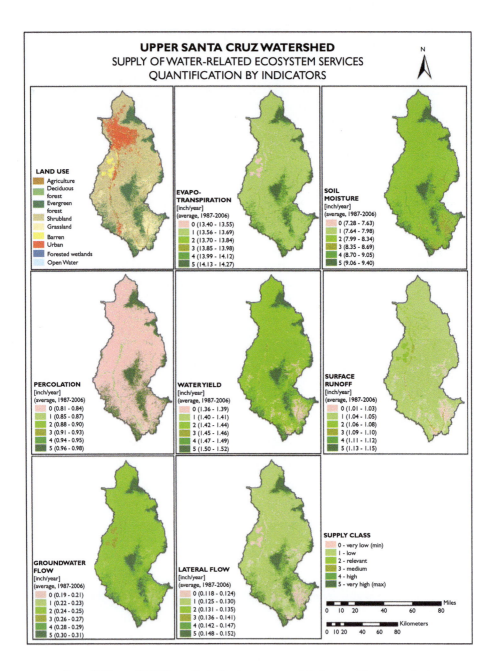

Figure 7 Maps of WRES supply by land use type and their quantities per indicator.

predominantly very low supply across the watershed due to the very low percolation supply of some of the most represented land use types, such as shrubland (59%), urban (12%) and grassland (9%). Meanwhile, evergreen forests show very high percolation supply and significant spatial representation (14.5%) in the highland areas.

Surface runoff shows generally low supply at the watershed scale due to the low surface runoff supply of the predominant land use types, such as shrubland, evergreen forests and urban areas and the very high supply only for open water, which has very low representation within the watershed (0.04%).

Evapotranspiration and lateral flow supply are low for shrub lands and urban areas, which have very high representation within the watershed. This is why the maps of evapotranspiration and lateral flow supply represent low overall supply within the watershed. Nevertheless, evergreen forests show very high supply of evapotranspiration and high supply of lateral flow, which is a major contributor to the overall supply within the watershed.

The highest overall supply within the watershed is of the indicator soil moisture, followed by water yield and groundwater flow, which have relevant supply for the most represented land use type in the watershed – shrub lands, as well as for the urban areas, and very high supply for evergreen forests.

2.3 WRES supply classes per service

The overall supply of WRES is calculated as the average supply of its indicators, with equal weight given to all indicators (*Table 4*). The calculation of average supply of ES based on multiple indicators leads to generalization and loss of information for the individual indicators. This generalization is also visible in the WRES maps (*Figure 8*). However, important aggregated information about the supply of ES can be conveyed and applied in decision-making.

- Evergreen forests and forested wetlands show very high supply of freshwater provision and high supply of water flow regulation and water purification. These

Table 4 WRES supply classes of the different land use types.

Land use	WRES supply classes		
	Freshwater Provision	Water flow regulation	Water purification
Agriculture	1.8	1.6	2.0
Deciduous forest	2.3	1.6	1.3
Evergreen forest	5.0	4.2	3.7
Shrubland	1.3	1.0	1.3
Grassland	1.0	0.4	0.3
Barren	1.0	1.0	1.7
Urban	1.3	1.0	1.3
Open water	1.3	2.2	3.3
Forested wetlands	5.0	4.2	3.7

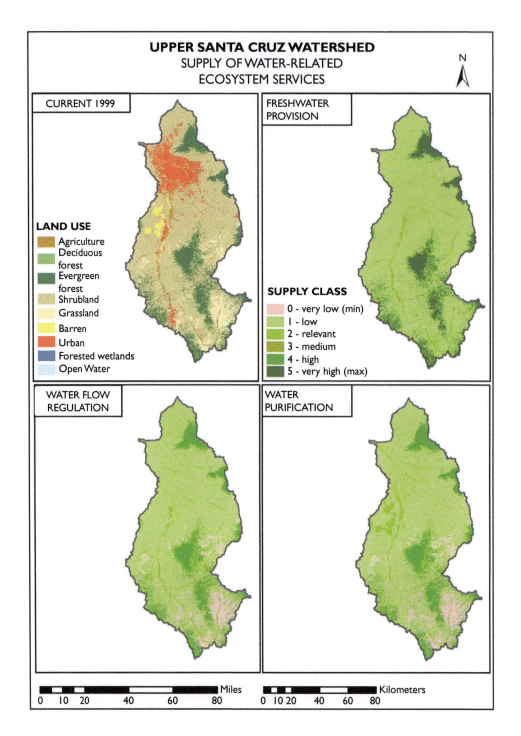

Figure 8 Maps of WRES supply by land use type.

results indicate that they have the highest potential to supply WRES within the watershed, but depending on their areas they have different contributions to the overall supply.
- Urban areas show low supply of all three WRES and agricultural areas – relevant supply of all services.
- Deciduous forests show relevant supply of freshwater provision and water flow regulation and low supply of water purification.
- Open water areas show low supply of freshwater provision, but relevant supply of water flow regulation and medium supply of water purification.
- Barren areas show low supply of freshwater provision and water flow regulation, but relevant supply of water purification.
- Grasslands show low supply of freshwater provision and very low supply of water flow regulation and water purification, making it the land use type within the watershed providing the least WRES.

Maps of the supplied WRES show their spatial representation (*Figure 8*). These maps indicate low overall supply of WRES within the watershed, with high and very high supply only from evergreen forests and forested wetlands. It is important to note that forested wetlands, even though having very small representation within the watershed (0.3%) have very high relevance for the supply of WRES, showing high and very high supply for most of the individual indicators and respective services. Furthermore, they are very important sites for habitat and biodiversity (*Norman et al., 2013*), which increases the importance of their conservation and restoration. Evergreen forests have fairly high representation within the watershed and provide high and very high supply of WRES, therefore are of very high importance for the overall provision of WRES within the watershed and their preservation is of very high relevance.

2.4 Bundles of WRES supplied by land use types

All land use types provide bundles of ES. Depending on land use structure and function, the supply of some services is very low, while of others is very high. Such an analysis provides information about ES trade-offs, which is important when decisions regarding land use change are to be taken, leading to the increase in the supply of some services and the decrease or loss of others.

Figure 9 presents, via a spider diagram, the bundles of chosen indicators of WRES supply and their classes for all land types within the Upper Santa Cruz watershed. The diagram shows that evergreen forests and forested wetlands have a very high supply of almost all indicators, while deciduous forests show a quite diverse supply, depending on the indicator. Urban areas and shrub lands show mostly low supply and open water shows a very high supply of some indicators and very low of others. The supply of all indicators from barren and agricultural areas is generally below medium, and grasslands show low supply of all indicators except for soil moisture (which is medium). The changes in supply of different indicators within the different land use types are also easily visible. The line for water yield stays close to the outer edge of the diagram for evergreen forests, forested wetlands and open water, which indicates its high supply in these areas, and goes closer to the center of the diagram for all other

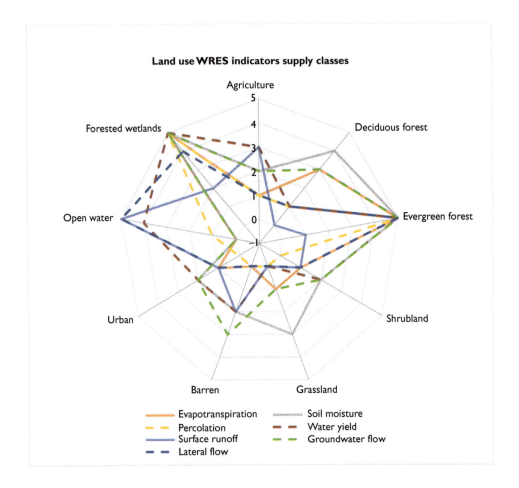

Figure 9 Bundles of WRES supply by land use type and their classes per indicator.

land use types, respectively indicating its lower supply by them. A similar observation is valid for lateral flow. Percolation has a high supply only from evergreen forests and forested wetlands, and low supply from all other land use types, which indicates the high relevance of these land use types for the natural groundwater recharge within the watershed. A similar pattern is observed also for evapotranspiration, with the difference being that it shows generally higher (medium) supply compared to deciduous forests. Soil moisture shows higher overall supply for most of the land use types, and surface runoff has lower overall supply than all other indicators.

Similarly, *Figure 10* presents a spider diagram of the supply of WRES by the different land use types per service. The very high supply of WRES by evergreen forests and forested wetlands is clearly visible in comparison to the moderate and below moderate supply of WRES by all other land use types. Water purification shows lowest overall supply in comparison to the other WRES. Open water shows the biggest diversity in supply of WRES (from low to medium), depending on the service of interest.

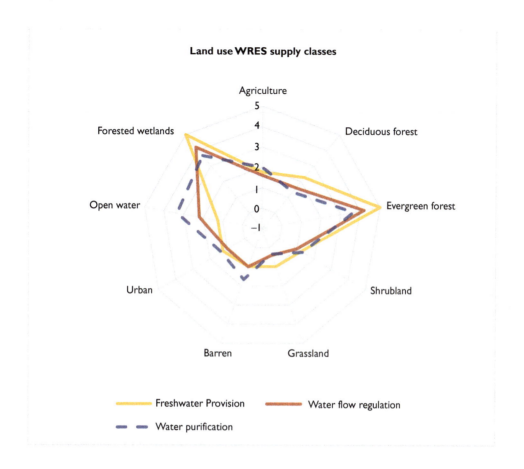

Figure 10 Bundles of WRES supply by land use type and their classes.

2.5 Hypothetical influence of land use change scenarios on the supply of WRES

The average area weighted supply index for the whole Upper Santa Cruz watershed indicates decreases in the supply of almost all WRES indicators in all three scenarios (*Figure 11*). Slight increases in supply are observed only for water yield in the conservation scenario and for surface runoff for all three scenarios. The smallest change in the indicators' supply is expected for the conservation scenario and the biggest decrease in supply is for the current trend scenario.

These results suggest that the "current trend" land use change scenario will have most significant impact on the supply of WRES as a result of the loss of land use types with high and very high potential to supply those services (evergreen forests and forested wetlands) and the significant increase of the urban areas which have low potential to supply WRES. This projection is of strong concern given the water-scarce character of the Upper Santa Cruz watershed, and taking into consideration the dependence of the area on groundwater resources and water allocated from the Colorado River (which is

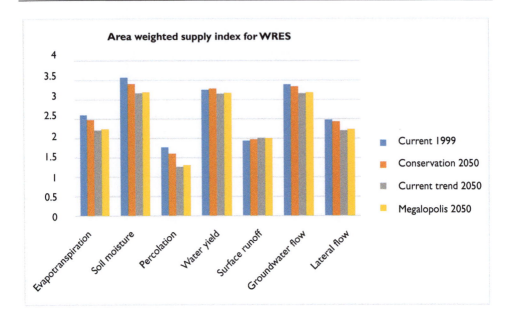

Figure 11 Area weighted supply index – indicating the contribution of the hydrological variables to the supply of WRES within the Upper Santa Cruz watershed, based on the land use areal representation.

currently experiencing an extended period of severe drought *(Cayan et al., 2010)*, over-exploitation and extremely vulnerable ecological status *(Triedman, 2012))*.

Considering the uncertainties related to the long-term continuation of the supply of CAP water to the Tucson AMA, a land use scenario that would lead to significant decreases in the water resources for the area should be very critically examined. Even though the Tucson AMA overlaps only partially with the Upper Santa Cruz watershed *(Figure 2)*, a comparison of the total quantity of water supplied for recharge by the CAP to the Tucson AMA against the quantities of natural groundwater recharge (percolation) within the Upper Santa Cruz watershed (despite the lack of full spatial overlap) indicates that the CAP water delivered to Tucson AMA for recharge[2] (approximately 31×10^3 AF/year; 38×10^6 m³/year) is on average equal to about 17% of the modeled natural groundwater recharge (approximately 187×10^3 AF/year; 230×10^6 m³/year) for the period 1993–2000. For the period 2001–2006 the volume of delivered CAP water for recharge increased four times (approximately 124×10^3 AF/year; 153×10^6 m³/year) while the natural groundwater recharge decreased two times (approximately 93×10^3 AF/year; 115×10^6 m³/year) due to the beginning of the dry period in the early 2000's. Therefore the delivered CAP water exceeded the natural recharge for that period by one third and is equal to 133% of it. These percentages highlight the significance of the CAP recharge water and the vulnerability of

---

2 The total CAP water delivered for recharge to Underground Storage Facilities (USF) and Groundwater Savings Facilities (GSF). Data available at: http://www.azwater.gov/AzDWR/Watermanagement/AMAs/TucsonAMA/TAMAOverview.htm#waterbudget.

natural groundwater recharge and generally freshwater resources after the beginning of the drought in Arizona, considering that the groundwater recharge equals the quantity of water available for abstraction in the long term (*Rushton and Ward, 1978*).

3 CONCLUSIONS

Application of the SWAT hydrological model for quantification and mapping of WRES provides significant information about the supply of those services by different land use types within the Upper Santa Cruz watershed. The very high supply of WRES by evergreen forests and forested wetlands is distinguishable in relation to all other present land use types. Their conservation and preservation is of significant relevance for the future supply of WRES within the area, especially when taking into consideration the ongoing drought and projected scenarios regarding land use change. Projected urban growth is expected to significantly increase the demand for WRES, while, at the same time, their natural supply within the watershed will decrease. Such land-use change scenarios should be viewed very critically give the semi-arid nature of the region, where water resources are extremely vulnerable and the water *'independence'* of the area is of high priority. Land and water are immutably connected and their management should be done in conjunction.

The methodology presented here provides new opportunities for the application of the SWAT hydrological model that unfold the potential for improved WRES analysis. The model simulations provide information that is not intuitive or measurable, and the interpretation of the results into ES provides enhanced possibilities for application into practice and management. The resulting tables, maps and diagrams can serve as tools to improve the dialogue with decision-makers and to bridge between specialized hydrological knowledge and a diverse group of stakeholders. Our methods and results are complimentary to ongoing research initiatives in the area, including the Targeted Watersheds Grant Program: Santa Cruz River, AZ and Mexico[3] (EPA), Santa Cruz River Initiative[4] (Sonoran Institute), the Santa Cruz Watershed Ecosystem Portfolio Model (SCWEPM)[5] (USGS), and can contribute to the efforts to *'close the water demand-supply gap in Arizona'* (Eden et al., 2015).

REFERENCES

Arizona Department of Water Resources (ADWR) (2015) Arizona's Next Century: A Strategic Vision for Water Supply Sustainability (Strategic Vision). January 2014. http://www.azwater.gov/AzDWR/Arizonas_Strategic_Vision/

Arnold, J.G. and Fohrer, N. (2005). SWAT2000: Current capabilities and research opportunities in applied watershed modeling, *Hydrological Processes*. 19(3): 563–572. doi:10.1002/hyp.5611

---

[3] http://water.epa.gov/grants_funding/twg/initiative_index.cfm.
[4] http://www.sonoraninstitute.org/where-we-work/southwest/santa-cruz-river.html.
[5] http://geography.wr.usgs.gov/science/ecoSevicesSCWatershed.html.

Arnold, J.G., Srinivasan, R., Muttiah, R.S. and Williams, J.R. (1998). Large area hydrologic modeling and assessment part 1: Model development. *The American Water Resources Association American Water Resources Association*, 34(1): 73–89. doi:10.1111/j.1752-1688.1998.tb05961.x.

Bagstad, K.J., Semmens, D.J., Waage, S. and Winthrop, R. (2013). A comparative assessment of decision-support tools for ecosystem services quantification and valuation. *Ecosystem Services*, 5: 27–39. doi:10.1016/j.ecoser.2013.07.004.

Bekele, E.G., Lant, C.L., Soman, S. and Misgna G. (2013). The evolution and empirical estimation of ecological-economic production possibilities frontiers. *Ecol. Econ.*, 90: 1–9. doi: 10.1016/j.ecolecon.2013.02.012.

Boyanova, K., Nedkov, S. and Burkhard, B. (2014). Quantification and mapping of flood regulating ecosystem services in different watersheds – case studies in Bulgaria and Arizona, USA. In: Bandrova, T., Konecny, M. and Zlatanova, S. (Eds.), *Thematic Cartography for the Society*, Dortrecht, Springer: 237–255.

Brauman, K.a., Daily, G.C., Duarte, T.K. and Mooney, H.A. (2007). The nature and value of ecosystem services: an overview highlighting hydrologic services. *Annual Review of Environment and Resources*, 32(1): 67–98. doi:10.1146/annurev.energy.32.031306.102758.

Burkhard, B., Kroll, F., Müller, F. and Windhorst, W. (2009). Landscapes' capacities to provide ecosystem services – A concept for land-cover based assessments. *Landscape Online*, 15(1): 1–22. doi:10.3097/LO.200915.

Burkhard, B., de Groot, R., Costanza, R., Seppelt, R., Jørgensen, S.E. and Potschin, M. (2012a). Solutions for sustaining natural capital and ecosystem services. *Ecological Indicators*, 21: 1–6. doi:10.1016/j.ecolind.2012.03.008.

Burkhard, B., Kroll, F., Nedkov, S. and Müller, F. (2012b). Mapping ecosystem service supply, demand and budgets. *Ecological Indicators*, 21: 17–29. doi:10.1016/j.ecolind.2011.06.019.

Burkhard, B., Kandziora, M., Hou, Y. and Müller, F. (2014). Ecosystem service potentials, flows and demands-concepts for spatial localisation, indication and quantification. *Landscape Online*, 34(1): 1–32. doi:10.3097/LO.201434.

Cayan, D.R., Das, T., Pierce, D.W., Barnett, T.P., Tyree, M. and Gershunov, A. (2010). Future dryness in the southwest US and the hydrology of the early 21st century drought. *Proceedings of the National Academy of Sciences of the United States of America*, 107(50): 21271–21276. doi:10.1073/pnas.0912391107.

Crossman, N.D., Burkhard, B., Nedkov, S., Willemen, L., Petz, K., Palomo, I. and Maes, J. (2013). A blueprint for mapping and modelling ecosystem services. *Ecosystem Services*, 4: 4–14. doi:10.1016/j.ecoser.2013.02.001.

Daily, G.C. and Matson, P.A. (2008). Ecosystem services: from theory to implementation. *Proceedings of the National Academy of Sciences of the United States of America*, 105(28): 9455–9456. doi:10.1073/pnas.0804960105.

Daily, G.C., Polasky, S., Goldstein, J., Kareiva, P.M., Mooney, H.a., Pejchar, L. and Shallenberger, R. (2009). Ecosystem services in decision making: Time to deliver. *Frontiers in Ecology and the Environment*, 7(1): 21:28. doi:10.1890/080025.

De Groot, R.S., Alkemade, R., Braat, L., Hein, L. and Willemen, L. (2010). Challenges in integrating the concept of ecosystem services and values in landscape planning, management and decision making. *Ecological Complexity*, 7(3): 260–272. doi:10.1016/j.ecocom.2009.10.006.

Dominguez, F., Cañon, J. and Valdes, J. (2010). IPCC-AR4 climate simulations for the Southwestern US: The importance of future ENSO projections. *Climatic Change*, 99(3): 499–514. doi:10.1007/s10584-009-9672-5.

Eden, S., Ryder, M. and Capehart, M.A. (2015). Closing the Water Demand-Supply Gap in Arizona, Arroyo, University of Arizona Water Resources Research Center, Tucson, AZ. http://wrrc.arizona.edu/publications/arroyo-newsletter/arroyo-2015-Closing-Demand-Supply-Gap.

Falkenmark, M. (2003). Water management and ecosystems: Living with change. *TEC Background Papers* n. 9, Global Water Partnership, Stockholm.

Falkenmark, M. and Rockström, J. (2006). The New Blue and Green Water Paradigm: Breaking New Ground for Water Resources Planning and Management. *Journal of Water Resources Planning and Management Editorial*, May/June 2006: 129–132.

Guswa, A., Brauman, K., Brown, C., Hamel, P., Keeler, B. and Sayre, S.S. (2014). Ecosystem services: Challenges and opportunities for hydrologic modeling to support decision making. *Water Resources Research*: 1–10. doi:10.1002/2014WR015497.

Haines-Young, R. and Potschin, M. (2013). CICES V4.3 – Report prepared following consultation on CICES Version 4, August-December 2012. EEA Framework Contract No EEA/IEA/09/003

Hoekstra, A.Y., Chapagain, A.K., Aldaya, M.M. and Mekonnen, M.M. (2011). *The Water Footprint Assessment Manual: Setting the Global Standard*. New York, Routledge, Earthscan.

Jenkins, W.A., Murray, B.C., Kramer A.R. and Faulkner, S.P. (2010). Valuing ecosystem services from wetlands restoration in the Mississippi Alluvial Valley. *Ecological Economics*, 59: 1051–1051.

Kandziora, M., Burkhard, B. and Müller, F. (2013). Interactions of ecosystem properties, ecosystem integrity and ecosystem service indicators: A theoretical matrix exercise. *Ecological Indicators*, 28: 54–78. doi:10.1016/j.ecolind.2012.09.006.

Logsdon, R.A. and Chaubey, I. (2013). A quantitative approach to evaluating ecosystem services. *Ecological Modelling*, 257: 57–65. doi:10.1016/j.ecolmodel.2013.02.009.

Maes, J., Teller, A., Erhard, M., Liquete, C., Braat, L. and Berry, P., et al. (2013). Mapping and Assessment of Ecosystems and their Services. An analytical framework for ecosystem assessments under action 5 of the EU biodiversity strategy to 2020. Publications office of the European Union, Luxembourg.

Maes, W.H., Heuvelmans, G. and Muys, B. (2009). Assessment of land use impact on water-related ecosystem services capturing the integrated terrestrial–aquatic system. *Environmental Science Andtechnology*, 43(19): 7324–7330.

MA (Millennium Ecosystem Assessment) (2005). Ecosystems and Human Well-being: Synthesis. Island Press/World Resources Institute, Washington, DC.

Martínez-Harms, M.J. and Balvanera, P. (2012). Methods for mapping ecosystem service supply: a review. *International Journal of Biodiversity Science, Ecosystem Services & Management*, 8(1–2): 17–25. doi:10.1080/21513732.2012.663792.

Michael A.M (1978). *Irrigation Theories and Practice*, Vikas Publishing House, India.

Nedkov, S. and Burkhard, B. (2012). Flood regulating ecosystem services – Mapping supply and demand, in the Etropole municipality, Bulgaria, *Ecological Indicators*, 21: 67–79. doi:10.1016/j.ecolind.2011.06.022.

Nedkov, S., Boyanova, K. and Burkhard, B. (2015). Quantifying, Modelling and Mapping Ecosystem Services in Watersheds. In: Chicharo, L., Müller, F., Fohrer, N. (Eds.), *Ecosystem Services and River Basin Ecohydrology*, Dordrecht, Springer: 133–149.

Neitsch, S., Arnold, J., Kiniry, J. and Williams, J. (2011). Soil & Water Assessment Tool Theoretical Documentation, Version 2009: 1–647.

Niemeijer, D. and de Groot, R. (2008). A conceptual framework for selecting environmental indicator sets, *Ecological Indicators*, 8:14–25.

Niraula, R., Norman, L.M., Meixner, T. and Callegary J.B. (2012a) Multi-gauge Calibration for modeling the Semi-Arid Santa Cruz watershed in Arizona-Mexico border area using SWAT, *Air, Soil and Water Research*, 5: 41–57. doi:10.4137/ASWR.S9410.

Niraula, R., Kalin, L., Wang, R. and Srivastava, P. (2012b) Determining nutrient and sediment critical source areas with SWAT model. effect of lumped calibration, *Transactions of the ASABE*, 55(1): 137–147.

Niraula, R., Kalin, L., Srivastava, P. and Anderson, C.J. (2013). Identifying critical source areas of nonpoint source pollution with SWAT and GWLF. *Ecological Modeling*, 268:123–133.

Niraula, R., Meixner, T. and Norman, L.M. (2015). Determining the importance of model calibration for forecasting absolute/relative changes in streamflow from LULC and climate changes. *Journal of Hydrology*, 522: 439–451. doi:10.1016/j.jhydrol.2015.01.007.

Norman, L.M., Feller, M. and Villarreal, M.L. (2012). Developing spatially explicit footprints of plausible land-use scenarios in the Santa Cruz Watershed, Arizona and Sonora. *Landscape and Urban Planning*, 107(3): 225–235. doi:10.1016/j.landurbplan.2012.06.015.

Norman, L.M., Villarreal, M.L., Niraula, R., Meixner, T., Frisvold, G. and Labiosa, W. (2013). Framing scenarios of binational water policy with a tool to visualize, quantify and valuate changes in ecosystem services. *Water (Switzerland)*, 5(3): 852–874. doi:10.3390/w5030852.

Notter, B., Hurni, H., Wiesmann, U. and Abbaspour, K.C. (2012). Modelling water provision as an ecosystem service in a large East African river basin. *Hydrology & Earth System Sciences*, 16: 69–86. doi:10.5194/hess-16-69-2012.

Rushton, K.R. and Ward, C. (1979). The estimation of groundwater recharge. *Journal of Hydrology*, 41(3–4): 345–351. doi:10.1015/0022-1594(79)90070-2.

Schilling, K.E., Jha, M.K., Zhang, Y.-K., Gassman, P.W. and Wolter, C.F. (2008). Impact of land use and land cover change on the water balance of a large agricultural watershed: Historical effects and future directions. *Water Resources Research*, 44: 1–12. doi:10.1029/2007WR006644.

Staub, C., Ott, W., Heusi, F., Klingler, G., Jenny, A., Häcki, M. and Hauser, A. (2011). Indicators for Ecosystem Goods and Services: Framework, methodology and recommendations for a welfare-related environmental reporting. Federal Office for the Environment, Bern. Environmental studies no. 1102: 17 S. Available at: ftp://ftp.wsl.ch/ALR/Papers/papers_ALR_2a/Indicators_Ecosyste_Goodand_Services.pdf.

Stone, M.C., Hotchkiss, R.H., Hubbard, C.M., Fontaine, T.a, Mearns, L. and Arnold, J.G. (2001). Impacts of climate change on Missouri River Basin water yield. *Journal of the American Water Resources Association*, 37(5): 1119–1129. doi:10.1111/j.1752-1688.2001.tb03626.x.

Stürck, J., Poortinga, A. and Verburg, P.H. (2014). Mapping ecosystem services: The supply and demand of flood regulation services in Europe. *Ecological Indicators*, 38: 198–211. doi:10.1016/j.ecolind.2013.11.010.

TEEB (2010). The Economics of Ecosystems and Biodiversity: mainstreaming the economics of nature: a synthesis of the approach, conclusions and recommendations of TEEB.

Triedman, B.N. (2012). Environment and Ecology of the Colorado River Basin. The 2012 State of the Rockies Report Card. The Colorado River Basin: Agenda for Use, Restoration, and Sustainability for the Next Generation. Available at: https://www.coloradocollege.edu/other/stateoftherockies/report-card/2012-report-card.dot.

van Oudenhoven, A.P.E., Petz, K., Alkemade, R., Hein, L. and de Groot, R.S. (2012). Framework for systematic indicator selection to assess effects of land management on ecosystem services. *Ecological Indicators*, 21: 110–122.

Vigerstol, K.L. and Aukema, J.E. (2011). A comparison of tools for modeling freshwater ecosystem services. *Journal of Environmental Management*, 92(10): 2403–2409. doi:10.1016/j.jenvman.2011.06.040.

Villamagna, A.M., Angermeier, P.L. and Bennett, E.M. (2013). Capacity, pressure, demand, and flow: A conceptual framework for analyzing ecosystem service provision and delivery. *Ecological Complexity*, 15: 114–121. doi:10.1016/j.ecocom.2013.07.004.

Wiggering, H. and Müller, F. (Eds.) (2004). *Umweltziele und Indikatoren*, Berlin/Heidelberg/New York, Springer.

WRDC (Water Resources Development Commission) (2010). Final Report, Volume I, October 1, 2011. Available at: http://www.azwater.gov/AzDWR/WaterManagement/WRDC_HB2661/Meetings_Schedule.htm.

Chapter 13

Qualitative assessment of the supply and demand of ecosystem services in the Pantano Wash watershed

Rositsa Yaneva
National Institute of Geophysics, Geodesy and Geography, Bulgarian Academy of Science

INTRODUCTION

The ecosystem services concept

The ecosystem services concept (*MA 2005; de Groot et al., 2010; Haines-Young, 2011*) has come to be acknowledged as an instrument that is useful in every step of the process of sustainable environmental management. The subject has been studied for decades (*Costanza et al., 1997; Daily, 1997*), but the term was introduces to the general public in 2005. Ecosystem services are variously defined as *"the benefits people obtain from the ecosystems"* (MA 2005) and *"the contributions of ecosystem structure and function—in combination with other inputs—to human well-being"* (Burkhard et al., 2012). A proper analysis of ecosystems services requires adoption of a transdisciplinary research approach that integrates across scientific disciplines and links environmental and socio-ecological concepts. The essence of this concept is in the emphasis placed on understanding the inter-relationships between the ecological variables while taking properly into account the presence of humans in the system. Without consideration of the anthropogenic factor (i.e., people who benefit from the ecosystem services and who are a major driver for environmental changes), the concept would not be able to sufficiently explain the functionality of an environmental system. Moreover, the concept of Ecosystem Services (ES) is pivotal to human well-being. Ultimately, it is *"stakeholder-driven concept"* (Koschke et al., 2014) and so it is essential for communication to occur among the relevant social, political, and environmental units.

Traditionally, the investigation of an ecosystem has often placed focus on the local and short-term dimensions of sustainability (considered to be the endurance of the systems' structure and processes), and high importance has been given to the system's sensitivity and ability to recover after a disturbance (such as a drought, over-exploitation of the resources, or a wildfire). These investigations provide only limited insights regarding the spatial and temporal changes in ecosystem structure *(Borisova, 2013; Burkhard et al., 2013)* and are unable to properly address the important challenge of accounting for anthropogenic presence as a primary agent for changes in system dynamics and stability, rather than an external factor of disturbance.

The holistic approach taken by the ES concept has been widely adopted by planners and decision makers, and is now integrated into many environmental policy acts at the national level, including the *UK National Ecosystem Assessment*[1] (UK NEA), *US Ecosystem Services Research Program*[2], the *World Bank WAVES 2013*[3], and the *European Union Biodiversity Strategy*[4] that is incorporated by all member states (*EU Biodiversity Strategy to 2020*). The approach requires that strong collaboration be established among users, planners, and decision-makers from the very beginning of the project, especially so when the involvement of stakeholders is critical to the assessment of geo-biophysical structure and function *(Hein et al., 2006; Malone et al., 2010; Burkhard et al., 2013; Koschke et al., 2014)*. Moreover, it is very important to consider the spatial scales of the valuation process, so as to correctly address the heterogeneity in biophysical and social aspects of a coupled human-environmental system when analyzing and valuing ecosystem demands and flows in support of environmental management planning *(Hein et al., 2006; Burkhard et al., 2013)*.

Expert opinions could be referred to an approach which synthesizes subjective opinions on a specific subject and/or problem. The experts were carefully selected to include participants with a good understanding of ecosystems and a familiarity with the ecosystem services concept. The expert opinion procedure is a core method when conducting surveys and obtaining reliable responses. It should be developed in such a way that minimizes inherent biases in subjective judgment in the elicited outcomes *(Knol, Sluijs, & Slottje, 2008)*.

Assessing and mapping supply of, and demand for ecosystem services

Before beginning the process of mapping supply of, and demand for, ecosystem services in the Pantano Wash watershed, we conducted an expert-based evaluation to obtain a qualitative assessment of the capacity of each type of land cover to provide such services. The criteria used for the selection of the experts was the diversity of their knowledge, and the group included participants familiar with the ES concept. The group consisted of three kinds of participants, classified according to their affiliation and understanding of the peculiarities of the case study—University of Arizona academics, SWAN project researchers, and University of Arizona students. Each of these potential "stakeholders" was both interviewed and asked to respond to a survey. Even though only some of the respondents identified themselves as being knowledgeable about the case study location and its local natural characteristics, their knowledge ranged broadly across many topics and disciplines and contributed to a holistic understanding of the structure and functions of the ecosystem. It is important to note that an opinion was not judged as *"right"* or *"wrong"* and all opinions were treated as providing knowledge regarding value of the ecosystem to human well-being. Although we anticipated that differences would arise among the survey responses, an examination of the experts' perceptions provided important insights

---

1 http://uknea.unep-wcmc.org/.
2 http://www2.epa.gov/eco-research/ecosystems-services.
3 http://documents.worldbank.org/curated/en/2013/04/17938923/wealth-accounting-valuation-ecosystem-services-global-partnership-2013-annual-report.
4 http://ec.europa.eu/environment/nature/biodiversity/comm2006/2020.htm.

regarding ecosystem capacity. An interesting aspect of this study is that it reflects the mindsets of different stakeholder groups in the sense of highlighting differences in their perceptions regarding this human-environmental system. Such data is important as it helps develop an understanding of factors underlying a complex transdisciplinary issue such as this, and helps to establish where a meeting point might exist between the opinions of inside and outside experts on land-water-people relationships.

The ES concept requires the development of maps that present the outcomes of the qualitative assessment for investigation, and that enables a spatially comparison of the supply of, and demand for, ecosystem services. The resulting maps developed in our study reflect the needs of humans regarding desired levels of a specific service, and thereby help to identify spatial mismatches across the landscape (*Paetzold et al., 2010*). This benchmark analysis serves as a solid base for ongoing research that seeks to track the provision of ecosystem services and changes in their spatial distribution over time. By quantifying the present capacities of different ecosystems to provide services, these maps provide useful tools for natural resource management.

1 BACKGROUND OF THE RESEARCH

1.1 Identification of the problem

Identification, quantification and evaluation of ecosystem services can contribute in significant ways to strategic decision making in two main areas: a) the design of new policies, and b) the strategic development of a landscape. Analysis of human-environmental system with particular attention to ecosystem integrity (*Kandziora et al., 2012*) can serve as an important tool in the communications between stakeholders in both public and private sectors.

Changes in the natural behavior of environmental processes, most often caused by the climate change and/or human-driven Land Use and Land Cover (LULC) change, can make a natural system less resilient to disturbance, more vulnerable to natural hazards, and lead to an eventual scarcity of the natural resources it provides. Recent studies have reported that since the 1950's, the southwestern United States has experienced a rather severe and extended drought *(Garfin et al., 2013)*. The scale of this phenomenon has a multidimensional impact on the natural and social aspects of the system, which must be taken into consideration. Further, the complexity of ecosystem functioning requires that both empirically accumulated knowledge and the results of past efforts to develop sustainable management be considered (*McGinnis & Ostrom, 2014*). Humans typically make considerable demands on the environment, and their comfort and security is directly dependent on the benefits derived from natural processes. Therefore, a social-ecological system framework is needed to properly address the complexity of the environmental system *(Ostrom, 2009; Binder et al., 2013; McGinnis & Ostrom, 2014)*, where ecosystem services represent the intermediate step between resource units (systems) and individual units (systems) defined by the social, economic, and political setting

Because ecosystems have different functions, they have different capacities to provide various ecosystem services. In particular, the ecosystem function and structure is strongly related to land uses and the resulting land cover changes.

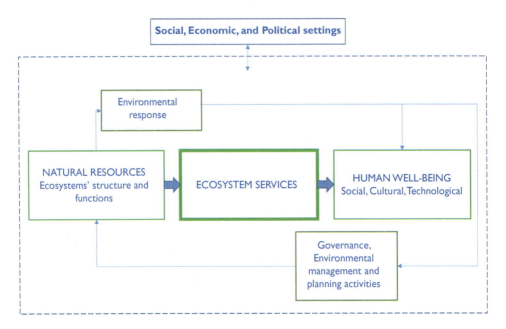

Figure 1 Human-environmental system as Ecosystem Services (ES) core-based social-ecological system (SES) framework (based on DPSIR model[5] and McGinnis & Ostrom, 2014).

Accordingly, if the supply of a particular ecosystem service is significantly changed, it is possible that the human demand for that service might go unfulfilled (*Burkhard et al., 2012*).

To manage a natural resource in a sustainable way requires implementation of a holistic approach that can bring together manifold factors within an interdisciplinary framework. This chapter discusses a comprehensive qualitative evaluation of the demand for, and supply of, ecosystem services relevant to the Tucson basin.

1.2 Case study general characteristics

The area investigated includes part of the Pantano Wash in the Rillito River Watershed, and encompasses the Upper Cienega Creek watershed (Santa Cruz basin) with an outlet in the confluence of the wash with Rincon Creek before it flows through the urbanized part of Tucson. The outlet point was specifically selected to exclude the urbanized portion of the system. Located in southwestern Arizona, the area investigated encompasses agricultural and naturally vegetated territories, and has a total area 555 mi² (1437 km²) and a mean elevation of 5905 ft/1800 m (*Figure 2*).

---

5 Smeets, E. and Weterings, R. 1999. Environmental indicators: Typology and overview. Technical report No 25. EEA, Copenhagen.

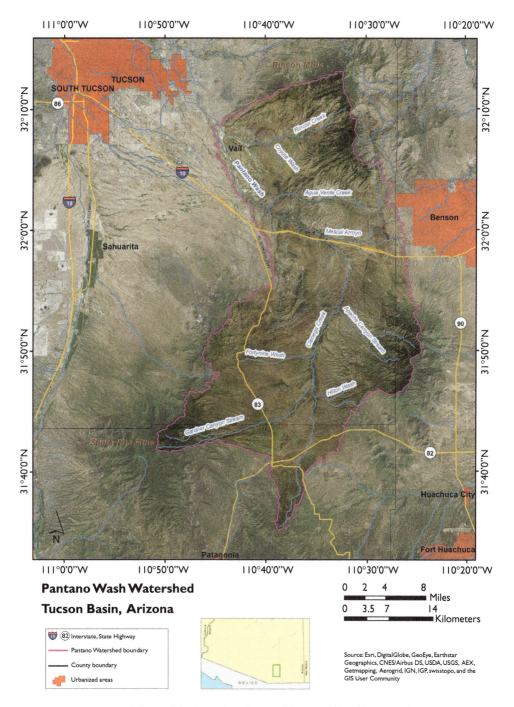

Figure 2 Map of the investigated area of Pantano Wash Watershed.

The Cienega Creek watershed has some unique landscape characteristics due to the perennial water flow. Because of its "*outstanding quality*," biodiversity, and richness of its natural resources, environmental reports proclaim the watershed to be: "Unique Water of Arizona" (*AZ Dept. of Environmental Quality*). The watershed contains about 488 mi^2/1264 km^2 of scrubland and 5 mi^2/13 km^2 of herbaceous rangeland. Evergreen forests cover about 50 mi^2/129.5 km^2 in the higher elevations. The riparian environment represents the original ecological nature of the Tucson area prior to 1900 (*McGann and Associates, Inc., 1994*). Current concerns regarding resources in the watershed include soil erosion, rangeland site stability, rangeland hydrologic cycle, excessive runoff (causing flooding or ponding), aquifer overdraft, excessive suspended sediment and turbidity in surface water, air quality on visibility and plant health, threatened or endangered plant and animal species, noxious and invasive plants, wildfire hazard, inadequate water for fish and wildlife, habitat fragmentation, and inadequate stock water for domestic animals (*USDA NRCS, 2007*).

The development pattern projected for Tucson for 2030 indicates that population and housing density in the urban fringe (areas outside the suburban areas) will increase at a faster rate than urban sprawl. The forecast is that the ex-urban areas (urban fringe) will expand from approximately 20% to 80% of the total area of the Pantano Wash—Rillito River Watershed. Because the urban fringe consumes much more land, it leaves a greater footprint on the general ecological health, and causes habitat fragmentation and other natural resource concerns (*Luck & Wu, 2002; Theobald, 2005; USDA NRCS, 2007*).

2 ASSESSMENT NETWORK/MATERIALS AND METHODS

2.1 Assessment approach

The objectives of the present research are twofold. First to investigate the ecosystem services relevant to the environmental conditions of the case study and second to identify the supply of and demand for ecosystem services, to develop assessment matrices of the ecosystem services, and to present spatially explicit maps of the distribution of ecosystems services. We follow the methodology for ecosystem services' quantification and supply and demand mapping presented in *Burkhard et al., 2009; Nedkov & Burkhard, 2012* and the ecosystem services budgeting approach described in *Burkhard et al., 2012*.

This assessment was conducted in several stages. The starting point for identifying the ecosystem services relevant to the Tucson Basin Case study was to conduct an expert assessment of the ecosystem services provided. This assessment was performed by interviewing different target groups, recognized here as *stakeholders*. The selected groups and selection criterion were as follows: *University of Arizona Academia* considered to provide expert opinions, *SWAN-team members* considered to have information based on their practical experience and research in the case study area, and *University of Arizona students* considered to have independent and unbiased opinions without in-depth knowledge. Further description of the applied qualitative assessment is presented in the following sections.

2.2 Identification of ecosystem services provided—importance, supply and demand capacities, and matrix model

Based on the ecosystem services list provided in *de Groot et al. (2002), MA (2005)* and *Kandziora et al. (2012)*, a list of 31 ecosystem services was developed and categorized in terms of provisioning, regulating and cultural services (*Appendix A*). The initial step was to prepare an online survey questionnaire that was distributed to the stakeholders. All the participants of that first survey were asked to answer the question: *"Which regulating/provisioning/cultural ecosystem services do you think are the most relevant for the TBCS?"* Each service was to be evaluated with a score from 0–5, where 0 = no relevance, 1 = low relevance, 2 = relevance, 3 = medium relevance, 4 = high relevance, 5 = very high relevance. The main idea was to collect information about the importance of ecosystem services in the case study. The survey was also designed to be completed in-person or by a teleconference interview, according to whichever the interviewee found more convenient. The results were collected via online collector and also by in-person interviews with scientists from the University of Arizona. The input they provided contributed significantly to the case study analysis and helped in the design of the following stage.

The second part of the survey was developed using the results obtained from the previous questionnaire. In this part, only the most representative results and ecosystem services rated with scores greater than 3 were considered as the relevant basis and data. The "matrix model" method *(Burkhard et al., 2009, 2012b; Nedkov & Burkhard, 2012; Jacobs et al., 2015)* was applied, as it successfully fulfills the needs of the ecosystem services expert-based qualitative assessment. The primary advantages of this approach were that the survey results could be easily linked with GIS, and ecosystem services data could be mapped and visualized in thematic maps. The matrix model provides an easy mapping approach that links the ecosystem services with the land cover and land use (LCLU) dataset (*Figure 3*). The model consists of a matrix with the ecosystem services as columns and geospatial units (LCLU classes) as rows. The LCLU classification that was used in the research follows the National Land Cover Data 2006 classification (USEPA[6]). Only land cover classes relevant to the case study were considered.

The core of the matrix model consists of estimates provided by expert perceptions regarding the unit's capacity to provide a particular ecosystem service at each intersection (*Figure 3*). This part of interviewing process again consists of a self-assessment by each interviewee (group member) of the ecosystem services' supply and demand within the specific land cover type on a 6 level scale (0 to 5) Likert range. This approach to scaling responses is based on the assumption that the interval between the answers can be anticipated to be equal. The questions that were asked in the online data collector included: *"What capacity do you think these land cover classes have to supply Regulating/Provisioning/Cultural ES?"* and *"What is the demand for Regulating/Provisioning/Cultural ES within different land cover classes?"* In the investigation process, the supply capacity of the ecosystem refers to *"the capacity of a particular area to provide a specific bundle of ecosystem goods and services within a*

---

[6] Unites States Environmental Protection Agency http://www.epa.gov/mrlc/nlcd-2006.html.

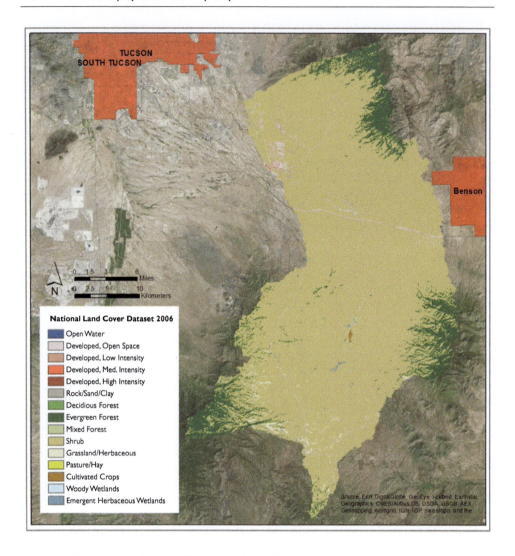

Figure 3 LCLU classes in Pantano Wash Watershed (NLCD 2006).

given time period" and demand was defined as the "*sum of all ecosystem goods and services currently consumed or used in a particular area over a given time period*" (Burkhard, Kroll, et al., 2012). Because environmental conditions change, and spatial heterogeneity is related to the ecological integrity and vulnerability to changes in territorial pattern, the provision and consumption of ecosystem services can be tracked and evaluated (*Figure 2*). While ecosystem services supply varies with spatial factors and landscape setting, ecosystem services demand varies in relation to the number of beneficiaries and their location, the socio-economic context, individual preferences and social practices *(Tardieu et al., 2013)*.

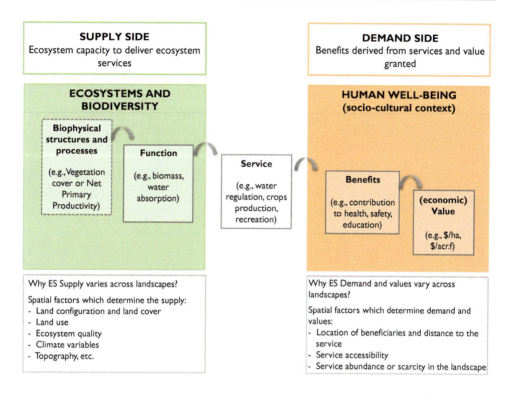

Figure 4 The spatial dimension of ecosystem services supply, demand and values, after Tardieu et al., 2013 (adapted from Haines-Young and Potschin, 2010).

2.3 Qualitative assessment

Given that this approach has previously been mainly applied in Europe[7], it was important to have in-person meetings and provide definitions and explanations of the investigation goals. Initially, the questionnaire was distributed to *University of Arizona (UofA) scientists* whose work was relevant to the ecosystem services concept and had scientific experience in the Tucson Basin area. This group of stakeholders consisted of researchers from the College of Agriculture and Life Science (CALS), College of Architecture, Planning and Landscape Architecture (CAPLA), College of Social Behavioral Sciences (SBS), College of Engineering, Udall Center for Studies in Public Policy, and United States Department of Agriculture (USDA). The stakeholders' subjective judgments on the subject provided strong input where insufficient quantitative data was available.

The second group of stakeholders was formed by *UofA graduate students* – master and PhD students from the Schools of Geography and Development, Natural Resources and the Environment, and the Department of Hydrology and Water Resources. Despite their relative unfamiliarity with the ecosystem services concept, the students were also able to provide the requested information.

---

7 To the knowledge of the author, the application of this methodology is novel for US environmental projects and ES assessments, since so far there is no information if it has been applied in the USA.

232 Water bankruptcy in the land of plenty

	regulating services								provisioning services				cultural services				
scale for assessing supply capacities		Local climate regulation	Air quality regulation	Water flow regulation	Water purification	Erosion regulation	Natural hazard protection	Pollination	Regulation of waste		Crops	Freshwater	Mineral resources	Abiotic energy sources		Recreation and tourism	Landscape aesthetic, amenity and inspiration	Knowledge systems	Cultural heritage and cultural diversity	Natural heritage and natural diversity
0 = no relevant capacity																				
1 = low relevant capacity																				
2 = relevant capacity																				
3 = medium relevant capacity																				
4 = high relevant capacity																				
5 = very high relevant capacity																				
NLCD 2006																				
Open Water		3	2	4	2	1	3	2	3		1	3	1	2		4	4	4	3	4
Developed, Open Space		2	1	1	1	1	1	1	1		1	1	1	3		3	2	3	4	2
Developed, Low Intensity		2	1	1	1	1	1	1	1		1	1	1	3		3	2	3	4	2
Developed, Med. Intensity		2	1	1	1	1	1	1	1		1	1	1	3		3	2	3	4	2
Developed, High Intensity		2	1	1	1	1	1	1	1		1	1	1	3		3	2	3	4	2
Rock/Sand/Clay		1	1	1	1	1	1	1	1		0	1	3	2		2	3	3	1	3
Deciduous Forest		3	3	4	3	4	3	3	2		1	3	1	1		4	4	3	2	4
Evergreen Forest		4	4	3	3	4	3	3	3		1	3	1	1		4	4	3	3	4
Mixed Forest		3	4	4	3	4	3	3	3		1	3	1	1		4	4	3	3	4
Shrub/Scrub		2	2	2	2	3	2	3	2		1	2	1	1		2	3	3	2	3
Grassland/Herbaceous		2	2	3	2	3	2	3	2		2	2	1	1		3	3	3	2	3
Pasture/Hay		2	2	2	1	2	1	2	1		2	1	1	1		1	2	2	2	2
Cultivated Crops		2	1	2	1	2	1	2	1		4	1	1	1		1	1	3	2	1
Wetlands		3	2	4	3	3	3	3	4		1	4	1	2		4	4	4	3	4
Emergent Herbaceous Wetlands		3	2	4	4	3	3	3	4		1	4	1	2		4	4	4	3	4

Figure 5 Assessment matrix of the supply capacities of the different land cover classes to provide ecosystem services.

The third group was formed by *SWAN project members*. SWAN researchers provided very subjective perceptions on the importance of ecosystem services being provided in the study area. The responses they provided were valuable because there was a variety of expertise that brought to the assessment a unique interdisciplinary perspective.

The second stage of valuation was performed post hoc in order to analyze the impact of the varying responses. As the research was focused on a very specific subject, the sample of individuals involved in this stage was relatively small. The stakeholder groups were formed only by members of the UofA Academia and the SWAN project team, who collectively have only a narrow range of backgrounds. The expert opinions provide the qualitative results of the ecosystem services supply and demand for an area that could be physically defined. For this reason, the main objective of this stage of interviewing was not to sample as many stakeholders as possible, but to capture quantifiable expert-based perceptions, opinions, and insights.

scale for assessing demands	regulating services								provisioning services					cultural services				
0 = no relevant demand 1 = low relevant demand 2 = relevant demand 3 = medium relevant demand 4 = high relevant demand 5 = very high relevant demand	Local climate regulation	Air quality regulation	Water flow regulation	Water purification	Erosion regulation	Natural hazard protection	Pollination	Regulation of waste		Crops	Freshwater	Mineral resources	Abiotic energy sources		Recreation and tourism	Landscape aesthetic, amenity and inspiration	Knowledge systems	Cultural heritage and cultural diversity	Natural heritage and natural diversity
NLCD 2006																			
Open Water	3	2	3	3	2	3	2	3		1	3	1	1		3	3	3	3	3
Developed, Open Space	3	3	3	3	3	3	2	3		2	3	2	4		3	2	3	3	2
Developed, Low Intensity	3	3	3	3	3	3	2	3		2	3	2	4		3	2	3	3	2
Developed, Med. Intensity	3	3	3	3	3	3	2	3		2	3	2	4		3	2	3	3	2
Developed, High Intensity	3	3	3	3	3	3	2	3		2	3	2	4		3	2	3	3	2
Rock/Sand/Clay	1	1	2	1	2	2	1	1		1	1	2	1		2	2	2	1	2
Deciduous Forest	3	3	3	3	3	3	3	2		1	3	1	1		3	3	3	2	3
Evergreen Forest	3	3	3	3	3	3	3	2		1	3	1	1		3	3	3	2	3
Mixed Forest	3	3	3	3	3	3	3	2		1	3	1	1		3	3	2	2	3
Shrub/Scrub	2	2	2	2	3	3	3	2		1	2	1	1		2	2	2	2	2
Grassland/Herbaceous	3	2	3	3	3	3	3	2		2	2	1	1		3	2	3	2	3
Pasture/Hay	2	2	2	2	3	2	3	2		2	2	1	1		1	2	2	1	2
Cultivated Crops	2	2	2	2	3	2	3	2		3	3	1	1		1	1	2	1	1
Wetlands	3	2	3	3	3	2	3	3		1	3	1	1		3	3	3	2	3
Emergent Herbaceous Wetlands	3	2	3	3	3	2	3	3		1	3	1	1		3	3	3	2	3

Figure 6 Assessment matrix of the demands of ecosystem services within the different land cover classes.

2.4 Data analysis

As described above, the data collected from experts falls under the category of purposeful sampling, and it is important to emphasize that samples were not taken randomly, especially during the second stage of interviewing where only participants with a good understanding of the ecosystem services concept were included. The first survey, entitled *"Importance of Ecosystem Services in Tucson Basin Case Study (TBCS)"* was completed by 55 participants out of 86 questionnaires sent, a 65% response rate. The 2nd survey entitled *"Supply And Demand of ES in Pantano Wash Watershed"* collected 18 responses (out of 57 stakeholders approached). Despite the initial goal to reach as many participants as possible (*"the more, the better"*), the objective was mainly to collect representative data that could be used to compare different stakeholder groups' perceptions on ecosystem service importance and supply-demand patterns. This relatively broad survey was needed to obtain a representative pool that was distinguished not by the number of participants, but by the diverse range of their scientific backgrounds (Jacobs et al., 2015).

234 Water bankruptcy in the land of plenty

NLCD 2006	Local climate regulation	Air quality regulation	Water flow regulation	Water purification	Erosion regulation	Natural hazard protection	Pollination	Regulation of waste	Crops	Freshwater	Mineral resources	Abiotic energy sources	Recreation and tourism	Landscape aesthetic, amenity and inspiration	Knowledge systems	Cultural heritage and cultural diversity	Natural heritage and natural diversity
Open Water	0	0	1	-1	-1	0	0	0	-1	0	0	0	1	1	1	0	1
Developed, Open Space	-1	-1	-2	-2	-2	-2	-2	-2	0	-2	-1	-1	0	0	0	0	0
Developed, Low Intensity	-1	-1	-2	-2	-2	-2	-2	-2	0	-2	-1	-1	0	0	0	0	0
Developed, Med. Intensity	-1	-1	-2	-2	-2	-2	-2	-2	0	-2	-1	-1	0	0	0	0	0
Developed, High Intensity	-1	-1	-2	-2	-2	-2	-2	-2	0	-2	-1	-1	0	0	0	0	0
Rock/Sand/Clay	0	0	0	-1	-1	0	-1	-1	0	0	1	1	1	0	1	0	1
Deciduous Forest	1	1	1	0	0	0	0	0	0	0	0	0	1	1	1	1	1
Evergreen Forest	0	1	1	0	0	0	0	0	0	0	0	0	1	1	0	1	1
Mixed Forest	0	1	1	0	0	0	0	1	0	0	0	0	1	1	1	1	1
Shrub/Scrub	0	0	0	0	0	0	0	0	0	-1	0	0	0	0	1	0	1
Grassland/Herbaceous	-1	0	0	0	0	-1	0	0	0	-1	0	1	0	0	0	0	0
Pasture/Hay	-1	0	0	-1	0	-1	-1	-1	0	-1	0	1	0	0	1	1	0
Cultivated Crops	-1	-1	-1	-1	-1	-1	-1	-1	1	-2	0	0	0	0	1	0	0
Wetlands	0	0	1	1	0	1	0	1	0	1	0	1	1	1	0	1	1
Emergent Herbaceous Wetlands	0	0	1	1	0	1	0	1	0	1	0	1	1	1	0	1	1

scale for ecosystem services balances:
-5
-4 demand exceeds
-3 supply
-2
-1
0 neutral balance
1
2
3 supply exceeds
4 demand
5

Figure 7 Assessment budget matrix of ES supply and demand within different NLCD classes.

An essential aspect of the ecosystem services concept, as related to the human-environmental system, is that a particular service cannot exist without having people (beneficiaries) of that service, that create a demand for a certain level of provision of that service. To assess demand, the data was derived via the survey entitled *"Supply And Demand of ES in Pantano Wash Watershed"*. The expert values were transferred into a matrix linking different land cover types with ecosystem supply capacities (*Figure 5*) and demands for ES (*Figure 6*).

The NLCD 2006 land cover classes are listed on the y-axis of the matrix. The regulating, provisioning, and cultural ecosystem services are shown along the x-axis. *Figure 5* clearly illustrates that the demands for ES are the highest in human-dominated land cover classes. The more urbanized and commercial a particular unit is, the higher the demand for ecosystem services. In contrast, the more natural and near-natural land cover types, characterized by low population and low human environmental impacts, show lower demand responses. More than 2/3 of the investigated territory is covered by scrub (*Figure 3*) and shows lower relevant demands for regulating ecosystem services.

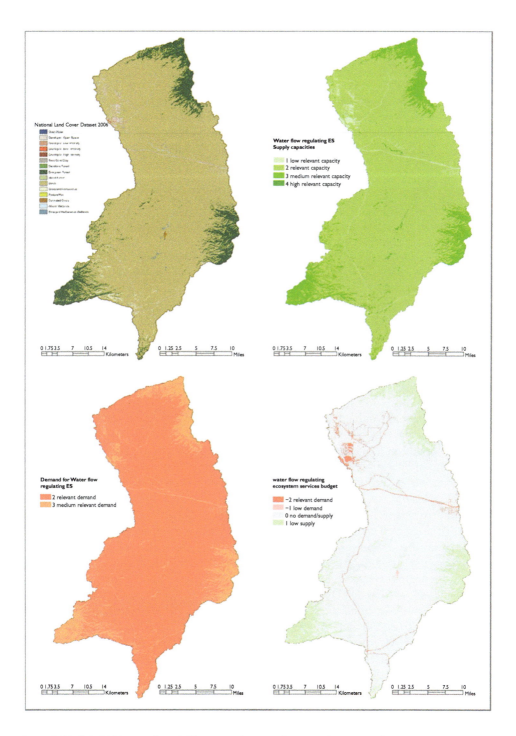

Figure 8 NLCD 2006 map (top left); maps of water flow regulation supply capacities (top right), demand (bottom left); budget map of water flow regulating ecosystem services (bottom right).

The complete qualitative assessment and analysis of ES supply and demand for each LCLU class requires identification of the ES budgets. This approach requires merging the data from the matrixes with the supply and demand maps (*Burkhard et al., 2012; Nedkov & Burkhard, 2012*). Each field in the budget matrix (*Figure 6*) was calculated as the difference between the supply matrix and the corresponding demand matrix. The scale for ES balance (budget) ranges from −5 = demand greatly exceeds supply (strong undersupply) to +5 = supply greatly exceeds demand (strong oversupply). The 0 values indicate a neutral balance (demand is equivalent to supply).

3 RESULTS

Ecosystems in the study area of the Tucson Active Management Area (TAMA) are vulnerable to the influences of industry and traffic, and also to rapid increases in housing at the expense of natural area. The anthropogenic activities undoubtedly impose a strong footprint on the environment and affect its ecological resilience. We have assessed the supply and demand of three of the most sensitive ecosystem services directly related to human well-being by developing spatially explicit maps of water flow regulation (regulating ES). The maps provide a holistic picture of stakeholders' perceptions regarding the temporal and spatial dynamics in the ecosystems.

Based on the information provided by the supply capacity and demand assessment matrixes, *Figure 8* shows maps of supply and demand of ecosystem services for the Pantano wash watershed. The relative scale values (0–5) corresponding to each spatial unit (LULC class), and representing the mean scores derived from the stakeholders' evaluation results, have been assigned to GIS layers. This assigns qualitative and statistical data to each spatial unit and enables and analysis of quantitatively different scale units (*Nedkov & Burkhard, 2012*). Accordingly, the supply-demand budget map indicates where qualitatively high demand coincides with low provisioning and vice versa. The map shows clearly that there is a balance between supply and demand of this particular ecosystem service within the "Shrub" land cover class. Ecological integrity is well maintained and the system is still resilient to external impacts (drivers of change). The mountain foothills in the periphery of the study area (covered by evergreen and deciduous forests) show medium to high capacity to provide water flow regulating ecosystem services. In contrast, in the urban and developed areas, the demand for these services exceeds the supply.

Of course, when assessing regulating ES, the patterns in the spatial landscape and the interactions between adjacent ES should also be taken into consideration (*Nedkov & Burkhard, 2012*). Further, the mapping approach allows further integration of biophysical, qualitative, and statistical data into a common spatial model that will help to locate and track areas where the ecosystem services are actually generated.

4 DISCUSSION

4.1 Relevance of the results

The mapping method allows comparison of the perceptions of different stakeholder groups (Academia and SWAN members) regarding ecosystem capacity, and facilitates tracking of spatial mismatches between evaluation results. The maps illustrate the spatial distribution of water flow regulation, fresh water (provisioning ES), and landscape aesthetic, amenity and inspiration (cultural ES) services. The results illustrate both the capacities of various land cover units to supply ecosystem services, and the human demand for such services. In general, the highest demand is found in urban and semi-urban areas, while more "wild" areas (mountain foothills) are characterized by relatively resilient ecological integrity and high supply capacity. By comparing the perceptions of different groups it is possible to further investigate the question: To what extent do perceptions vary regarding the importance of the ES and socio-environmental system that we work on?

4.2 Many participants, many opinions

In this investigation, we found that the perceptions of different stakeholder groups depended strongly on the strength of their scientific background in relation to the specific environmental conditions of the study area. When comparing expert judgments, there is no straightforward way of deciding whose opinion is more accurate, but practical experience would seem to be of key importance. Further, the results cannot be "weighed" in terms of the "accuracy" of an experts' opinions. So, when assessing human-ecological systems it is difficult to rely on holistic judgment alone (*Smith et al., 2012*), and uncertainties will increase and limit the outcomes of the assessment.

Certainly, we anticipated that differences would arise among the survey responses. Academic experts, having practical experience in the Tucson basin area, provided very detailed and carefully considered responses, along with additional information about the environmental and socio-cultural conditions in the study area. They assessed evergreen forests land cover classes as having medium supply capacity of water flow regulating ecosystem services, in contrast with the SWAN project members' corresponding assessment of "high" and "very high" capacity (*Figure 9*). Grassland land cover class was assessed a value of 4 by the Academic experts but a value of 2 by the SWAN group. A final objective result is difficult to provide, because each group could provide either and overestimation or underestimation.

As another example, very contradictory responses were obtained for the demand for water flow regulating ecosystem service. The Academic experts assessed evergreen forests as having very high (5-value) demand for a particular ecosystem service (see *Figure 10*), whereas the SWAN project members' assessed this land cover class as having no relevant demand (0-value). Similar discrepancies can be seen for the grasslands class.

Note that all of the respondents in the second interview stage were physical scientists, with most having a strong background in human-environmental systems in general, and ecosystem services in particular. What is of key importance here is their practical familiarity with the ecosystem services methodology. We did, however, observed variances in the answers within each group. Responders that were interviewed "in-person" provided different evaluations of the supply and demand of ecosystem services from those that provided answers online. Interestingly, the in-person scores tend to have relative similarity with those of SWAN project participants having

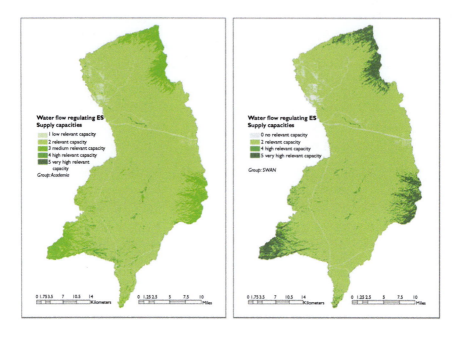

Figure 9 Maps of Water flow regulating ecosystem service supply based on Academia group (left) and SWAN group (right) evaluations.

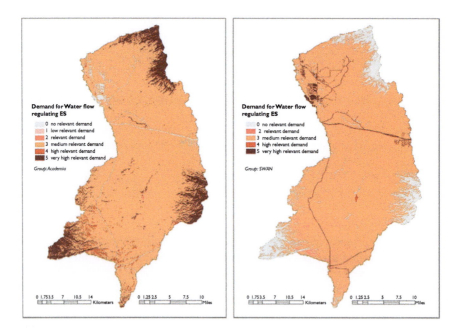

Figure 10 Water flow regulating ecosystem service demand maps based on Academia group (left) and SWAN group (right) evaluations.

a background in landscape science. Other possible reasons for such contrasting perceptions could be the lack of sufficient information regarding ecosystem integrity and the relative unfamiliarity of different stakeholder groups with the ecosystem services concept. Variation in the scores can also be based on factors related to the process of acquiring expert knowledge (i.e. the respondent has long-term experience with the topic or is currently a relative novice).

These problems should be viewed in the context that the number of studies that have investigated the supply of ecosystem services by far exceeds those concerning ecosystem demand, and that the former are easier to support with accurate and precise quantitative data. Improving the analysis would require comparing the assessment results with appropriate quantitative information. Further, the survey-based assessment could be improved by including detailed characteristics and examples of each ecosystem service relevant to the case study, so as to raise the socio-ecological awareness of the participants, and to streamline the survey process.

4.3 Identifying opportunities associated with a policy, plan or decision

Ecosystems, being complex adaptive systems, differ in their ecological integrity, stability, and dynamics. Referring to their intermediate place in the socio-ecological framework (*Figure 1*), the prerequisites for adaptive capacity of an ecosystem are the institutional, social, and industrial context and their present state. When environmental assessment initiatives are concerned with pristine ecosystems and wilderness areas, and when they try to take into consideration human demands for consumption, conflicts can arise between groups concerned variously with environmental management, political ecology, and human well-being.

In such situations, maps can be used as a valuable communication tool to initiate discussions between stakeholders, by indicating the locations where valuable ecosystem services are produced (or consumed) and by explaining the relevance of ecosystem services to the general public (*Maes et al., 2012b*). By helping to visualize the assessments of experts, maps can provide key support to the implementation of the ecosystem service's concept within decision making in regards to conservation, biophysical analysis, and ecological monitoring. Further, the mapping approach enables the integration of temporal data, and can be used to indicate trends (*Burkhard et al., 2009; Maes et al., 2012a; Grêt-Regamey et al., 2013*). From the perspective of effective environmental management, this approach can serve as a basis for further socio-economic and socio-cultural valuation of ecosystem services.

4.4 Uncertainties and limitation of the results

While the mapping approach has many advantages and provides support for sustainable environmental management, it also suffers from limitations and drawbacks that lend a certain level of uncertainty to the results. In regards to the expert opinion process, the cognitive capacity of the human brain has limits (*Smith et al., 2012*), and judgments are typically based on a range of heuristics that can introduce bias into the outcome. Influenced by moral and professional responsibility, the assumptions made by an expert can lead to either "under confidence" or "over confidence" and affect the quality of the assessment (*Knol et al., 2008*). Expert-based assessments are strongly dependent on scientific and practical experience of the respondents and, when the assessments are not well informed by quantitative data, these responses will typically reflect personal judgments and perceptions. On the other side, the use of expert knowledge can help to improve the results and significantly reduce uncertainties (*Grêt-Regamey et al., 2013; Presnall, López-Hoffman, & Miller, 2014*). Ultimately, when assimilated into a comprehensive human-environmental model, expert judgment can help to initiate ecological and institutional change in the process of natural resource management (*Redpath et al., 2013; Cote & Nightingale, 2015*).

Another factor hampering reliability of the approach is poor availability of relevant data. Bringing additional quantitative information, indicators of trends, and statistical information to bear can help to improve the temporal and spatial resolution of the assessment. Examples of such data include in-depth landscape investigations, physical and socioeconomic characteristics, and land cover data validation.

5 CONCLUSION

Map-based assessments of ecosystem service supply and demand can be extremely valuable to the decision-making process. Expert-based assessments can be used to obtain qualitative results regarding the capacity of each land cover type to provide ecosystem services. The approach helps in the identification of supply-demand mismatches and changes across the landscape. This study constitutes an application of a European-based and tested methodology to the southwestern United States. By bridging gaps in the integration of the available empirical and spatial data, this study has shown that the methodology can be successfully implemented at the local scale. The results of such an expert-based assessment can contribute to improved water planning and the development of optimal strategies for sustainable management, by highlighting the need for balance between ecosystem services supply and demand in various areas of interest. By highlighting the roles that ecosystems play in human well-being, the approach can lead to improved environmental benefits for citizens and help in the establishment of their socio-economic priorities.

REFERENCES

Arizona Water Atlas. Volume 8: Active management area planning area. Section 8.0 Overview. Arizona Department of water resources. Retrieved from http://www.azwater.gov/azdwr/StatewidePlanning/WaterAtlas/ActiveManagementAreas/default.htm.

Binder, C.R., Hinkel, J., Bots, P.W.G. and Pahl-Wostl, C. (2013). Comparison of frameworks for analyzing social-ecological systems. *Ecology and Society*, 18(4): 26–44.

Burkhard, B., Crossman, N., Nedkov, S., Petz, K. and Alkemade, R. (2013). Mapping and modelling ecosystem services for science, policy and practice. *Ecosystem Services*, 4: 1–3.

Burkhard, B., De Groot, R., Costanza, R., Seppelt, R., Jørgensen, S.E. and Potschin, M. (2012). Solutions for sustaining natural capital and ecosystem services. *Ecological Indicators*, 21: 1–6.

Burkhard, B., Kroll, F. and Müller, F. (2009). Landscapes' Capacities to provide ecosystem services—a concept for land-cover based assessments. *Landscape Online*, 15(1): 1–22.

Burkhard, B., Kroll, F., Nedkov, S. and Müller, F. (2012). Mapping ecosystem service supply, demand and budgets. *Ecological Indicators*, 21: 17–29.

Costanza, R., d'Arge, R., de Groot, R., Farber, S., Grasso, M., Hannon, B., Van den Belt, M. (1997). The value of the world's ecosystem services and natural capital. *Nature*, 387(6630): 253–260.

Cote, M. and Nightingale, A.J. (2015). Resilience thinking meets social theory: Situating social change in socio-ecological systems (SES) research, 36(4): 475–489.

Daily, G. (1997). Nature's services: Societal dependence on natural ecosystems. Washington, DC: Island Press.

De Groot, R.S., Alkemade, R., Braat, L., Hein, L. and Willemen, L. (2010). Challenges in integrating the concept of ecosystem services and values in landscape planning, management and decision making. *Ecological Complexity*, 7(3): 260–272.

De Groot, R.S., Wilson, M. a. and Boumans, R.M. (2002). A typology for the classification, description and valuation of ecosystem functions, goods and services. *Ecological Economics*, 41(3): 393–408.

Grêt-Regamey, A., Brunner, S., Altwegg, J., Christen, M., Bebi, P., Dynamics, E. and Bebi, P. (2013). Integrating expert knowledge into mapping ecosystem services trade- offs for sustainable forest management. *Society and Ecology*, 18(3): 34–54.

Haines-Young, R. (2011). Exploring ecosystem service issues across diverse knowledge domains using Bayesian belief networks. *Progress in Physical Geography*, 35(5): 681–699.

Hein, L., van Koppen, K., de Groot, R.S. and Van Ierland, E.C. (2006). Spatial scales, stakeholders and the valuation of ecosystem services. *Ecological Economics*, 57(2): 209–228.

Jacobs, S., Burkhard, B., Daele, T. Van, Staes, J. and Schneiders, A. (2015). "The Matrix Reloaded": A review of expert knowledge use for mapping ecosystem services. *Ecological Modelling*, 295: 21–30.

Kandziora, M., Burkhard, B. and Müller, F. (2012). Interactions of ecosystem properties, ecosystem integrity and ecosystem service indicators—A theoretical matrix exercise. *Ecological Indicators*, 28: 54–78.

Knol, A., Sluijs, J.P. Van Der and Slottje, P. (2008). Expert Elicitation: Methodological suggestions for its use in environmental health impact assessments. *RIVM Letter Report 630004001/2008*, 56.

Koschke, L., Van der Meulen, S., Frank, S., Schneidergruber, A., Kruse, M., Fürst, C. and Bastian, O. (2014). Do You Have 5 Minutes To Spare? – The challenges of stakeholder processes In ecosystem services studies. *Landscape Online*, 25: 1–25.

Maes, J., Egoh, B., Willemen, L., Liquete, C., Vihervaara, P., Schägner, J.P. and Bidoglio, G. (2012). Mapping ecosystem services for policy support and decision making in the European Union. *Ecosystem Services*, 1(1): 31–39.

Maes, J., Teller, A., Erhard, M., Liquete, C., Braat, L., Berry, P. and Bidoglio, G. (2013). Mapping and assessment of ecosystems and their services – an analytical framework for ecosystem assessment under action 5 of the eu biodiversity strategy to 2020. *Environment*, (3). European Union.

Malone, E.L., Dooley, J.J. and Bradbury, J.A. (2010). Moving from misinformation derived from public attitude surveys on carbon dioxide capture and storage towards realistic stakeholder involvement. *International Journal of Greenhouse Gas Control*, 4(2): 419–425.

McGinnis, M.D. and Ostrom, E. (2014). Social-ecological system framework: Initial changes and continuing challenges. *Ecology and Society*, 19(2): 30–41.

Millennium Ecosystem Assessment [MA] (2005). *Ecosystems and Human Well-being: Synthesis*. Washington, D.C.: Island Press/World Resources Institute.

Nedkov, S. and Burkhard, B. (2012). Flood regulating ecosystem services—Mapping supply and demand, in the Etropole municipality, Bulgaria. *Ecological Indicators*, 21: 67–79.

Ostrom, E. (2009). A general framework for analyzing sustainability of social-ecological systems. *Science (New York, N.Y.)*, 325: 419–422.

Presnall, C., López-Hoffman, L. and Miller, M. (2014). Adding ecosystem services to environmental impact analyses: More sequins on a "bloated Elvis" or rockin'idea? *Ecological Economics*, 115: 29–38.

Redpath, S.M., Young, J., Evely, A., Adams, W.M., Sutherland, W.J., Whitehouse, A. and Gutiérrez, R.J. (2013). Understanding and managing conservation conflicts. *Trends in Ecology and Evolution*, 28 (2): 100–109.

Smith, R., Dick, J., Trench, H. and Van Oijen, M. (2012). Extending a bayesian belief network for ecosystem evaluation. *Conference Paper of the Human Dimensions of Global Environmental Change on "Evidence for Sustainable Development,"* 12. Retrieved from http://nora.nerc.ac.uk/id/eprint/501468.

Tardieu, L., Roussel, S. and Salles, J.M. (2013). Assessing and mapping global climate regulation service loss induced by Terrestrial Transport Infrastructure construction. *Ecosystem Services*, 4: 73–81.

Theobald, D. (2005). Landscape patterns of exurban growth in the USA from 1980 to 2020. *Ecology and Society*, 10(1): 32–65.

USDA Natural Resource Conservation Service – AZ. (2007). Pantano Wash – Rillito River Watershed. Arizona Rapid Watershed Assessment. University of Arizona: Water Resources Research Center, Retrieved from http://www.nrcs.usda.gov/Internet/FSE_DOCUMENTS/nrcs144p2_064621.pdf.

World Bank (2013). Wealth accounting and the valuation of ecosystem services: global partnership – 2013 annual report. Washington DC: World Bank, Retrieved from http://www-wds.worldbank.org/external/default/WDSContentServer/WDSP/IB/2013/06/28/000445729_20130628122618/Rendered/PDF/768110REPLACEMENT0FILE0same0box00PUBLIC0.pdf.

APPENDIX A. Ecosystem services and definitions

No.	Regulating service	Definition
1	Global climate regulation	Long-term storage of greenhouse gases in ecosystems.
2	Local climate regulation	Changes in local climate components like wind, precipitation, temperature, radiation due to ecosystem properties.
3	Air quality regulation	Capturing/filtering of dust, chemicals and gases.
4	Water flow regulation	Maintaining of water cycle features (e.g. water storage and buffer, natural drainage, irrigation and drought prevention).
5	Water purification	The capacity of an ecosystem to purify water, e.g. from sediments, pesticides, disease-causing microbes and pathogens.
6	Nutrient regulation	The capacity of an ecosystem to recycle nutrients, e.g. N, P.
7	Erosion regulation	Soil retention and the capacity to prevent and mitigate soil erosion and landslides.
8	Natural hazard protection	Protection and mitigation of floods, dryness, storms (hurricanes, typhoons ...), fires and avalanches.
9	Pollination	Bees, birds, bats, moths, flies, wind, non-flying animals contribute to the dispersal of seeds and the reproduction of lots of plants.
10	Pest and disease control	The capacity of an ecosystem to control pests and diseases due to genetic variations of plants and animals making them less disease-prone and by actions of predators and parasites.
11	Regulation of waste	The capacity of an ecosystem to filter and decompose organic material in water and soils.

No.	Provisioning service	Definition
12	Crops	Cultivation of edible plants and harvest of these plants on agricultural fields and gardens which are used for human nutrition.
13	Biomass for energy	Plants used for energy conversion (e.g. sugar cane, maize).
14	Fodder	Cultivation and harvest of fodder for domestic animals.
15	Livestock (domestic)	Production and utilization of domestic animals for nutrition and use of related products (e.g. dairy, wool).
16	Fiber	Cultivation and harvest of natural fiber (e.g. cotton, jute sisal, silk, cellulose) for, e.g. cloths, fabric, paper.
17	Timber	Wood used for construction purposes.
18	Wood fuel	Wood used for energy conversion and/or heat production.
19	Fish, seafood and edible algae	Catch of seafood/algae for food, fish meal and fish oil.
20	Aquaculture	Harvest of seafood/algae from marine and terrestrial aquaculture farms.
21	Wild food, semi-domestic livestock and ornamental resources	Harvest of berries, mushrooms, (edible) plants, hunted wild animals, fish catch from recreational fishing, semi-domestic animal husbandry and collection of natural ornaments (e.g. seashells, leaves and twigs for ornamental or religious purposes).
22	Biochemicals and medicine	Natural products used as biochemicals, medicine and/or cosmetics.
23	Freshwater	Used freshwater (e.g. for drinking, domestic use, industrial use, irrigation).
24	Mineral resources	Minerals excavated close from surface or above surface (e.g. sand for construction, lignite, gold).
25	Abiotic energy sources	Sources used for energy conversion (e.g. solar power, wind power, water power and geothermic power).

No.	Cultural service	Definition
26	Recreation and tourism	Outdoor activities and tourism relating to the local environment or landscape, including forms of sports, leisure and outdoor pursuit.
27	Landscape aesthetic, amenity and inspiration	Visual quality of the landscape/ecosystems or parts of them which influences human well-being and the need to create something, esp. in art, music and literature. The sense of beauty people obtain from looking at landscapes/ecosystems as ecosystems provide a rich source of inspiration for art, folklore, national symbols, architecture, advertising and technology.
28	Knowledge systems	Environmental education based on ecosystem/landscape, i.e. out of a formal schools context, and knowledge in terms of traditional knowledge and specialist expertise arising from living in this particular environment.
29	Religious and spiritual experience	Spiritual or emotional values that people or religions attach to local environments or landscapes due to religious and/or spiritual experience.
30	Cultural heritage and cultural diversity	Values that humans place on the maintenance of historically important (cultural) landscapes and forms of land use (cultural heritage).
31	Natural heritage and natural diversity	The existence value of nature and species themselves, beyond economic or human benefits.

APPENDIX B

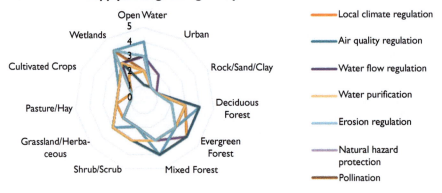

Figure 11

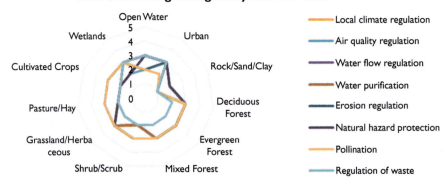

Figure 12

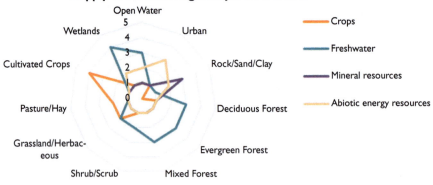

Figure 13

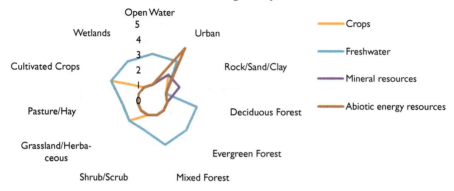

Figure 14

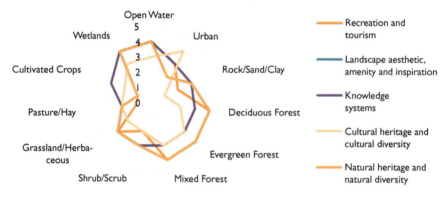

Figure 15

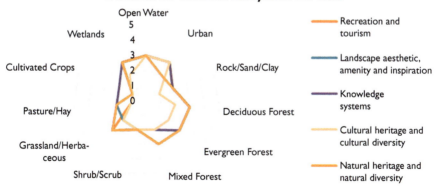

Figure 16

Figure 11-16 Spider web diagrams showing the assessed (0-5) ecosystems' supply capacities and human demands for all regulating, provisioning, and cultural services in relation to all NLCD 2006 land cover/land use classes Pantano Wash watershed.

APPENDIX C.

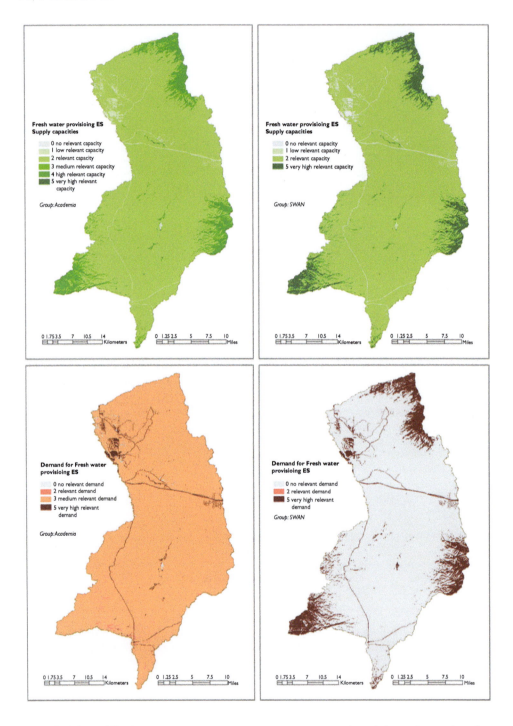

Figure 17 Maps of Fresh water provisioning ecosystem service supply and demand maps based on Academia group (left) and SWAN group (right) evaluations.

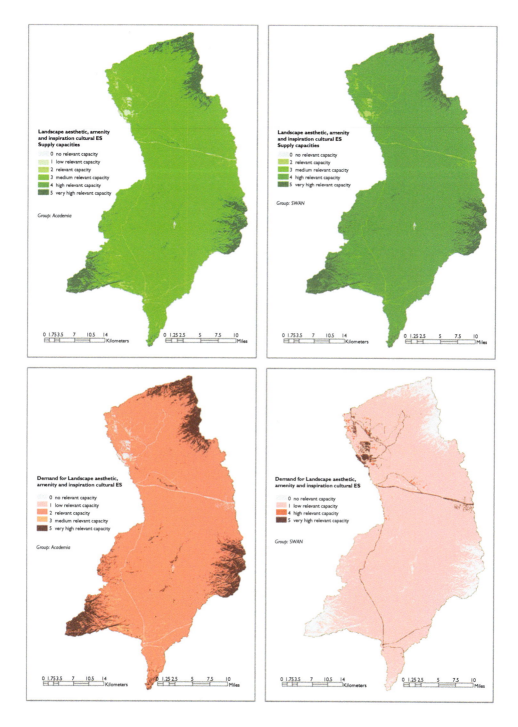

Figure 18 Maps of Landscape aesthetic, amenity and inspiration ecosystem service supply and demand maps based on Academia group (left) and SWAN group (right) evaluations.

Chapter 14

The role of biodiversity in the hydrological cycle

The case of the American Southwest

Maria A. Sans-Fuentes
Biosphere 2, University of Arizona, USA

Thomas Meixner
Department of Hydrology and Atmospheric Sciences,
University of Arizona, USA

INTRODUCTION

Human population growth and the associated economic activities have a major influence on the health and evolution of the natural environment. Humans have influenced natural environments in two ways: by reducing the extents of natural areas, and consequently their biodiversity, due to overexploitation of resources; and by creating protected areas where biodiversity is both regulated and conserved with a view to preserving the services provided to humans, or with the intent to repair previous damage. Throughout human history, the tendency to reduce the extents of natural habitats has dominated over the impulse to restoration and, consequently, humans are now facing a rapidly increasing rate of loss of biodiversity. This trend is aggravated by climate change, particularly in semi-arid areas such as the southwestern United States. Dryland ecosystems[1] are particularly susceptible to climate variation and, in such regions, the (lack of) availability of water (surface water, groundwater and air moisture) acts as the main constraint on biological activity.

Drylands constitute 41% of the land surface of the Earth (*Reynolds et al., 2007*). In North America, they form a block that includes five deserts (see next section for more detail), most of which are in the American Southwest. The definition of what actually constitutes the American Southwest varies from source to source, being mostly a general agreement that the core of the region consists of Arizona and New Mexico. Other states that are often included in the various definitions are California, Nevada, Utah, Colorado, Texas and Oklahoma. In this chapter, we will take the American Southwest to be the region formed by the states of Arizona, California, Colorado, Nevada, New Mexico, Texas and Utah (*Figure 1*).

One of the main challenges that is facing the deserts of the American Southwest, and more generally drylands, is the onset and progression of desertification; i.e., the degradation of land quality in arid, semi-arid, and dry sub-humid areas, resulting from

---

[1] According to the United Nations Environment Programme (UNEP), drylands are defined as regions with an aridity index (AI) less than 0.65, where AI is the ratio of mean annual precipitation to mean annual potential evapotranspiration. Arid lands include hyper-arid (AI < 0.05), arid (0.05 <AI <0.20), semi-arid (0.20 <AI <0.50) and dry sub-humid (0.50 <AI <0.65) areas (*Middleton & Thomas, 1997*).

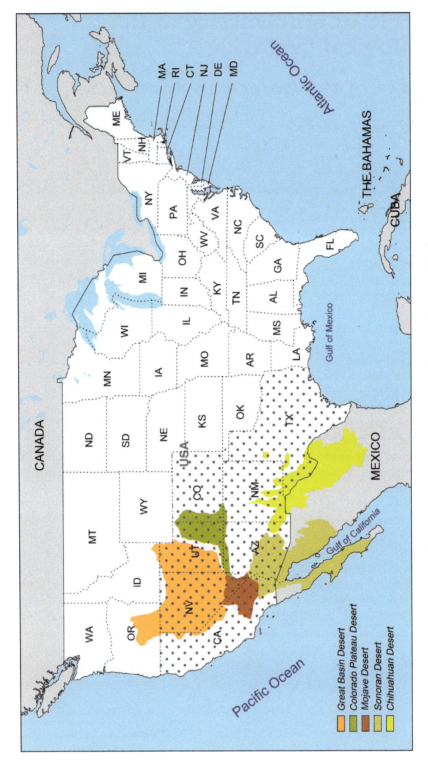

Figure 1 Main deserts of North America. The dotted area shows the states that form the American Southwest.

various factors including climate variability and human activity (as defined by the United Nations Convention to Combat Desertification). Examples of serious problems that affect these areas include overgrazing, unsustainable farming, ground water pumping, and diversion of water. These problems have been aggravated by significant development pressures caused by exponential population growth; for example, the growth rate in almost all states of the American Southwest was higher than the US average (3.3%) for the period 2010–2014[2] (Arizona 5.3%, California 4.2%, Nevada 5.1%, Utah 6.5%, Colorado 6.5%, New Mexico 1.3%, and Texas 7.2%). The fastest growing cities in the region are Las Vegas, Phoenix, Tucson, Albuquerque, and El Paso (*Malloy et al., 2013*).

Anthropogenic activities associated with population growth have both direct and indirect impacts on the environment. By altering *in situ* the area occupied (by increasing air and water pollution and decreasing the water availability for ecosystem use), growth contributes to fragmentation of habitats and degradation of the surrounding environments. To secure adequate water, water must then be brought to the developed areas by artificial means, diverted from major rivers or pumped from groundwater aquifers. Such diversions disrupt the natural drainage systems and reduce the amount and quality of water available for ecosystem use. Projections for the American Southwest indicate that this situation will likely be aggravated by increased aridity caused by changes to climate that are expected to bring about substantial declines in average annual rainfall and increased frequency and intensity of drought (*Seager et al., 2007*). Riparian areas are of special concern because they provide support for the regional biodiversity (*Sabo & Soykan, 2006; Sabo et al., 2005*). With increasing aridity, the availability of surface water will decline and the depths to groundwater will increase (*Stromberg et al., 2013*).

Arid regions are particularly sensitive to changes in climate, because the animals and plants that live there operate very close to their physiological limits. Therefore, changes in water availability and temperature can rapidly have significant impacts on species composition, distribution and density. In turn, the number and types of species that inhabit a region determine the organismal traits that influence the functioning of local ecosystem processes (*Chapin et al., 2000*), including the hydrological cycle. Since one of the most important factors affecting the cycle of water is the composition and properties of the soil, it follows that the composition and activity of subterranean fauna play an important role in modifying soil structure, which in turn affects the pathways and rates of water infiltration from the surface to the subsurface.

Most research related to the impacts of ecosystems on availability and quality of water has focused on flora conservation and the role of plants in water retention (e.g., *Stromberg et al., 2013; Stromberg et al., 2007*). When focused on fauna, the literature has mainly looked at aquatic animals. Consequently, little is yet known about the roles that subterranean fauna and soil-dwelling and burrowing species play in influencing the hydrological cycle. Such fauna, by creating large numbers of burrows, can be expected to play a key role in water dynamics, because these activities change the overall perme-

---

[2] According to U.S. Census Bureau (2010). State & County Quick Facts: http://quickfacts.census.gov/qfd/index.html.

ability of the soil. In turn, such changes in permeability can further impact the patterns of surface runoff and movement and storage of water within the soil.

In this chapter, we review what is known about the impacts of biodiversity on the hydrological cycle, with particular focus on the burrowing rodents of the southwestern desert regions. We identify gaps in the research conducted in this area, and analyze the potential for such knowledge to impact both water management and programs for biodiversity conservation.

I THE IMPORTANCE OF SOIL BIODIVERSITY IN THE ARID LANDS OF THE AMERICAN SOUTHWEST

North American deserts extend through the western United States and northern Mexico between the latitudes 44°N and 22°N. The five major deserts are the *Sonoran* (Arizona, California in USA, and Sonora, Sinaloa, Baja California in Mexico), *Chihuahuan* (Arizona, New Mexico, and Texas in USA, and Chihuahua, Coahuila, Durango, Zacatecas, and Nuevo León in Mexico), *Mojave* (Arizona, California, and Nevada in USA), *Great Basin* (Idaho, Nevada, Oregon, and Utah in USA), and *Colorado Plateau* (Utah, Colorado, New Mexico, and Arizona in USA) (Shreve, 1942). While the *Colorado Plateau* is often considered part of the *Great Basin Desert*, and the *Mojave Desert* is considered to be a transition zone between the *Great Basin* and *Sonoran* deserts, this chapter will treat them as separate regions (*Figure 1*). The *Great Basin, Colorado Plateau* and *Mojave* deserts lie entirely within the US, while the *Sonoran* and *Chihuahuan* deserts stretch across the southwestern US and into Mexico. While the *Chihuahuan* desert is the largest in North America, only a small fraction lies within the US. The largest and coldest desert in the US is the *Great Basin*, whereas the smallest and driest is the *Mojave* desert, and the hottest is the *Sonoran* desert (*Table 1*).

The aridity of these deserts is mainly caused by two factors: i) Orographic barriers block the delivery of moisture from the Pacific in winter and from the Gulf of Mexico in summer (due to the rain shadow effect); and ii) Subtropical high-pressure cells tend to limit the flow of atmospheric water vapor to these regions. Aridity in the *Great Basin, Colorado Plateau* and *Mojave* deserts is mainly caused by the rain shadow effect, while the *Sonoran* and *Chihuahuan* deserts, although also affected by orographic rain shadows, owe their aridity mainly to their latitudinal position in the subtropical belts. Due to physiographic differences, not all the deserts in North America have the same climatic characteristics. The *Great Basin* and *Colorado Plateau* are considered "cold" deserts (having winter temperatures mainly below 32 °F / 0°C), being located in a northerly position with high average elevations and receiving 60% of their moisture as winter snow. The *Colorado Plateau* has two regions, north and south, with the *Northern Colorado Plateau* being more similar to the *Great Basin*, and the *Southern Colorado Plateau* having characteristics transitional between the *Northern Plateau* and the warmer southern deserts. The *Mojave, Sonoran* and *Chihuahuan* deserts are considered "warm" deserts where most of the precipitation occurs as rain. In particular, the *Sonoran* and *Southern Colorado Plateau* deserts have a bimodal rainfall regime, with a rainfall maximum during the summer (monsoon thunderstorms) and a second peak in winter. The *Chihuahuan* desert receives most of its annual rain in summer during the monsoon season. In contrast, most of the precipitation in the *Mojave* desert comes from winter rainfall, as is the case for the *Great Basin* and *Northern Colorado Plateau* deserts *(Table 1)* *(Dregne, 1984; Laity, 2009; Sayre, 1998)*.

Due to the extreme nature of desert environments, species living in the arid lands of North America have evolved to be capable of surviving under conditions characterized by high temperatures and scarce water supplies *(Chaves et al., 2003; Walsberg, 2000)*. In spite of this, the number of species living in arid lands is surprisingly high. The amount of biodiversity ranges from *low* in the *Great Basin* to *high* in the *Sonoran* and *Chihuahuan* deserts, with the latter being one of the most diverse deserts in the world *(Jones, 1986)*.

Where the diversity in arid lands is low, this is mainly due to lower primary productivity (biomass production) of plants, which is mainly constrained by water availability and the amounts of rainwater stored in the soil to be available during dry periods. This depends, in turn, on the properties of the soil, which are influenced by a) the climate, b) the organisms that inhabit the soil, c) the nature of the parent material, d) topography and e) time *(Jenny, 1994)*. Therefore, the soil is one of the most important components influencing the nature of arid lands, forming an important part of a complex network of interactions with other factors such as water, climate and organisms.

In this regard, the soils of arid lands are *'poorly developed'* and contain a large percentage of primary rock minerals, whose composition determines the degree of fertility of the soil. Soils developed from parent rocks with high iron and magnesium content tend to be fine-textured, rich in clay minerals, and more fertile. In contrast, soils formed from silica-rich parent materials that are high in quartz and feldspar, tend to have coarse textures and low fertility.

While the fertility and texture of the soil determine the kinds of organisms (macrofauna, microfauna, vascular plants) that inhabit the soil, the organisms are in turn important drivers of soil formation. Macrofauna such as burrowing mammals, earthworms, ants and termites bring deeper materials to the surface, contributing to the distribution of the nutrients and water in the soil, and reducing organic matter to particle sizes. On the other hand, microfauna such as fungi and bacteria convert nutrients from organic to inorganic form and make them available to other organisms such as plants.

When microfauna form biological soil crusts (BSC)[3], this contributes to carbon sequestration, nitrogen fixation, decreased soil albedo, reduced erosion by wind and water, and changes in the hydrology of the soil by either increasing or decreasing water infiltration *(Belnap et al., 2001; Brotherson & Rushforth, 1983)*. In some studies, BSCs have been related to enhancement of the vascular plant cover and higher levels of nutrients in vascular plants. In turn, vascular plants can aid the survival of crustal components by generating conducive microclimate conditions *(Belnap et al., 2001)*. BSCs are dependent on water, being active only when wet *(Lange at al., 1998; Tuba et al., 1996)*. Vascular plants, besides providing organic material to the soil,

---

[3] Biological soil crust is an association of cyanobacteria, green algae, lichens, mosses, and microfungi that grows on or just below the soil surface creating a crust of soil particles bound together. They cover the soil spaces not occupied by green plants.

Table 1 Physiographic characteristics of the deserts of North America (Note: Separate values of Area and Elevation for the North and South Colorado Plateaus could not be found).

	Area (miles²)	Elevation range (feet)	Average annual Temperature range (°F)	Average summer Temperature range (°F)	Average winter Temperature range (°F)	Annual precipitation range (inches)	Summer precipitation range (inches)	Winter precipitation range (inches)
Great Basin[1,7,8]	154,441	1,063–14,246	45.0–58.0	50.0–90.0	10.0–40.0	3.3–15.8	1.0–3.2	1.50–10.0
Colorado Plateau North[2,7,9]	148,263	1,200–12,700	32.0–68.0	55.0–100.0	20.0–55.0	7.9–37.4	3.1–5.6	2.0–8.7
Colorado Plateau South[3,7]			44.8–69.1	61.2–87.3	22.5–54.7	6.4–21.3	2.1–7.9	2.6–9.3
Sonoran desert[4,7,10]	106,178	200–4,000	50.0–91.4	67.3–98.4	38.0–65.0	3.0–19.7	5.3–7.7	3.3–6.4
Mojave desert[5,7]	54,054	−282–11,000	33.8–68	55.4–86	33.8–64.4	1.3–12.2	0.02–4.9	1.0–9.8
Chihuahuan desert[6,7,11]	200,000	3,000–10,000	57.2–73.4	73.4–86	39.2–50.0	3.9–19.7	4.9–9.2	1.9–3.1

Data from: 1. Comstock & Ehleringer, 1992; Nachlinger et al., 2001; Warner, 2004; 2. Davey et al., 2006; Witwicki, 2013; 3. Thomas et al., 2006; 4. Davey et al., 2007c; 5. Davey et al., 2007b; Hereford et al., 2004; 6. Davey et al., 2007a; Schmidt, 1986; 7. Laity, 2009; 8. Nachlinger et al., 2001; 9. Hereford et al., 2002; 10. Phillips & Wentworth Comus, 2000; 11. Chihuahuan Desert Ecoregional Assessment Team, 2004.

also contribute to the cycling of water (e.g., evaporation, transport of water to deeper layer of the soil through dead roots).

The spatial distribution of desert plants is usually patchy, providing microhabitats for the wildlife, whose diversity therefore depends on the distribution and abundance of these microhabitats. Alteration of the nature of these microhabitats (either by climate change or anthropogenic effects) can significantly affect wildlife diversity by altering the delicate soil-water-climate-organism equilibrium (*Jones, 1986*). Special attention therefore needs to be given to the water, given its role as the primary limiting resource for many organisms in drylands, and given that changes in water distribution or availability will lead to major changes in habitat nature and distribution (*Huennke & Noble, 1996*).

In the next sections we investigate aspects of this complex network of interactions between burrowing rodents and the soil water cycle, keeping in mind its relationship with the bigger picture, the ecosystem, with a view to identifying knowledge gaps.

2 RODENTS AS ECOSYSTEM ENGINEERS, AND THEIR IMPORTANCE TO THE WATER CYCLE

The water content in soil plays a key role in the functioning of ecosystems, with a clear bidirectional relationship between water and biodiversity: while water is necessary to maintain biodiversity, fauna and flora influence the water cycle. The water content of the soil depends upon a variety of hydrological processes including precipitation, infiltration, runoff, and evapotranspiration. Upon reaching the land surface as precipitation, water can either runoff or infiltrate, depending on the structure and water content of the soil. If the soil is already saturated with water, or if the soil has low permeability, most of the precipitation will flow over the surface in the form of runoff. However, if the properties of the soil allow, the water will infiltrate into the soil where it is redistributed through the soil and its layers by gravity and capillary forces. A portion of this water will be absorbed by plants and returned to the atmosphere by the process of evapotranspiration, whereas a portion will recharge the groundwater system and a portion will move downslope (as interflow) to eventually exit the system as stream flow (*Figure 2*). For the joint management of water and ecosystems, it is important to understand the mechanisms by which rainwater moves through the soil surface, the factors that affect this movement, and the factors that can act to minimize runoff (because it can cause erosion), flooding and degradation of water quality.

The porosity of soil is intimately related to its capacity for infiltration, which thereby determines its hydraulic conductivity, i.e., the capacity of the porous medium to transmit water under a hydraulic gradient. Moreover, hydraulic conductivity is interrelated with the infiltration capacity of the medium; i.e., the maximum rate at which water can enter into or be absorbed by the soil (*Ward, 1975*).

Fauna and flora can actively modify the structure and hydraulic properties of the soil by altering its porosity, thereby altering its water conductivity, and ultimately affecting its water content. Organisms that do this can justifiably be called *ecosystem engineers*, defined as "*organisms that directly or indirectly modulate the availability of resources (other than themselves) to other species, by causing physical state*

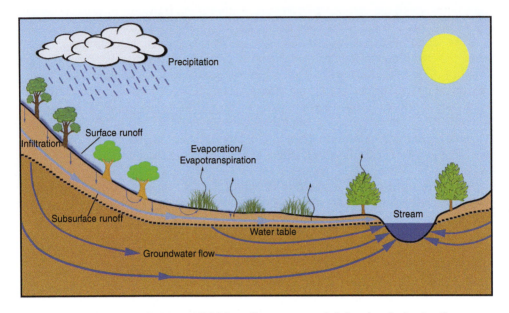

Figure 2 Water cycle. Modified from USGS (http://water.usgs.gov/edu/earthgwdecline.html).

changes in biotic or abiotic materials. In so doing they modify, maintain and/or create habitats" (*Jones, Lawton, & Shachak, 1994*). The process is called bioturbation, defined by *Paton et al. (1995)* as reactions between animals, plants and soil material, during which soil fabric is altered, by detachment, transport, sorting, and deposition of material, both within the soil mantle and on its surface. Bioturbation plays a major role in increasing soil heterogeneity, which in turn enhances biodiversity by affecting productivity, nutrient cycling, sediment transport, and soil erosion. It is, therefore, an important variable to account for in the analysis of ecosystem function (*Burnett et al., 1998; Gabet, 2000; Whitford & Kay, 1999; Williams & Houseman, 2014*). Preservation of biodiversity is especially important in arid lands since biodiversity provides the basis for ecosystem services in this biome, and its preservation can mitigate desertification (*Al-Eisawi, 2003*).

Small burrowing rodents such as prairie dogs, ground squirrels, kangaroo rats, and pocket gophers are the among most important bioturbators in the desert landscape (*Kelt, 2011; Whitford & Kay, 1999*). They are known as *allogenic* engineers, since they use mechanical or other means to transform living or non-living material from one physical state to another, in contrast with *autogenic* engineers (such as plants) that can change the environment using their own physical structure (*Hole, 1981*). Burrowing rodents affect the infiltration of water by modifying the porosity of the soil through mechanical work (mounding, mixing, forming voids, back filling of voids or *krotovina*, and forming and destroying peds[4]). In particular, soil fauna contribute actively to the formation and maintenance of *macropores* (pores with diameter

---

[4] Natural units of soil structure such as crumbs, granules, blocks, and prisms.

>3,000 μm) (*Beven & Germann, 1982*). The pores formed by burrowing rodents are primarily tubular in shape, can be over 2 in (5 cm) in diameter, and are usually concentrated close to the soil surface, from a few inches to more than 33 feet (10 m) in depth (*Beven & Germann, 1982*). Although such pores may represent only a small proportion of the total porosity of the soil, they can have a very important influence on its saturated hydraulic conductivity (*Beven & Germann, 1982*). In general, bioturbated areas have higher infiltration rates than non-bioturbated ones. Moreover, burrows may act as subsurface pipes, redistributing the water throughout the horizon. When they are connected downslope to the ground surface or to a seepage face, they can act as local conduits for lateral flow and, thereby, provide an important control on runoff generation. When the deeper soil horizons saturate during the storm season, macropore flow provides an active and rapid pathway for surface discharge (*Montgomery & Dietrich, 1995*). The geomorphic effect of the burrowing animals is greater in slopes because the bioturbated materials tend to move downslope (*Price, 1971*).

The relationship between burrowing animals and the structure and moisture content of soil is complex. While the composition of fauna in the soil will depend on the soil characteristics and local environment (climate, vegetation, soil moisture, soil texture, and pH), the fauna modifies the soil it inhabits by bioturbation. The structure, depth and length of rodent burrows will depend on the species and the properties of the soil (*Laundré & Reynolds, 1993; Wilkinson et al., 2009*). For example, some burrow systems have been found to be quite complex, having several entrances, more than one main tunnel, multiple minor branches used for foraging, and different chambers used for nesting, food storage and fecal mater deposition. Nevertheless, other burrow systems can be quite simple, having just two entrances and a main tunnel (*Baker et al., 2003; Hoogland, 1995; Reichman & Seabloom, 2002*). Surrounding the burrow openings it is common to find mounds (i.e., accumulated soil deposited as a result of the burrowing activity), whose sizes vary according to the size of the burrow.

All of these burrow characteristics act to influence the amount of water infiltrated and the depth of water penetration (*Whitford & Kay, 1999*). Moreover, there can also be indirect impacts of the burrow system on water infiltration. On the one hand, burrows of fossorial mammals that provide chambers for storage of food and for defecation result in an increase in the mineral nutrient content of the soil, which in turn may increase vegetation density, thereby further modifying the infiltration characteristics of the soil (*Titus et al., 2002; Whitford & Kay, 1999*). On the other hand, animal burrows can provide plants with a direct source of water by maintaining higher water conductivity in the undercanopies and by promoting deeper percolation of water (*Shafer et al., 2007*). In this manner, the burrows made by fossorial animals contribute to the productivity and spatial heterogeneity of the ecosystems. This landscape heterogeneity can increase to a greater extent when more than one burrowing species coexist in the same area (*Davidson & Lightfoot, 2006; 2008*). It has also been observed that burrowing activity doubles the infiltration rate in open areas with no shrubs when compared to areas with shrubs, while burrowing had only a small effect under shrubs (*Soholt et al., 1975; Soholt & Irwin, 1976*).

The magnitude of the effects that ecosystem engineer species can cause in the environment will depend on: i) the population density; ii) spatial distribution; iii) the lifetime per capita of individual organisms; iv) the length of time the population has been present at the site; v) the durability of the perturbation caused in the absence of the original engineer; and vi) the number and types of resource flows that are modulated by the constructs and artifacts, and the number of species dependent upon these flows (*Jones et al., 1994*).

Although there is clear evidence that underground systems built by burrowing mammals provide an important ecosystem service by facilitating water infiltration (e.g. *Grant, French, & Folse, 1980; Laundré, 1993*), only limited research has been conducted on this topic, and the relevance of these organisms to the hydrological cycle has not been studied in detail. As pointed out by *Westbrook et al. (2013)*, the field of ecohydrology (study of the interaction between water and ecosystems) has a definite bias towards plant based-research. In that study, which included a literature review of articles from 2000 to 2011 indexed in Science Direct®, 17% of the articles related fauna with hydrology, but less than 7% (23 articles) analyzed fauna as a driver of hydrological patterns. Even fewer studies have addressed the role of burrowing rodents in hydrology, with only a very small number focused on drylands. In fact, most of the studies on fauna have been focused on the role of soil invertebrates in water infiltration, soil moisture and runoff (e.g., *Fischer et al., 2014; James et al., 2008; Léonard & Rajot, 2001; Schaik et al., 2014*), and research of the effects on burrowing rodents on hydrological processes is extremely scarce.

3 THE MAIN ECOSYSTEM ENGINEER RODENTS IN DESERTS OF THE SOUTHWEST

The following discussion is based on a review of studies regarding the most relevant burrowing rodents in the southwestern US, where climate change and human overexploitation are expected to result in even more severe drought than has historically been experienced. Specifically, we summarize the literature related to five groups of ecosystem engineers (ground squirrels, marmots, prairie dogs, pocket gophers, and kangaroo rats) and their relationship to the water cycle.

3.1 Ecosystem Engineers and water cycle

Ground squirrels include four genera: *Spermophilus* (true ground squirrels), *Ammospermophilus* (antelope ground squirrels), *Marmota* (marmots) and *Cynomys* (prairie dogs). In this section we will cover the two first genera (*Spermophilus* and *Ammospermophilus*), while marmots and prairie dogs are treated later. In the drylands of the Southwest we can find 9 species of true ground squirrels (out of the 26 living in North America), and 3 of antelope ground squirrels (out of the 5 in NA) (*Yensen & Sherman, 2003*) (*Figure 3*).

Ground squirrels are diurnal, and although most species can climb, they live mainly on and in the ground. They are primarily, but not exclusively, generalist herbivorous; they can eat grasses, flowers, forbs, seeds, berries, fruits, insects and small vertebrates. They live in a wide range of densities, usually <8 individuals/ac (<20 individuals/ha). The densities can be very high near rich food resources: 132 individuals/ac (331 individuals/ha).

Ground squirrels excavate their own burrows, but occasionally they can use burrows excavated by other species. Many ground squirrel burrows are in soil containing

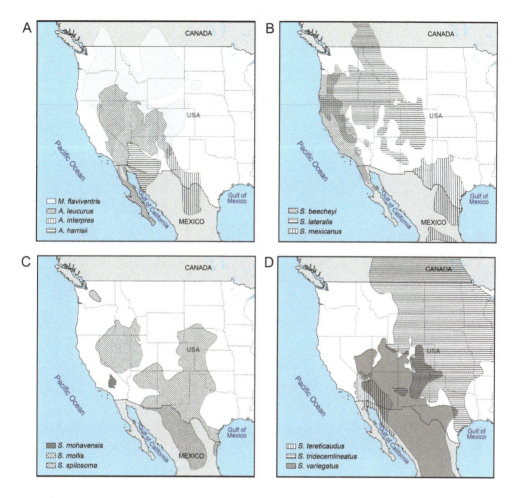

Figure 3 Distribution of *Marmota flaviventris, Ammospermophilus leucurus, A. interpres, A. harrisii* (A); *Spermophilus beecheyi, S. lateralis, S. mexicanus* (B); *S. mohavensis, S. mollis, S. spilosoma* (C); *S. tereticaudus, S. tridecemlineatus, S. variegatus* (D). Adapted from *IUCN (2015)*.

boulders or in rock fissures or ledges with soil accumulations. They use a variety of burrows for different purposes: nesting, refuge and hibernation (*Yensen & Sherman, 2003*). The burrows used for nesting usually are more complex, longer and deeper and have more entrances and chambers than burrows used for other purposes. Nesting burrows usually have several entrances. *Berentsen & Salmon (2001)* have reported between 2 and 5 entrances for *S. beecheyi*. The entrances are frequently plugged to make them inconspicuous to predators and are excavated under trees rocks or shrubs. The tunnels are often parallel to the surface, although there is a tendency to burrow more horizontally on sloped ground. The diameter of the burrow depends on the size, age and species of the animal and can have a wide range of variation (*Table 2*). For example, *S. lateralis* excavates burrows that have an entrance with a diameter of 2.32–2.56 in (5.9–6.5 cm) and the tunnel has a diameter of 1.97–2.2 in (5–5.6 cm), while

Table 2 Burrows/mound measures for the different groups of burrowing rodents' species inhabiting the deserts of the Southwest (for the species composition see table 4).

Species Groups	Species	Soil texture	Tunnel morphology	Burrow diameter (inches)	Burrow cross section area (inches²)*	Tunnel length (feet)	Tunnel depth (inches)	Mound Diameter (inches)	Mound height (inches)	Mound/burrow density (mound/acre) (burrow/acre)**	Mound volume (inches³)	Excavation Rate (feet³/acre·year)	(feet³/acre)	Surface area covered (%)
Ground squirrels[1]		Clay sand	Horizontal	1.97–4.02	3.05–12.69	0.33–138	1.97–72	36–590.55	6–39.37	NA	NA	NA	NA	40–80
Marmots[2]	M. flaviventris	Sandy loam	Inclined	NA	26.23–68.44	12.47–14.44	15.75–64	NA	NA	0.24–17.4 **	NA	0.11	NA	NA
Prairie dogs[3]		Clay loam	Inclined	3.15–8.86	7.79–61.65	9.35–108.27	7.87–196.85	24–240	6–39.37	23.9–50 20–120 **	5184–9153.56	NA	52.31–554.41	5
Pocket gopher[4]		Sandy loam to loam	Horizontal	1.97–5.00	3.05–19.63	31–341	3.11–51.97	7.87–60	1–12.50	133–19426	67.13–670.95	40.01–2800.84	NA	2–25.9
Kangaroo rats[5]	D. ordii	Sandy loam	inclined	1.56–5.20	1.91–21.24	1.64–17.72	0.66–134.19	NA	NA	NA	NA	NA	NA	NA
	D spectabilis	Sandy loam	Complex network of tunnels	3.15–4.33	7.79–14.73	104.99	20–35.4	2.6–248	3.94–48	0.12–38.06	12,200.68–280,708.42	NA	1.37–256.63	1.6–3

1 Berentsen & Salmon, 2001; Burt, 1936; Criddle, 1939; Davis & Schmidly, 1994; Desha, 1966; Grinnell & Dixon, 1918; Howell, 1938; G. E. Johnson, 1917; Juelson, 1970; Price, 1971; Ryckman, 1971; Wallace, 1991.
2 Plaster, 2003; Svendsen, 1976.
3 Butler, 2006; Ceballos et al., 1999; Clark, 1971; Cooke & Swiecki, 1992; Egoscue & Frank, 1984; Gedeon et al., 2012; J. L. Hoogland, 1995; 1996; Koford, 1958; Thorp, 1949; Verdolin et al., 2008; Wilcomb, 1954.
4 Black & Montgomery, 1991; Grant et al., 1980; Hickman, 1977; Hobbs & Mooney, 1985; 1991; Laycock, 1958; Miller, 1957; Reed, 2013; Reichman et al., 1982; Thorn, 1978; Vleck, 1981; Yurkewycz et al., 2014.
5 Andersen & Kay, 1999; Bancroft et al., 2004; Best, 1988; Eldridge et al., 2009; Holdenried, 1957; Moorhead et al., 1988; Moroka et al., 1982; Mun & Whitford, 1990; Oliver & Wright, 2010; Reichman et al., 1985; Schroder & Geluso, 1975; Taylor, 1930; Vorhies & Taylor, 1922.
(*) The burrow cross-section area has been calculated from the diameter of the burrows in all the species except for marmots, assuming the burrows have a circular cross-section shape. In the case of marmots, the area has been estimated from the length of the sides provided by Plaster (2003).

the burrow entrances of *S. variegatus* have a diameter of 3–12 in (7.6–30.5 cm) and the tunnels have a mean diameter of 4.02 in (10.2 cm) (*Bihr & Smith, 1998; Hatt, 1927; Juelson, 1970*). The number of chambers per burrow is 2.1, with one or more exits. In the burrow chamber wood, grass, seeds, feces and human trash can be found. In general, for *S. lateralis*, food caches are rare (*Bihr & Smith, 1998*). The auxiliary burrows are used as refuge, and are built at different distances from the nest burrow. These burrows can be just short holes under rocks or shrubs, a steep tunnel that ends in a chamber or can be burrows built by other species (*Yensen & Sherman, 2003*). Hibernation burrows are only used for prolonged dormancy. The chambers are usually spherical and the tunnels can be from less than 3.3 ft (1 m) to more than 6.6 ft (2 m) long. No food has been found in these tunnels (*Yensen & Sherman, 2003; Young, 1990*).

Only a few species have burrows surrounded by mounds from excavated soil (*Yensen & Sherman, 2003*). For example, *S. beecheyi* mounds have been described as having a diameter of 36–590.55 in (91.44–1500 cm), and a height of 6–39.37 in (15.4–100 cm). Sometimes, elongated mounds can be found which are composed of coalesced individual mounds. The dimensions of these mounds can be as large as 32.81 ft (10 m) width by 164.04–328.08 ft (50–100 m) long, and in extreme cases the length can be more than 656.17 ft (200 m) (*Wallace, 1991*). Mounds can represent 40–80% of large areas of land (*Grinnell & Dixon, 1918; Wallace, 1991*). Mounds have also been described for *S. lateralis* (*Bihr & Smith, 1998*).

The construction of these burrows represents an important bioturbation and a large amount of mixed soil. Excavation rates for ground squirrels species inhabiting the southwestern deserts are not available. Nevertheless, these rates have been measured in other species. For example, arctic ground squirrel (*Spermophilus parryii*) has an excavation rate of 197.05 ft^3/ac.yr (13.79 m^3/ha.yr) (study conducted in Ruby Range, Yukon, Canada; *Price, 1971*). In the Front Ranges of the Canadian Rocky Mountains, the average rates of sediment transport to the ground surface were 12.32–14.78 ft^3/ac.yr (0.86–1.03 m^3/ha.yr) for *S. columbianus*. The average volume of sediment added to each burrow mound was 8.33–15.54 ft^3/yr (0.25–0.44 m^3/yr), suggesting that every year a section of 13–23 ft (4–7 m) was added to the tunnel system (*Smith & Gardner, 1985*).

These tunnel systems have been proven to increase water infiltration. An experiment conducted in the Big Canyon (Idaho) showed that burrows built by ground squirrels (Townsned's ground squirrel, *S. townsendii* and Wyoming ground squirrel, *S. elegans*) increase water infiltration in the soil (up to 34% of the amount of winter snow melt that enters into the soil in spring). The recharge amount was also higher in burrows areas than non-burrow areas, and is positively related to burrow density. Moreover, burrows also add water to the deeper portions of the soil profile (*Laundré, 1993*). This increase in soil moisture has been related to an increase of plant productivity (*Laundré, 1998*).

Marmots. The Yellow-bellied marmot (*Marmota flaviventris*) is the only marmot that inhabits the deserts of the Southwest (*Figure 3A*). This species is diurnal and live in burrows excavated in grass-forbs meadows on well-drained slopes ranging from 2° to 33° with a mean of 18.6°, and with rocky outcrops or scree. They live as members of a colony, singly or in pairs (*Armitage, 2003; Frase & Hoffmann, 1980; Plaster, 2003; Yensen & Sherman, 2003*). They prefer sandy loam soils containing rocks (*Svendsen, 1976*).

The density of the burrows can range from 0.24/ac (0.6/ha) to 17.4/ac (43/ha), depending on the area (*Blumstein et al., 2001*). There are three types of burrows: home burrow or nest; auxiliary, scape or flight burrows, and hibernation. The yellow-bellied marmots do not dig new burrows, but use the same burrow year after year. Flight burrows usually have one entrance, while the home burrows typically have several entrances. In this case, there is one main entrance to the burrow, and up to 3 entrances can join the main tunnel. No data is available on the diameter of the burrows, although we found data regarding the dimensions of the entrances to the tunnels. Since the entrances can be either oval, rectangular or triangular, we estimated the area of the entrances from the length of the sides provided by *Plaster (2003)* (*Table 2*). The nest chamber is usually higher than the entrance, due to the slope of the terrain. Short lateral tunnels may be present, and blind burrows might extend from the nest chambers. Marmots use plugs in the entrances of the hibernacula (*Armitage, 1991; 2003; Plaster, 2003; Svendsen, 1976*). *M. flaviventris* burrows have mounds from the excavated soil, but descriptive data about mounds is not available. The amount of soil displaced by *M flaviventris* in Colorado has been estimated to be 0.11 ft^3/ac.yr (0.027 m^3/ha.yr) (*Plaster, 2003*).

No research has been conducted on the influence of the burrow systems of *M. flaviventris* on the soil water cycle.

Prairie dogs are considered to be a keystone species[5] and the ecosystem engineer species of the grasslands. They live underground in burrows that they dig for shelter, refuge from predators, flood control, and reproduction among other uses. Prairie dogs are endemic to North America and their distribution extends from Canada to Mexico. Prairie dogs include five species: black-tailed (*Cynomys ludovicianus*), Gunnison's (*C. gunnisoni*), white-tailed (*C. leucurus*), Mexican (*C. mexicanus*), and Utah (*C. parvidens*). All of them can be found in the deserts of the Southwest, except the Mexican prairie dog, which is endemic to north-central Mexico (*Figure 4*). Among all the species, the black-tailed prairie dog has the greatest range (from south-central Canada to the northern part of Mexico).

The burrows are not very complex. They are usually 4 to 12 in (10 to 30 cm) in diameter at the entrances and slightly narrower underground. The length and depth of the tunnels have a wide range of variation (*Table 2*). Burrows may contain different kinds of chambers: nesting chambers that are packed with dry grass, small burrow enlargements used as turnaround point or temporary refuges, side pockets that are used to store food (*Sheets et al., 1971*). Some prairie dog burrows harbor fecal pellets while others do not (*Hoogland, 2003*). In some cases a fecal chamber dedicated to the accumulation of fecal pellets has been described (*Cooke & Swiecki, 1992*).

The burrows usually have one or two entrances, although burrows with more than 15 entrances have been observed. The entrances can be flat holes in the ground or be surrounded by mounds of soil, with diameters of 78.74–118.11 in (2–3 m) and as high as 7.9–11.8 in (20–30 cm; dome craters) or diameters of 39.37–59.06 in (1–1.5 m) and as high as 39.37 in (1 m; rim crater) (*Hoogland, 1995*). While the dome mounds are built with subterranean soil, the rim mounds are built using surface materials (*Cincotta, 1999*). For *C. gunnisoni*, Gedeon et al. (2012) have estimated a mound

---

5 A keystone species is one that has a large overall effect on the ecosystem structure and function.

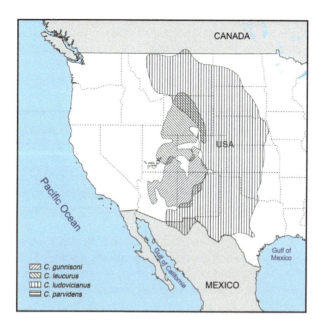

Figure 4 Distribution of *Cynomys gunnisoni, C. leucurus, C. ludovicianus,* and *C. parvidens.* Adapted from IUCN (2015).

density 27.8 mounds/ac (68.7 mounds/ha), while *Thorp (1949)* counted 50 mounds/ac (123 mound/ha). The area occupied by mounds is 5% (Hole, 1981; Thorp, 1949). The prairie dog mounds are the result of the work of many generations (*Egoscue & Frank, 1984; Thorp, 1949*). Taking into account that prairie dogs live in colonies or dog towns and the land cover by burrows can be high (the number of entrances may range 20–120/ac or 50–300/ha; *Whicker & Detling, 1988*), the volume of soil that they move can be large. For example, in Northern Colorado, the volume of a normal mound is 5,184 in^3 for *C. ludovicianus* (*Koford, 1958*), while in Northern Arizona, the volume of a mound is 9,154 in^3 for *C. gunnisoni* (*Gedeon et al., 2012*). In the case of the black tailed prairie dog, for example, for 12 burrows with a total of 25 entrances, prairie dogs move to the surface about 4 tons of soil (*Gedeon et al., 2012; Koford, 1958*). *Ceballos et al. (1999)* have calculated that *Cynomys ludovicianus* mixes 52.31–106.97 ft^3/ac (3.66–7.18 m^3/ha) in Chihuahua, Mexico; while *Thorp (1949)* calculated for the same species a mixing rate of 554.41 ft^3/ac (38.80 m^3/ha; assuming a soil density of 1.3 g/cm^3), in Akron, Colorado. *Gedeon et al. (2012)* estimated a mixing rate of 158.59–252.99 ft^3/ac (11.1–17.70 m^3/ha) in *C. gunnisoni*, in Flagstaff, Arizona.

Research conducted on how prairie dogs bioturbation directly affects water infiltration is scarce but relevant. Estimates from *Koford, (1958)* show that one acre (0.4 ha) of a dog town with 25 holes would contain 700 gallons (2,650 liters) of water that could run off or penetrate into the ground. A recent study conducted on black-tailed prairie dog in Janos Biosphere Reserve (Chihuahuan desert, Mexico) showed that water infiltration rates in prairie dog grasslands were higher (14.1 ± 11.36 in/hr; 357 ± 288.45 mm/hr) than grasslands (11.14 ± 7.65 in/hr; 283 ± 194.36 mm/hr), and mesquite (3.82 ± 3.06 in/hr; 97 ± 77.82 mm/hr). This increased infiltration is because soils within the prairie dog grasslands are less compacted than the soils of the areas dominated by mesquites, and have more pore spaces that allow the movement of water, increasing the water infiltration (*Martínez-Estévez et al., 2013*). Also, *Barth et al., (2014)* have compared the infiltration in on-mound areas (11.81 in or 30 cm from the burrow center) versus off-mound (78.74 in or 200 cm from the burrow center) and control areas in South Dakota. The authors found that infiltration rates in on-mound areas (0.53–2.66 in/hr; 13.5–67.5 mm/h) were higher than in off-mound (0.12–0.71 in/hr; 3–18 mm/hr) and control areas (0.20–1.46 in/hr; 5–37 mm/hr).

Prairie dogs burrows might affect runoff dynamics because they avoid building tunnels on level ground choosing slopes that ranges from 2% to 30% in order to avoid flooding (*Koford, 1958*). Moreover, the mounds around the burrow entrances, besides serving as lookout stations or to enhance ventilation of the tunnels, they act as a barrier preventing water from entering the tunnels after torrential rainstorms (*Hoogland, 1995*).

Pocket gophers range from Canada to Panama. In the USA only three genera (*Cratogemys, Geomys* and *Thomomys*) occur, including a total of 18 species. Only six species are present in the drylands of the Southwest (*Cratogeomys castanops, Geomys arenarius, Thomomys bottae, T. talpoides, T. umbrinus, T. townsendii*, Figure 5). They live preferentially in soils of fine sand to loam, porous, well drained and with high nutrient content, avoiding in general soils with high clay content that do not drain well. Pocket gophers are fossorial and generalist herbivores. They feed on roots or entire plants and their diet changes with the season (*Feldhamer et al., 2003*).

Pocket gophers are important bioengineers of the desert and their impact on the soil is significant. Although their burrowing effects on the soil can differ between species and location (*Ellison, 1946; Grant et al., 1980; Kerley et al., 2004*), in general, the structure of burrows is complex. Usually, they have one or two main tunnels, and several secondary branches used for foraging. The burrows are parallel to the soil surface with a wide range of depths, depending on the species and the soil (*Table 2*). For example, *Best (1973)* reported mean burrow depth of 5.20 in (13.20 cm) for *T. bottae* and 6.5 in (16.51 cm) for *C. castanops*. Nevertheless, *Hickman (1977)* reported depths of 3.94–51.97 in (10–132 cm) for *C. castanops*. For *T. talpoides, Thorn (1978)* reported mean values of 3.11–3.58 in (7.9–9.1 cm); while *Miller (1957)* reported depths of 4–28 in (10–71 cm). The tunnel cross-section area also varies according to the species, and, together with the length and depth of the burrow, can determine a burrow's effectiveness in thwarting predators (*Laundré, 1989*) (*Table 2*).

The tunnels used for foraging constitute 80% of the tunnels and are located along the root zone. *Vleck (1981)* has reported depths of 2–8 in (15–35 cm) and diameters of 1.97–2.91 in (5–7.4 cm) for foraging tunnels. The burrows can have several chambers for nesting, food storage and fecal deposition (*Baker et al., 2003; Reichman et al., 1982*). The rate of soil disturbance of pocket gopher burrows can be as high as 25.9%, although interannual variability is likely to be an important component (*Hobbs & Mooney, 1991*). Most of the soil extracted from the excavation is deposited on the surface forming the mounds, changing the topography of the soil. The diameter and size of the mound also depends on the species and location (*Table 2*). The soil density of the mounds is 10–40% of the soil density of the underlying soil (*Inouye et al., 1997*). Regarding mound density, several studies have shown differ-

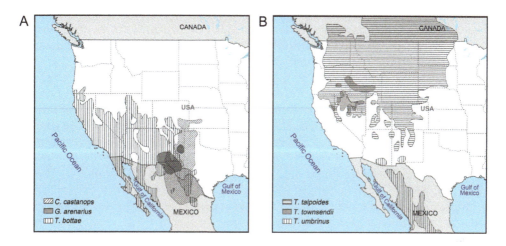

Figure 5 Distribution of *Cratogeomys castanops, Geomys arenarius, Thomomys bottae* (A); *T. talpoides, T. townsendii, T. umbrinus* (B). Adapted from IUCN (2015).

ent values for different species. For *Geomys arenarius*, it has been reported a mound density of 19,433.20 mounds/ac (48,000 mounds/ha), while for *Thomomys bottae* the density ranges from 647.77 mounds/ac (1600 mounds/ha) in interfluves to 3,643.72 mounds/ac (9000 mounds/ha) in areas from toe slopes to lower slopes (*Black & Montgomery, 1991; Kerley et al., 2004*). The average rate of excavation of one gopher is quite variable, with a range of 40–2800 ft^3/ac.yr (*Kerley et al., 2004; Reed, 2013*).

Over the years, pocket gophers move more soil than prairie dogs because prairie dogs dig little after their burrows are established, while gophers burrow continuously because of their feeding (*Koford, 1958*). These modifications of the soil might contribute to decrease runoff and erosion since gopher mounds cause ponding and meandering that decrease runoff velocities (*Grant et al., 1980; Hakonson, 1999*). Because the soil of the mounds is less compact, it absorbs more water and has higher water holding capacity, increasing water infiltration and soil moisture, although in the presence of plants the soil moisture might be reduced due to evapotranspiration (*Grant et al., 1980; Hakonson, 1999; Mielke, 1977*). For example, mound soils of Black Mesa (Colorado) populated by *T. talpoides* contained 196% more soil moisture than the inter-mound soils due to less soil compaction (*Mielke, 1977*). Nevertheless, in a study conducted in the volcano Mount St Helens (Washington State), the authors found that water infiltration was one third less on mounds than on undisturbed soils. The mounds presented a crust probably formed by pumiceous material derived from the last eruption that presumably was the cause of the reduced infiltration (*Yurkewycz et al., 2014*). Likewise, *Grant et al. (1980)* although detecting an increase of infiltration rates in mounds by pocket gophers in comparison to undisturbed areas, the soil water content did not differ. As later was pointed out by *Laundré (1993)*, the study by *Grant et al. (1980)* did not specify the amount of water added in his experiment and the depth of the sampling.

The tunnels excavated by pocket gophers also increase infiltration and water penetration. Two studies conducted in landfill covers[6] have shown that pocket gopher burrows increase water infiltration and water penetration into the soil, moving large amount of pounded water (*Breshears & Nyhan, 2005; Hakonson, 1999*). However, a study conducted on *T. talpoides* in Alberta, Canada, showed no effect of burrowing activity on infiltration (*Zaitlin et al., 2007*). On alpine slopes of the Colorado Front Range, abandoned tunnels systems can evolve in areas of miniature terracettes (*Thorn, 1978*). Terracettes might temporarily retain or store runoff, increase infiltration and reduce erosion (*Greenwood et al., 2015*). Nevertheless, it has also been observed that tunnels can canalize runoff and by subsurface erosion the tunnel can collapse leading to a gully, increasing erosion (*Swanson et al., 1989; Thorn, 1978*).

Kangaroo rats are small nocturnal rodents of genus *Dipodomys* that are native to arid and semiarid regions of western North America. There are 21 species described, and only 6 of them inhabit the deserts of the Southwest (*Figure 6*): Desert Kangaroo Rat (*Dipodomys deserti*), Merriam's Kangaroo Rat (*Dipodomys merriami*), Houserock Chisel-toothed Kangaroo Rat (*Dipodomys microps*), Ord's Kangaroo Rat (*Dipodomys ordii*), Panamint Kangaroo Rat (*Dipodomys panamintinus*), and Banner-tailed Kangaroo Rat (*Dipodomys spectabilis*).

Kangaroo rats spend most of the daily cycle in burrows. Burrows are usually occupied by a single individual. The complexity of the burrows mainly depends on the species and can be very simple, such as the case of *D. ordii*, or quite complex such as the burrows built by *D. spectabilis*. *D. ordii* excavate burrows with one to three entrances and one or two main tunnels with either one or two shorter side tunnels or an enlarged area off the main tunnel. The diameter of the entrance to the tunnels is slightly smaller than the diameters of the tunnels. The length of the tunnels is variable and they are quite superficial (*Table 2*). The volume of the burrow was estimated to be 0.25–0.29 ft³ (0.0071–0.0082 m³). Length and volume of the soil excavated seems to be directly related to the percentage of the silt and clay in the soil. These burrows can contain nest chambers. The food is stored in the evaginations or along the main or side tunnels. Although most of the entrances are surrounded by mounds, neither data about the dimensions of the mounds or related to excavation rates have been found (*Laundré & Reynolds, 1993; Reynolds & Wakkinen, 1987; White, 2009*).

Burrows built by *D. spectabilis* are more complex and consist of a labyrinth of tunnels and chambers. Each mound is inhabited by an adult, and the animals occupy non-overlapping home ranges. Each mound opening can group from 6 to 12 entrances (slightly higher than the tunnel diameter), and the number of entrances can be modified according to the season. In summer, with higher temperatures and higher precipitation, the number of entrances is higher than in winter (*Edelman, 2011*). Each burrow can have up to four levels of tunnels. We have not found data about the total length of the tunnels, probably due to their complexity. Nevertheless, we estimated the length from *Figure 2* in *Vorhies & Taylor (1922)* (*Table 2*). The depth of the tunnel usually is 20 in (50 cm) below ground, although the tunnel containing the nest is

---

6 Structures created to dispose of industrial or household waste, isolated from the ground to prevent the waste from infiltrating to the ground water.

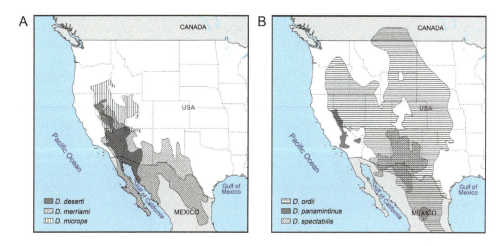

Figure 6 Distribution of *Dipodomys deserti, D. merriami, D. microps* (A); *D. ordii, D. panamintinus, D. spectabilis* (B). Adapted from IUCN (2015).

usually deeper (25.59 in/65 cm), and some branches can reach even deeper (+35.43 in; +90 cm; *Table 2; Vorhies & Taylor, 1922*). There are evaginations to store food (seeds, grasses and forbs), and the amount of food stored can be large (13–20 gal; 50–75 l). The seeds are often infected by molds (*Vorhies & Taylor, 1922*). Apparently, *Dipodomys*, burrows do not have fecal pockets, and feces are distributed along the tunnels of the den *(Best et al., 1988; Reichman et al., 1985; Vorhies & Taylor, 1922)*.

The mounds of *D. spectabilis* can be very big and occupy a considerable area (*Table 2*). *Eldridge et al. (2009)* have reported mound volumes of 12,200.68–280,708.42 in^3 (200,000–4,600,000 cm^3). The average area per mound ranges from 215 to 700 ft^2 (20–65 m^2) (*Moroka et al., 1982; Schooley et al., 2000*). Nevertheless, the mounds of other species of kangaroo rats are smaller. *Hornbeck et al. (2011)* have reported mound volumes of 0.61–915.97 in^3 (10–15010 cm^3) for *D. microps*.

The area occupied by mounds of *D. spectabilis* ranges between 1.6 and 3% of the habitat. The distribution of the mounds can be random or in aggregates (*Moroka et al., 1982; Taylor, 1930*). *D. spectabilis* can displace 1.37–256.63 ft^3/ac (0.096–17.96 m^3/ha) (*Eldridge et al., 2009*). Mound development takes a long time (23–30 months) and they deteriorate in about one year when they are abandoned. Nevertheless, mounds are often passed from generation to generation and can persist more than 50 years (*Schooley et al., 2000*). In a transect done in the Chihuahuan Desert watershed, Kangaroo rats (*D. ordii, D. merriami*, and *D. spectabilis*) produced more surface disturbance (burrows, burrow system, and foraging pits) than the pocket gopher *T. bottae*: 443.47 ft^2/ac (101.8 m^2/ha) *vs.* 298.4 ft^2/ac (68.5 m^2/ha) (*Whitford & Kay, 1999*).

Burrows of kangaroo rats (*D. spectabilis*) are characterized by higher values for soil moisture and water holding capacity of the surface soil, and this is accompanied by an increased proportion of clay and finer sized soil particles (*Greene & Murphy, 1932*). *Chew & Whitford (1992)* showed that mound soil could achieve more water storage during the rainy season than intermound soil that was calcium impregnated in an area dominated by creosotebush (*Larrea tridentata*). However, *Titus et al. (2002)*, in a study conducted in the Mojave desert, did not find differences in infiltration when two microsites were compared based of a mixed *L. tridentata*–deciduous shrub clump with and without small mammal burrows of *Dipodomys spp*, although the values were slightly higher in the microsite with burrows: 1.86 ± 0.34 fl. oz./min (55 ± 10 ml/min) *vs.* 1.52 ± 0.68 fl. oz./min (45 ± 20 ml/min). The shrubs in the microsite with burrows were larger than in the microsite without burrows, probably due to the higher levels of nitrogen, phosphorus and potassium. In an another study, *Mun & Whitford (1990)* showed that the soil water potential of mound soils was consistently lower than in intermound soils. This result was due to the higher evaporation and high infiltration rates from the top soil of the mound where soil bulk is usually lower. Moreover, there is an additional water loss by plant transpiration; the above- ground biomass and the root biomass on the mounds were greater than in intermounds areas. Similar results have been found in other studies (*Moorhead et al., 1988*).

3.2 Discussion

As shown in the previous section, the concept of burrowing fauna as modifiers of the hydrological cycle is not new, and some early studies already mentioned this relationship (*Koford, 1958*). For most of the species groups inhabiting the deserts of the Southwest, the burrows systems have been well characterized, although not always in the desert areas. Nevertheless, research focused on the direct effect of burrowing rodents on dynamics of the water through building tunnels, mixing the soil (decreasing its density in the surface), or changing the micro-geography as a by-product of the mounds is scarce, and a great part of this research has not been conducted in deserts (e.g., *Barth et al., 2014; Price, 1971; Yurkewycz et al., 2014*).

Furthermore, there is an indirect impact on the hydrologic cycle by modification of the vegetation. Usually, mound areas are denuded of vegetation due to the mixing and moving of the soil, but in the immediate areas the vegetation is denser. This last effect might be due to two main factors: availability of nutrients and availability of water. Burrows usually contain food and fecal material that produce an increase of nutrients such as nitrogen, phosphorus and potassium (*Kerley et al., 2004*). Vegetation can also modify water input into the soil in two ways: i) roots of dead plants can act as pipes to bring water into the soil (*Gabet et al., 2003*); ii) roots of living plant can act as a pump sucking water from the soil as a result of evapotranspiration (*Cejas et al., 2014*).

The relative impact of burrowing rodents on the hydrological system in the desert is difficult to estimate, because several physical aspects of the burrow system can determine the magnitude of this impact: complexity of the burrows, depth and diameter of the tunnels, size of the mounds and extent of mounds across the landscape. These four variables vary by species and the characteristics of the soil. Moreover, the density of the species is positively correlated to the magnitude of the bioturbation. It is difficult to know which group of species would have more impact on the hydrological system as a whole, and most likely each species will have a different kind of impact on infiltration, runoff or percolation.

When focused on infiltration, deeper tunnels will conduct the water into lower layers of the soil. Moreover, mounds around the burrow entrances can act as a barrier to water penetration into the burrows, or can facilitate water infiltration due to the lower density or the soil. *Whitford & Kay (1999)* have suggested that water infiltration is higher on short-lived, low bulk density soil ejecta[7] mounds and in areas with numerous small burrows than on long-lived mounds. According to this reasoning, we can hypothesize that from the species groups described previously, pocket gophers will have more effect on infiltration than prairie dogs, and prairie dogs more than kangaroo rats. Unfortunately, we do not have data for mound density and mound volumes for ground squirrels and marmots to infer their relative affect on infiltration.

Pocket gophers have shorter-lived (1–3 years) and smaller mounds (*Table 2*) with higher densities, while kangaroo rats have the biggest mounds, followed by prairie dogs. The burrows of kangaroo rats are inherited through generations and can be established for more than 50 years. It is believed that prairie dog burrows are also maintained through several generations (*Hoogland, 2003; Schooley et al., 2000*). We can hypothesize that the older and bigger a mound is, the more compact the soil and the greater the likelihood that the soil crust on the surface will be thicker. Therefore, the water infiltration rate will be lower. Moreover, higher mounds will act as dikes to protect the burrows from being flooded. Thus, under a heavy rain, small mounds will not block enough water to prevent access to the burrows and to infiltrate to the low levels of the soil. Although a positive correlation between bulk density and runoff (less infiltration) has been shown (*Greenwood et al., 2015*), non comparable data about the bulk density of the mound soil has been found for the different species because only few studies have been done on different types of soils, and those studies only include some species.

Burrowing on hillslopes may also impact the runoff by channeling the water. Species that build their burrows in steeper slopes such as ground squirrels (0–30°), marmots (2–33°) and pocket gophers (2–30°) will have higher impact on runoff than those that prefer gentle slopes, such as prairie dogs (<6°) and kangaroo rats (0–2°) (*Best, 1972; Bihr & Smith, 1998; Ortega, 1987; Seabloom et al., 2000; Wagner & Drickamer, 2004; Whicker & Detling, 1988*). Moreover, ground squirrels and pocket gopher burrows can be very shallow, and erosion caused by subsurface runoff can collapse the burrows creating gullies, which will increase erosion and decrease water quality.

A rough calculation of the maximum volume of water that can be held in the burrows per unit area in case of flooding can provide context for the potential impact of burrowing mammals on hydrologic response. By combining the mid value of the diameter and the mid value of the burrow length (*Table 3*) we can calculate the total volume of water that can be held by a set of burrows. Using these values and assuming a cylindrical burrow shape, we calculated the burrow volume. Multiplying these volumes by the density of burrows, we obtained the total volume occupied by the burrows per unit area. According to these calculations the water held in the burrows system can be as low as 0.80 ft^3/ac (55.65 l/ha) in the case of the pocket gophers (*D. ordii*) or as high as 589.77 ft^3/ac (41,240.57 l/ha) for prairie dogs. This water has the potential to infiltrate or contribute to subsurface runoff, but its behavior will depend on several factors that would require a site-specific assessment.

---

[7] Soil ejecta is the soil that is ejected from the burrow as a result of the excavation activity.

Table 3 Total volume of burrows per area.

Species	Mid-value of burrow diameter (inches)	Mid value of burrow length (feet)	Burrow volume (feet³)	Burrow density (number/acre) min	Burrow density (number/acre) max	Total Volume per area (feet³/acre) min	Total Volume per area (feet³/acre) max	References for burrow density
Ground squirrels	2.99	69.16	3.36	3.64	12.75	12.22	42.80	Streubel & Fitzgerald, 1978
Marmots[1]	–	13.45	4.42	0.24	17.40	1.06	76.91	Blumstein et al, 2001
Prairie dogs	6.01	58.81	11.47	15.78	51.40	181.06	589.78	Hoogland et al., 1987; Long, 2002; Sheets et al., 1971
Pocket gophers	3.48	17.20	1.13	31.64	48.16	35.77	54.45	Reichman et al, 1982
Kangaroo rats (D. ordii)[2]	3.38	9.68	0.60	1.33	5.46	0.80	3.27	Sullivan, 2015

1 For marmots, we used the mid-value of the cross section (47.33 in²) to calculate the volume of the burrow.
2 Not enough data available for D. spectabilis.

These are just some examples of how different rodents can impact the water soil cycle according to the degree of bioturbation. However, to more precisely know how burrowing rodents influence the water dynamics in the desert soils of the Southwest, additional research is necessary to:

- *Quantify the population density of these species in the deserts of the Southwest.*
- *Better characterize burrow systems and quantify burrow densities in the deserts of the Southwest* (including burrow length, depth, diameter, volume, mound characteristics, etc). Most of these species have a wider range than deserts, and many studies have been conducted in areas other than deserts. Knowing the structure of burrow systems in the desert soils is necessary to understand how water flows into the soil.
- *Conduct research on water dynamics in burrow systems, to determine how burrows influence variability of soil moisture in space and time.*
- *Study the population trends and conservation needs of these species,* because declines in burrowing rodents can have cascading effects throughout the ecosystems in which they occur (see next section). For example, in the context of the soil water cycle, it is important to better understand the interactions of burrowing systems with vegetation, as both have an impact on the water cycle.

4 INTERDEPENDENCE OF ANIMAL ECOSYSTEM ENGINEERS AND OTHER COMPONENTS OF THE ECOSYSTEM, AND IMPLICATIONS FOR THE SOIL WATER CYCLE

Burrowing rodents play a key role in shaping ecosystems by increasing, through herbivory and ecosystem engineering, biodiversity and habitat heterogeneity across the landscape. Although there has been a recent increase in consciousness about the importance of the native fauna as geomorphic agents capable of modifying the ecosystems, humans have historically altered the functioning of ecosystems directly by eliminating native species (e.g., poisoning), or indirectly through the introduction of non-native species that have displaced the native ones.

Besides being important to the water cycle by increasing water infiltration and affecting water runoff, burrowing rodents provide a host of other ecosystem services (*Davidson et al., 2012; Guo, 1996; Martínez-Estévez et al., 2013*), including:

- Increasing soil organic matter and inorganic nutrients by soil mixing and urine and fecal deposition around mounds.
- Increasing soil stability.
- Increasing soil productivity.
- Increasing storage of atmospheric carbon in grasslands.
- Providing habitats for other animals
- Serving as prey for predators such as raptors, canids, felids, herpestids, mustelids, and some snakes
- Enhancing seed capture, seedling germination, recruitment, and plant diversity via foraging tunnels and food storage chambers.

- Redistribution of soil surface contaminants in the soils, such as lead (Pb). Burrowing rodents decrease Pb concentrations in surface soils, reducing the potential for erosional redistribution of Pb, and decrease Pb transport time through the soil profile as a result of soil mixing. Therefore, it is possible that they reduce the amount of Pb that arrives to the groundwater (*Mace et al., 1997*).

Thus, burrowing rodents are important drivers in the maintenance and improvement of ecosystem services. To maintain these services in the long term, it is necessary to preserve the ecological functions of their populations. Several impacts related to the decline of burrowing mammals populations have been observed, such as woody plant invasions in grasslands, decline of animals populations that rely on their colonies for nesting, decline of predators, loss of habitat heterogeneity, alteration of plant cover, and modification of soil water cycle.

Despite most of the burrowing rodents mentioned in this chapter being negatively impacted by humans in the Southwest, only few species have been listed as threatened, vulnerable or endangered by the Endangered Species Act (ESA) and/or the International Union for Conservation of Nature (IUCN) Red List of Threatened Species (*IUCN, 2015*) (*Table 4*). A very interesting example is the black-tailed prairie dog (*Cynomys ludovicianus*) that is not listed as threatened in any of the lists, although the range of this species across North America has been reduced by 98% during the last 200 years; in Arizona, the black-tailed prairie dog was last seen in 1960. The factors contributing to this decline are: sylvatic plague (*Yersinia pestis*), chemical control, shooting, and conversion of rangeland to cropland (*Hoogland, 2006*). Moreover, many of the remaining prairie dog towns are affected by urbanization and habitat fragmentation (*Johnson & Collinge, 2004; Lomolino & Smith, 2001*). Several petitions to include this species under the ESA have been filed, and it was considered a candidate to be listed under the ESA only until 2004. Given this situation, the Arizona Game and Fish Department initiated a reintroduction program of the black-tailed prairie dog, with a first release occurring in 2008 in New Mexico and in Las Cienegas National Conservation Area (located between Sonoran and Chihuahuan deserts) (*Underwood & Van Pelt, 2008*).

The primary threats impacting burrowing rodents are direct extermination of the individuals (trapping, shooting or poisoning), habitat loss, introduced species, disease, cattle grazing, plague, and climate change (*Davidson et al., 2012*). Frequently, there are campaigns to "control" burrowing rodents populations in order to benefit the livestock industry, despite research showing that poisoning affects other wildlife. Although high densities of burrowing rodents can compete with cattle for forage, low and moderate densities can increase land productivity (*Martínez-Estévez et al., 2013*). In fact, livestock mass has been observed to decline during the same period that burrowing mammals are poisoned, and livestock weight is not affected when burrowing mammal colonies cover less than 30% of the landscape (*Davidson et al., 2012*).

Habitat loss and habitat fragmentation have led to important reductions of suitable habitat for burrowing rodents and consequently the populations of many species have declined. Moreover, habitat loss/fragmentation has led to a decline in apex predators (predators at the top level of the food chain) and an increase of

Table 4 List of burrowing rodent's species present in each one of the deserts of the America Southwest. The conservation status is also listed.

			Presence in the America Southwest deserts					Conservation lists		
	Common name	Scientific name	Great Basin Desert	Colorado Plateau	Mojave Desert	Sonoran Desert	Chihuahuan Desert	IUCN Red list	ESA	Observations
Ground squirrels	White-tailed Antelope Squirrel	Ammospermophilus leucurus	x	x	x			Least concern	—	The populations of this species may be threatened on the two islands in the Gulf of California by predation from feral cats and by human activities; no known conservation measures specific to this species; there are several protected areas within its range.
	Texas Antelope Squirrel	Ammospermophilus interpres					x	Least concern	—	No known conservation measures specific to this species; there are several protected areas within its range.
	Harris's Antelope Squirrel	Ammospermophilus harrisii		x	x	x		Least concern	—	No known conservation measures specific to this species; there are several protected areas within its range.
	California Ground Squirrel	Spermophilus beecheyi			x			Least concern	—	No known conservation measures specific to this species; there are several protected areas within its range.

(Continued)

Table 4 (Continued)

Common name	Scientific name	Presence in the America Southwest deserts					Conservation lists			Observations
		Great Basin Desert	Colorado Plateau	Mojave Desert	Sonoran Desert	Chihuahuan Desert	IUCN Red list	ESA	
Golden Mantled Ground Squirrel	Spermophilus lateralis	x	x	x	?		Least concern	—	No known conservation measures specific to this species; there are several protected areas within its range.
Mexican Ground Squirrel	Spermophilus mexicanus					x	Least concern	—	No known conservation measures specific to this species; there are several protected areas within its range.
Mohave Ground Squirrel	Spermophilus mohavensis			x			Vulnerable	Not listed	Its occurrence range is approximately 20,000 km^2, its range is severely fragmented, and there is ongoing decline in the extent and quality of its habitat.
Piute Ground Squirrel	Spermophilus mollis	x					Least concern	—	There was a decline in the late 1980s in the Snake River Birds of Prey Area in southwestern Idaho, due to a conversion of vegetation from desert shrublands to exotic annual-dominated communities by wildfires.
Spotted Ground Squirrel	Spermophilus spilosoma		x		x	x	Least concern	—	No known conservation measures specific to this species; there are several protected areas within its range.

	Round-tailed Ground Squirrel	*Spermophilus tereticaudus*		x	Least concern	No listed	No known conservation measures specific to this species; there are several protected areas within its range.
	Thirteen-lined Ground Squirrel	*Spermophilus tridecemlineatus*	x		Least concern	Not listed	No known major threats; there are several protected areas within its range.
	Rock Squirrel	*Spermophilus variegatus*	x	x	Least concern	–	No known conservation measures specific to this species; there are several protected areas within its range.
Marmots	Yellow-bellied Marmot	*Marmota flaviventris*	x		Least concern	Not listed	No known major threats; there are several protected areas within its range.
Prairie dogs	Black-tailed prairie dog	*Cynomys ludovicianus*		x	Least concern	Not listed	It was considered a candidate to list under ESA until 2004. In 2009 its inclusion to ESA was not warranted.
	Gunnison's Prairie Dog	*Cynomys gunnisoni*	x		Least concern	Not listed	Candidate to be included in the future.
	White-tailed Prairie Dog	*Cynomys leucurus*	x		Least concern	Not listed	It has been petitioned to be included in ESA, but in 2010 its inclusion was not warranted.
	Utah Prairie Dog	*Cynomys parvidens*	x		Endangered	Threatened	Its population has been reduced from 95,000 adults in 1920 to less than 10,000 and its range has decreased from 1800 km^2 to 28 km^2 today.

(*Continued*)

Table 4 (Continued)

	Common name	Scientific name	Presence in the America Southwest deserts					Conservation lists		Observations
			Great Basin Desert	Colorado Plateau	Mojave Desert	Sonoran Desert	Chihuahuan Desert	IUCN Red list	ESA	
Pocket gophers	Yellow-faced pocket gopher	*Cratogeomys castanops*					x	Least concern	—	Since this species is considered an agricultural pest, populations are often reduced by trapping and poisoning, or by managing for the presence of large predatory birds including hawks and owls. There are several protected areas within its range.
	Desert pocket gopher	*Geomys arenarius*					x	Near Threatened	Not listed	The main threat is habitat fragmentation produced by a reduction and expanse of grassland. A protected population occurs on White Sands National Monument that is vulnerable to droughts.
	Botta's pocket gopher	*Thomomys bottae*	x	x	x	x	x	Least concern	—	It is considered an agricultural pest. Nevertheless, there are several protected areas within its range.
	Northern pocket gopher	*Thomomys talpoides*	x	x	?			Least concern	Not listed	There are several protected areas within its range.
	Townsend's pocket gopher	*Thomomys townsendii*	x					Least concern	—	There are several protected areas within its range.

Southern pocket gopher	*Thomomys umbrinus*	?		x	Least concern	Not listed	It is very limited in its distribution, but not immediate threats t its survival have been identified. It is not know if there are protected areas in its range.
Kangaroo rats Desert Kangaroo Rat	*Dipodomys deserti*		x		Least concern	—	No known major threats; there are several protected areas within its range.
Merriam's Kangaroo Rat	*Dipodomys merriami*		x	x	Least concern	Not listed	Only two subspecies (*D. m. parvus, D. m. Collinus*) have conservation measures and none of them inhabit in the deserts of the Southwest.
Houserock Chisel-toothed Kangaroo Rat	*Dipodomys microps*	x	x		Least concern	Not listed	Two of the subspecies are federal candidate taxa for conservation: *D. m. alfredi* and *D. m. Leucotis*; *D. m. leucotis* is also a candidate taxon for listing in Arizona.
Ord's Kangaroo Rat	*Dipodomys ordii*	x	?	x	Least concern	Not listed	Very abundant.
Panamint Kangaroo Rat	*Dipodomys panamintinus*		x		Least concern	—	No known major threats; there are several protected areas within its range.
Banner-tailed Kangaroo Rat	*Dipodomys spectabilis*	x	x	x	Near threatened	—	It is in significant decline (at a rate of less than 30% over ten years) due to widespread degradation of its desert grassland habitat through much of its range.

mesopredators[8] (mesopredator release[9]). This outbreak of mesopredators can place strains on other smaller species (*Prugh et al., 2009*). Moreover, introduced domestic animals (such as pet cats) can also contribute to the outbreak of mesopredators of the prey animals such as rodents, birds and lizards, causing declines in the populations of preys (*Malloy et al., 2013*).

Cattle grazing can also affect the populations of small mammals. The effect of livestock on small mammals populations can be direct (trampling burrows) or indirect (compacting the soil, reducing the seed stock in plants, changing the vegetation cover and structure). Plague, a disease caused by the bacterium *Yersinia pestis* was introduced from Asia to North America and has induced important declines in the populations of prairie dogs. Climate change also affects densities of burrowing animals, especially in the Southwest where drought is projected to be more severe over the oncoming years and temperature is expected to raise (*Davidson et al., 2012*). Moreover, desert rodent populations are limited by food availability, displaying dramatic increases in abundance during wet periods regardless of predator levels. For example, initiation of reproduction in Merriam's kangaroo rat (*Dipodomys merriami*) is highly dependent upon production of green vegetation in response to winter rainfall (*Moses et al., 2012; Van de Graaf & Balda, 1973*).

Although research shows evidence of the positive impact of these animals on the health and functioning of the ecosystem (*Martínez-Estévez et al., 2013*), burrowing rodents continue to be heavily prosecuted as the demand for food production increases. Educational plans on the diverse ecological roles of these species are need in order to change public opinion, address misconceptions, and reverse government policies that continue eradication programs. In the cases that densities of rodents really need to be controlled, managers may be able to reduce burrowing mammal populations by non-invasive methods such as reducing livestock grazing and allowing grass to grow tall (*Davidson et al., 2012*).

The introduction of several species, such *as C. ludovicianus,* on conservation lists (e.g., ESA, IUCN Red List) is needed to put in place conservation plans focused on increasing the population densities when possible, by including the creation of protected areas, commitment of local communities, and financial compensation of landowners for supporting burrowing mammals and the ecosystem services they provide. Conservation plans are preferable to reintroduction plans, as the latter are intensive, costly, and small-scale (*Davidson et al., 2014*).

To understand how plans for conservation of burrowing rodents might affect the water cycle, it is important to connect all the known factors that interact directly or indirectly with both water and burrowing rodents. *Figure 7* shows the roles that burrowing rodents play in deserts, with special attention to the soil water cycle, and their interaction with anthropogenic effects. The direction of these interactions is not easy to determine. For example, in the case of the apex predator removal, the strength and structure of the interactions among apex predators, mesopredators, and prey species

---

8 Mesopredator is any midranking predator in a food web, regardless of its size or taxonomy (*Prugh et al., 2009*).
9 Mesopredator release, is the expansion in density or distribution, or the change in behavior of a middle-rank predator, resulting from a decline in the density or distribution of an apex predator.

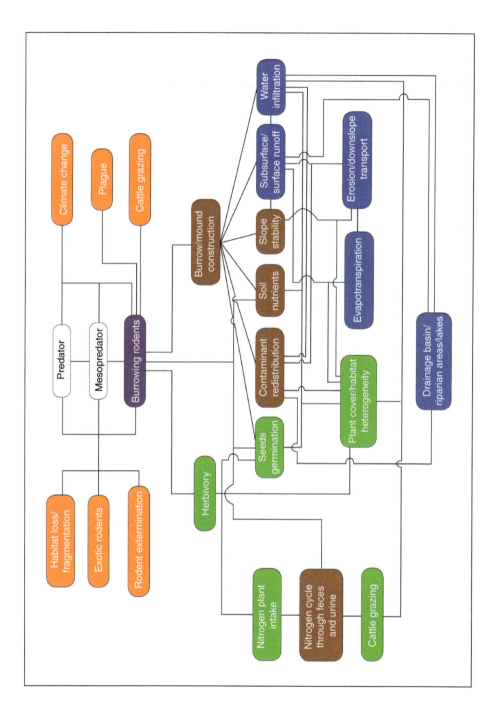

Figure 7 Conceptual diagram of anthropogenic effects on burrowing rodents and their geomorphic effects on the water cycle. Modified from *Butler (2006), and Davidson et al. (2012)*.

need to be first identified. Not always does the increase in densities of a mesopredator translate into a decrease of the prey (*Prugh et al., 2009*). Likewise, as we have seen previously, the impacts on water infiltration and water runoff of the construction of a burrow will depend on several factors, such as soil properties and burrowing species involved. Therefore, intensive research is needed in order to understand the biotic and abiotic factors that interact in the desert landscape, and the manner in which these interactions directly or indirectly affect the water cycle.

REFERENCES

Al-Eisawi, D. (2003). Effect of biodiversity conservation on arid ecosystem with a special emphasis on Bahrain. *Journal of Arid Environments*, 54(1): 10–10.

Andersen, M.C. and Kay, F.R. (1999). Banner-tailed kangaroo rat burrow mounds and desert grassland habitats. *Journal of Arid Environments*, 41: 147–160.

Armitage, K.B. (1991). Social and population dynamics of yellow-bellied marmots: results from long-term research. *Annual Review of Ecology and Systematics*, 22: 379–407.

Armitage, K.B. (2003). Marmots. In Feldhamer, A.G., Thompson, B.C. and Chapman J.A., (Eds.), *Wild mammals of North America: biology, management, and conservation*. Batimore, Maryland: The Johns Hopkins University Press.

Baker, R.J., Bradley, R.D. and McAliley, L.R., Jr. (2003). Pocket Gophers. In Feldhamer, A.G., Thompson, B.C. and Chapman J.A., (Eds.), *Wild mammals of North America: biology, management, and conservation*. Batimore, Maryland: The Johns Hopkins University Press.

Bancroft, W.J., Hill, D. and Roberts, J.D. (2004). A new method for calculating volume of excavated burrows: the geomorphic impact of wedge-tailed shearwater burrows on Rottnest Island. *Functional Ecology*, 18(5): 752–759.

Barth, C.J., Liebig, M.A., Hendrickson, J.R., Sedivec, K.K. and Halvorson, G. (2014). Soil change induced by prairie dogs across three ecological sites. *Soil Science Society of America Journal*, 78(6): 2054–2060.

Belnap, J., Rosentreter, R., Leonard, S., Kaltenecker, J.H., Williams, J. and Eldridge, D. (2001). *Biological soil crust: ecology and management* (pp. 1–118). Service, U S Department of the Interior; Bureau of Land Management; U.S. Geological Service.

Berentsen, A.R. and Salmon, T.P. (2001). The structure of California ground squirrel burrows: control implications. *Transactions of the Western Section of the Wildlife Society*, 37: 66–70.

Best, T.L. (1972). Mound development by a pioneer population of the banner-tailed kangaroo rat, *Dipodomys spectabilis baileyi* Goldman, in eastern New Mexico. *American Midland Naturalist*, 87(1): 201–206.

Best, T.L. (1973). Ecological separation of three genera of pocket gophers (Geomyidae). *Ecology*, 54(6): 1311–1319.

Best, T.L. (1988). *Dipodoms spectabilis*. The American Society of Mammalogists, 311: 1–10.

Best, T.L., Intress, C. and Shull, K.D. (1988). Mound structure in three taxa of mexican kangaroo rats (*Dipodomys spectabilis cratodon, D. s. zygomaticus* and *D. nelsoni*). *American Midland Naturalist*, 119(1): 216–220.

Beven, K. and Germann, P. (1982). Macropores and water flow in soils. *Water Resources Research*, 18(5): 1311–1325.

Bihr, K.J. and Smith, R.J. (1998). Location, structure, and contents of burrows of *Spermophilus lateralis* and *Tamias minimus*, two ground-dwelling sciurids. *The Southwestern Naturalist*, 43(3): 352–362.

Black, T.A. and Montgomery, D.R. (1991). Sediment transport by burrowing mammals, Marin County, California. *Earth Surface Processes and Landforms*, 16, 163–172.

Blumstein, D.T., Daniel, J.C. and Bryant, A.A. (2001). Anti-predator behavior of Vancouver Island marmots: using congeners to evaluate abilities of a critically endangered mammal. *Ethology*, 107, 1–14.

Breshears, D.D. and Nyhan, J.W. (2005). Ecohydrology monitoring and excavation of semiarid landfill covers a decade after installation. *Vadose Zone Journal*, 4, 798–810.

Brotherson, J.D. and Rushforth, S.R. (1983). Influence of cryptogamic crusts on moisture relationships of soils in Navajo National Monument, Arizona. *The Great Basin Naturalist*, 43(1): 73–78.

Burnett, M.R., August, P.V., Brown, J.H. Jr. and Killingbeck, K.T. (1998). The influence of geomorphological heterogeneity on biodiversity: I. A patch-scale perspective. *Conservation Biology*, 12(2): 363–370.

Burt, W.H. (1936). Notes on the habits of the mohave ground squirrel. *Journal of Mammalogy*, 17(3): 221–224.

Butler, D.R. (2006). Human-induced changes in animal populations and distributions, and the subsequent effects on fluvial systems. *Geomorphology*, 79(3–4): 448–459.

Ceballos, G., Pacheco, J. and List, R. (1999). Influence of prairie dogs (*Cynomys ludovicianus*) on habitat heterogeneity and mammalian diversity in Mexico. *Journal of Arid Environments*, 41, 161–172.

Cejas, C.M., Hough, L.A., Castaing, J.C., Fretigny, C. and Dreyfus, R. (2014). Simple analytical model of evapotranspiration in the presence of roots. *Physical Review E*, 90, 042716

Chapin, F.S., Zavaleta, E.S., Eviner, V.T., Naylor, R.L., Vitousek, P.M., Reynolds, H.L., et al. (2000). Consequences of changing biodiversity. *Nature*, 405(6783): 234–242.

Chaves, M.M., Maroco, J.P. and Pereira, J.S. (2003). Understanding plant responses to drought — from genes to the whole plant. *Functional Plant Biology*, 30(3): 239–264.

Chew, R.W. and Whitford, W.G. (1992). A long-term positive effect of kangaroo rats (Dipodomys spectabilis) on creosotebushes (*Larrea tridentata*). *Journal of Arid Environments*, 22, 375–386.

Chihuahuan Desert Ecoregional Assessment Team. (2004). *Ecoregional Conservation Assessment of the Chihuahuan Desert for Pronatura Noreste* (pp. 1–102). Second Edition, Revised July 2004 Copyright©, 2002, 2004 Pronatura Noreste, The Nature Conservancy and World Wildlife Fund.

Cincotta, R.P. (1999). Note on mound architecture of the black-tailed prairie dog. *Great Basin Naturalist*, 49, 621–623.

Clark, T.W. (1971). Notes on white-tailed prarie dog (*Cynomys leucurus*) burrows. *The Great Basin Naturalist*, 31(3): 115–124.

Comstock, J.P. and Ehleringer, J.R. (1992). Plant adaptation in the Great Basin and Colorado Plateau. *Great Basin Naturalist Memoirs*, 52(3): 195–215.

Cooke, L.A. and Swiecki, S.R. (1992). Structure of a white-tailed prairie dog burrow. *Western North American Naturalist*, 52(3): 288–289.

Criddle, S. (1939). The thirteen-striped ground squirrel in Manitoba. *The Canadian Field-Naturalist*, 53, 1–6.

Davey, C.A., Redmond, K.T. and Simeral, D.B. (2006). Weather and climate inventory, National Park Service, Northern Colorado Plateau Network. Natural Resource Technical Report NPS/NCPN/NRTR— 2006/002. National Park Service, Fort Collins, Colorado.

Davey, C.A., Redmond, K.T. and Simeral, D.B. (2007a). *Weather and climate inventory, National Park Service, Chihuahuan Desert Network*. Natural Resource Technical Report NPS/CHDN/NRTR—2007/034. National Park Service, Fort Collins, Colorado.

Davey, C.A., Redmond, K.T. and Simeral, D.B. (2007b). *Weather and climate inventory, National Park Service, Mojave Desert Network*. Natural Resource Technical Report NPS/MOJN/NRTR—2007/007. National Park Service, Fort Collins, Colorado.

Davey, C.A., Redmond, K.T. and Simeral, D.B. (2007c). *Weather and climate inventory National Park Service Sonoran Desert Network*. Natural Resource Technical Report NPS/SODN/NRTR—2007/044. National Park Service, Fort Collins, Colorado.

Davidson, A.D. and Lightfoot, D.C. (2006). Keystone Rodent Interactions: Prairie Dogs and Kangaroo Rats Structure the Biotic Composition of a Desertified Grassland. *Ecography*, 29(5): 755–765.

Davidson, A.D. and Lightfoot, D.C. (2008). Burrowing rodents increase landscape heterogeneity in a desert grassland. *Journal of Arid Environments*, 72(7): 1133–1145.

Davidson, A.D., Detling, J.K. and Brown, J.H. (2012). Ecological roles and conservation challenges of social, burrowing, herbivorous mammals in the world's grasslands. *Frontiers in Ecology and the Environment*, 10(9): 477–486.

Davidson, A.D., Friggens, M.T., Shoemaker, K.T., Hayes, C.L., Erz, J. and Duran, R. (2014). Population dynamics of reintroduced gunnison's prairie dogs in the southern portion of their range. *The Journal of Wildlife Management*, 78(3): 429–439.

Davis, W.B. and Schmidly, D.J. (1994). The Mammals of Texas—online edition. University of Texas Press. Retrieved from http://www.nsrl.ttu.edu/tmot1/Default.htm.

Desha, P.G. (1966). Observations on the burrow utilization of the thirteen-lined ground squirrel. *The Southwestern Naturalist*, 11(3): 408–410.

Dregne, H.E. (1984). North American deserts. In F. El-Baz (Ed.), *Deserts and Arid Lands* (pp. 1–12). Lexington, Mass., U.S.A.: Litton industries, inc.

Edelman, A.J. (2011). Kangaroo rats remodel burrows in response to seasonal changes in environmental conditions. *Ethology*, 117, 430–439.

Egoscue, H.J. and Frank, E.S. (1984). Burrowing and denning habits of a captive colony of the Utah prairie dog. *The Great Basin Naturalist*, 44(3): 495–498.

Eldridge, D.J., Whitford, W.G. and Duval, B.D. (2009). Animal Disturbances Promote Shrub Maintenance in a Desertified Grassland. *Journal of Ecology*, 97(6): 1302–1310.

Ellison, L. (1946). The pocket gopher in relation to soil erosion on Moutain Range. *Ecology*, 27(2): 101–114.

Feldhamer, G., Thompson, B. and Chapman, J. (Eds.). (2003). *Wild mammals of North America: Biology, management, and conservation* (2nd ed.). Baltimore, Md.: Johns Hopkins University Press.

Fischer, C., Roscher, C., Jensen, B., Eisenhauer, N., Baade, J., Attinger, S., et al. (2014). How do earthworms, soil texture and plant composition affect infiltration along an experimental plant diversity gradient in grassland? *PLoS ONE*, 9(2): e98987.

Frase, B.A. and Hoffmann, R.S. (1980). *Marmota flaviventris*. The American Society of Mammalogists, 135, 1–8.

Gabet, E.J. (2000). Gopher bioturbation: field evidence for non-linear hillslope diffusion. *Earth Surface Processes and Landforms*, (13): 1419–1428.

Gabet, E.J., Reichman, O.J. and Seabloom, E.W. (2003). The effects of bioturbation on soil processes and sediment transport. *Annual Review of Earth and Planetary Sciences*, 31, 249–273.

Gedeon, C.I., Drickamer, L.C. and Sanchez-meador, A.J. (2012). Importance of burrow-entrance mounds of Gunnison's prairie dogs (*Cynomys gunnisoni*) for vigilance and mixing of soil. *The Southwestern Naturalist*, 57(1): 100–104.

Grant, W.E., French, N.R. and L J Folse, S.J.S. (1980). Effects of pocket gopher mounds on plant production in shortgrass prairie ecosystems. *The Southwestern Naturalist*, 25(2): 215–224.

Greene, R.A. and Murphy, G.H. (1932). The Influence of two burrowing rodents, *Dipodomys spectabilis spectabilis* (kangaroo rat) and *Neotoma albigula albigula* (pack rat), on desert doils in Arizona. II Physical effects. *Ecology*, 13(4): 359–363.

Greenwood, P., Kuonen, S., Fister, W. and Kuhn, N.J. (2015). The influence of terracettes on the surface hydrology of steep-sloping and subalpine environments: some preliminary findings. *Geographica Helvetica* 70, 63–73.

Grinnell, J. and Dixon, J. (1918). *Natural history of the ground squirrels of California*. Sacramento: California State Print. Office.

Guo, Q. (1996). Effects of bannertail kangaroo rat mounds on small-scale plant community structure. *Oecologia*, 106(2): 247–256.

Hakonson, T.E. (1999). The effects of pocket gopher burrowing on water balance and erosion from landfill covers. *Journal of Environmental Quality*, 28, 659–665.

Hatt, R.T. (1927). Notes on the ground-squirrel, Callospermophilus. *Occasional Papers of the Museum of Zoology*, 185, 1–24.

Hereford, R., Webb, R.H. and Graham, S. (2002). *Precipitation History of the Colorado Plateau Region, 1900–2000*. (Hendley J.W., II and Stauffer, P.H., Eds.) (pp. 1–4). U.S. Department of the Interior; U.S. Geological Survey.

Hereford, R., Webb, R.H. and Longpre, C.I. (2004). *Precipitation History of the Mojave Desert Region, 1893–2001*. (Hendley A.W. II, Ed.) (pp. 1–4). Service, U S Department of the Interior; U.S. Geological Survey.

Hickman, G.C. (1977). Burrow system structure of *Pappogeomys castanops* (Geomyidae) in Lubbock County, Texas. *American Midland Naturalist*, 97(1): 50–58.

Hobbs, R.J. and Mooney, H.A. (1985). Community and population dynamics of serpentine grassland annuals in relation to gopher disturbance. *Oecologia*, 67(3): 342–351.

Hobbs, R.J. and Mooney, H.A. (1991). Effects of rainfall variability and gopher disturbance on serpentine annual grassland dynamics. *Ecology*, 72(1): 59–68.

Holdenried, R. (1957). Natural history of the bannertail kangaroo rat in New Mexico. *Journal of Mammalogy*, 38(3): 330–350.

Hole, F.D. (1981). Effects of animals on soil. *Geoderma*, 25, 75–112.

Hoogland, J.L. (1995). The Black-Tailed Prairie Dog. Chicago & London: University of Chicago Press.

Hoogland, J.L. (1996). *Cynomys ludovicianus*. *Mammalian Species*, (535): 1–10.

Hoogland, J.L. (2003). Black-tailed Prairie Dog. In Feldhamer, A.G., Thompson, B.C. and Chapman J.A., (Eds.), *Wild Mammals of North America: Biology, Management, and Conservation*. Batimore, Maryland: The Johns Hopkins University Press.

Hoogland, J.L., Angell, D.K., Daley, J.G. and Radcliffe, M.C. (1987). Demography and population dynamics of prairie dogs. *Great Plains Wildlife Damage Control Workshop Proceedings*, 18–22.

Hoogland, L.J. (2006). Demography and population dynamics of prairie dogs. In Hoogland J.L., (Ed.), *Conservation of the black- tailed prairie dog: saving North America's western grasslands* (pp. 27–52). Washington, D.C., USA: Island Press.

Hornbeck, H., Crokus, B., Gladding, E. and Childs, A. (2011). *Field sampling of biotic turbation of soils at the Clive Site, Tooele County, Utah* (pp. 1–39). SWCA Environmental Consultants.

Howell, A.H. (1938). Revision of the North American ground squirrels, with a Classification of the North American Sciuridae (Vol. 56). Washington, D.C.: North America Fauna Series.

Huennke, L.F. and Noble, I. (1996). Ecosystem function of biodiversity in arid ecosystems. In Mooney, H.A., Cushman, J.H., Medina, E., Sala, O.E. and Schulze E.-D., (Eds.), Functional Roles of Biodiversity: A Global Perspective (pp. 99–128). Chichester & New York: Wiley-Blackwell.

Inouye, R.S., Huntly, N. and Wasley, G.A. (1997). Effects of pocket gophers (*Geomys bursarius*) on microtopographic variation. *Journal of Mammalogy*, 78(4): 1144–1148.

IUCN (2015). *The IUCN red list of threatened species*. Version 2015–3. http://www.iucnredlist.org. Downloaded on 12 October 2015.

James, A.I., Eldridge, D.J., Koen, T.B. and Whitford, W.G. (2008). Landscape position moderates how ant nests affect hydrology and soil chemistry across a Chihuahuan Desert watershed. *Landscape Ecology*, 961–975.

Jenny, H. (1994). Factors of soil formation. A system of quantitative pedology (pp. 1–191). New York: Dover Publications.

Johnson, G.E. (1917). The habits of the thirteen lined ground squirrel (*Citellus tridecemlineatus*), with special reference to the burrows. *The Quarterly Journal of the University of North Dakota*, 7, 261–271.

Johnson, W.C. and Collinge, S.K. (2004). Landscape effects on black-tailed prairie dog colonies. *Biological Conservation*, 115, 487–497.

Jones, C.G., Lawton, J.H. and Shachak, M. (1994). Organisms as ecosystem engineers. *Oikos*, 69(3): 373–386.

Jones, K.B. (1986). Deserts. In Cooperrider, A.Y., Boyd, R.J. and Stuart H.R., (Eds.), Inventory and Monitoring of Wildlife Habitat (p. 858). Washington D.C.: U.S. Department of the Interior, Bureau of Land Management.

Juelson, T.C. (1970). *A study of the ecology and ethology of the rock squirrel, Spermophilus variegatus (Erxleben) in northern Utah* (Doctoral dissertation). University of Utah, Salt Lake City.

Kelt, D.A. (2011). Comparative ecology of desert small mammals: a selective review of the past 30 years. *Journal of Mammalogy*, 92(6): 1158–1178.

Kerley, G., Whitford, W.G. and Kay, F.R. (2004). Effects of pocket gophers on desert soils and vegetation. *Journal of Arid Environments*, 58, 155–166.

Koford, C.B. (1958). Prairie dogs, whitefaces, and blue grama. *Wildlife Monographs*, (3): 3–78.

Laity, J. (2009). Deserts of the world. In Deserts and Desert Environments (1st ed., pp. 14–43). Hoboken, NJ, USA: Wiley-Blackwell

Lange, O.L., Belnap, J. and Reichenberger, H. (1998). Photosynthesis of the cyanobacterial soil-crust lichen *Collema tenax* from arid lands in Southern Utah, USA: role of water content on light and temperature responses of CO2 exchange. *Functional Ecology*, 12(2): 195–202.

Laundré, J.W. (1993). Effects of small mammal burrows on water infiltration in a cool desert environment. *Oecologia*, 94(1): 43–48.

Laundré, J.W. and Reynolds, T.D. (1993). Effects of soil structure on burrow characteristics of five small mammal species. *The Great Basin Naturalist*, 53(4): 358–366.

Laundré, J.W. (1989). Horitzontal and vertical diameter of burrows of 5 small mammal speceis in southeastern Idaho. *Great Basin Naturalist*, 49(4): 646–649.

Laundré, J.W. (1998). Effect of ground squirrel burrows on plant productivity in a cool desert environment. *Journal of Range Management*, 51(6): 638–643.

Laycock, W.A. (1958). The initial pattern of revegetation of pocket gopher mounds. *Ecology*, 39(2): 346–351.

Léonard, J. and Rajot, J.L. (2001). Influence of termites on runoff and infiltration: quantification and analysis. *Geoderma*, 104, 17–40.

Lomolino, M.V. and Smith, G.A. (2001). Dynamic biogeography of prairie dog (*Cynomys ludovicianus*) towns near the edge of their range. *Journal of Mammalogy*, 82(4): 937–945.

Long, K. (2002). Prairie Dogs: A Wildlife Handbook. Boulder, CO: Johnson Publishing Company.

Mace, J.E., Graham, R.C. and Amrhein, C. (1997). Anthropogenic lead distribution in rodent-affected and undisturbed soils in Southern California. *Soil Science*, 162(1): 46.

Malloy, R., Brock, J., Floyd, A., Livingston, M. and Webb, R.H. (Eds.). (2013). Design with the desert: conservation and sustainable development. Boca Raton, London & New York: CRC Press.

Martínez-Estévez, L., Balvanera, P., Pacheco, J. and Ceballos, G. (2013). Prairie dog decline reduces the supply of ecosystem services and leads to desertification of semiarid grasslands. *PLoS ONE*, 8(10): e75229.

Middleton, N. and Thomas, D. (1997). World atlas of desertfication. United Nations Environment Programme (2nd ed.). New York, NY: John Willey & Sons.

Mielke, H.W. (1977). Mound building by pocket gophers (Geomyidae): their impact on soils and vegetation in North America. *Journal of Biogeography*, 4(2): 171–180.

Miller, M.A. (1957). Burrows of the Sacramento Valley pocket gopher in flood-irrigated alfalfa fields. *Hilgardia*, 26(8): 431–452.

Montgomery, D.R. and Dietrich, W.E. (1995). Hydrologic processes in a low-gradient source area. *Water Resources Research*, 31(1): 1–10.

Moorhead, D.L., Fisher, F.M. and Whitford, W.G. (1988). Cover of spring annuals on nitrogen-rich kangaroo rat mounds in a Chihuahuan Desert grassland. *American Midland Naturalist*, 120(2): 443–447.

Moroka, N., Beck, R.F. and Pieper, R.D. (1982). Impact of burrowing activity of the bannertail kangaroo rat on Southern New Mexico desert rangelands. *Journal of Range Management*, 35(6): 707–710.

Moses, M.R., Frey, J.K. and Roemer, G.W. (2012). Elevated surface temperature depresses survival of banner-tailed kangaroo rats: will climate change cook a desert icon? *Oecologia*, 168(1): 257–268.

Mun, H.T. and Whitford, W.G. (1990). Factors affecting annual plants assemblages on banner-tailed kangaroo rat mounds. *Journal of Arid Environments*, 18, 165–173.

Nachlinger, J., Sochi, K., Comer, P., Kittel, G. and Dorfman, D. (2001). *Great Basin: An Ecoregion-based Conservation Blueprint. The Nature Conservancy, Reno, NV* (pp. 1–397).

Oliver, G.V. and Wright, A.L. (2010). The banner-tailed kangaroo rat, *Dipodomys spectabilis* (Rodentia: Heteromyidae), in Utah. *Western North American Naturalist*, 70(4): 562–566.

Ortega, J.C. (1987). Den site selection by the rock squirrel (*Spermophilus variegatus*) in southeastern Arizona. *Journal of Mammalogy*, 68(4): 792–798.

Paton, T.R., Humphreys, G.S. and Mitchell, P.B. (1995). Soils: a new global view. New Haven & London: Yale University Press.

Phillips, S. and Wentworth Comus, P. (Eds.). (2000). *A natural history of the Sonoran Desert* (1st ed.). Tucson, Arizona: Arizona-Sonora Desert Museum Press & University of California Press.

Plaster, B.R. (2003). *The geomorphic impacts of marmots in Gothic, Colorado*. Master's Thesis, Texas State University, San Marcos, Texas.

Price, L.W. (1971). Geomorphic Effect of the Arctic Ground Squirrel in an Alpine Environment. *Geografiska Annaler. Series A Physical Geography*, 53(2): 100–106.

Prugh, L.R., Stoner, C.J., Epps, C.W., Bean, W.T., Ripple, W.J., Laliberte, A.S. and Brashares, J.S. (2009). The rise of the mesopredator. *BioScience*, 59(9): 779–791.

Reed, S.E. (2013, March 31). *Pedologic-biologic feedbacks on the Merced River chronosequence: The role of pocket gophers (Thomomys bottae) in Mima mound-vernal pool ecosystems of the San Joaquin Valley* (Doctoral dissertation). University of California, Berkeley.

Reichman, O.J. and Seabloom, E.W. (2002). The role of pocket gophers as subterranean ecosystem engineers. *Trends in Ecology & Evolution*, 17(1): 44–49.

Reichman, O.J., Whitham, T.G. and Ruffner, G.A. (1982). Adaptive geometry of burrow spacing in two pocket gopher populations. *Ecology*, 63(3): 687–695.

Reichman, O.J., Wicklow, D.T. and Rebar, C. (1985). Ecological and mycological characteristics of caches in the mounds of *Dipodomys spectabilis*. *Journal of Mammalogy*, 66(4): 643–651.

Reynolds, J.F., Smith, D.M.S., Lambin, E.F., B.L.T.S.I., Mortimore, M., Batterbury, S.P.J., et al. (2007). Global desertification: building a science for dryland development. *Science*, 316, 847–851.

Reynolds, T.D. and Wakkinen, W.L. (1987). Characteristics of the burrows of four species of rodents in undisturbed soils in southeastern Idaho. *American Midland Naturalist*, 118(2): 245–250.

Ryckman, R.E. (1971). Plague vector studies. Part II. The role of climatic factors in determining seasonal fluctuations of flea species associated with the Caifornia Ground Squirrel. *Journal of Medical Entomology*, 8(5): 541–549.

Sabo, J.L. and Soykan, C.U. (2006). Riparian zones increase regional richness by supporting different, not more, species: reply. *Ecology*, 87(8): 2128–2131.

Sabo, J.L., Sponseller, R., Dixon, M., Gade, K., Harms, T., Heffernan, J., et al. (2005). Riparian zones increase regional species richness by harboring different, not more, species. *Ecology*, 86(1): 56–62.

Sayre, A.P. (1998). North America. Brookfield, Connecticut: Twenty-First Century Books.

Schaik, L., Palm, J., Klaus, J., Zehe, E. and Schröder, B. (2014). Linking spatial earthworm distribution to macropore numbers and hydrological effectiveness. *Ecohydrology*, 7: 401–408.

Schmidt, R.J. (1986). Chihuahuan climate. *Second Symposium on Resources of the Chihuahuan Desert Region. United States and Mexico. 20–21 October 1983*, 40–63.

Schooley, R.L., Bestelmeyer, B.T. and Kelly, J.F. (2000). Influence of small-scale disturbances by kangaroo rats on Chihuahuan Desert ants. *Oecologia*, 125(1): 142–149

Schroder, G.D. and Geluso, K.N. (1975). Spatial distribution of *Dipodomys spectabilis* mounds. *Journal of Mammalogy*, 56(2): 363–368.

Seabloom, E.W., Reichman, O.J. and Gabet, E.J. (2000). The effect of hillslope angle on pocket gopher (*Thomomys bottae*) burrow geometry. *Oecologia*, 125(1): 26–34.

Seager, R., Ting, M., Held, I., Kushnir, Y., Lu, J., Vecchi, G., et al. (2007). Model projections of an imminent transition to a more arid climate in southwestern North America. *Science*, 316, 1181–1184.

Shafer, D.S., Young, M.H., Zitzer, S.F. and Caldwell, T.G. (2007). Impacts of interrelated biotic and abiotic processes during the past 125000 years of landscape evolution in the Northern Mojave Desert, Nevada, USA. *Journal of Arid Environments*, 69, 633–657.

Sheets, R.G., Linder, R.L. and Dahlgren, R.B. (1971). Burrow Systems of Prairie Dogs in South Dakota. *Journal of Mammalogy*, 52(2): 451–453.

Shreve, F. (1942). The Desert Vegetation of North America. *Botanical Review*, 8(4): 195–246.

Smith, D.J. and Gardner, J.S. (1985). Geomorphic Effects of Ground Squirrels in the Mount Rae Area, Canadian Rocky Mountains. *Arctic and Alpine Research*, 17(2): 205–210.

Soholt, L., Griffin, M.R. and Byers, L. (1975). The influence of digging rodents on primary production in Rock Valley. *US International Biological Program, Desert Biome, Utah State University, Logan, Utah Reports of 1974 Progress, Volume 3: Process Studies, RM 75–19*.

Soholt, L.F. and Irwin, W.K. 1976. The Influence of Digging Rodents on Primary Production in Rock Valley. U.S. International Biological Program, Desert Biome, Utah State University, Logan, Utah. Reports of 1975 Progress, Volume 3: Process Studies, RM 76–18

Streubel, D.P. and Fitzgerald, J.P. (1978). *Spermophilus spilosoma*. *Mammalian Species*, 101, 1–4.

Stromberg, J.C., Beauchamp, V.B., Dixon, M.D., Lite, S.J. and Paradzick, C. (2007). Importance of low-flow and high-flow characteristics to restoration of riparian vegetation along rivers in arid south-western United States. *Freshwater Biology*, 52(4): 651–679.

Stromberg, J.C., McCluney, K.E., Dixon, M.D. and Meixner, T. (2013). Dryland riparian ecosystems in the American Southwest: sensitivity and resilience to climatic extremes. *Ecosystems*, 16(3): 411–415.

Sullivan, J. (2015). *Dipodomys ordii*. Retrieved October 9, 2015, from http://www.fs.fed.us/database/feis/ [2015, October 9]

Svendsen, G.E. (1976). Structure and Location of burrows of yellow-bellied marmot. *The Southwestern Naturalist*, 20(4): 487–494.

Swanson, M.L., Kondolf, G.M. and Boison, P.J. (1989). An example of rapid gully initiation and extension by subsurface erosion: coastal San Mateo County, California. *Geomorphology*, 2, 393–403.

Taylor, W.P. (1930). Methods of determining rodent pressure on the range. *Ecology*, 11(3): 523–542.

Thomas, L.P., Hendrie, M.N., Lauver, C.L., Monroe, S.A., Tancreto, N.J., Garman, S.L. and Miller, M.E.. (2006). *Vital Signs Monitoring Plan for the Southern Colorado Plateau Network*. Natural Resource Report NPS/SCPN/NRR-2006/002. National Park Service, Fort Collins, Colorado.

Thorn, C.E. (1978). A preliminary assessment of the geomorphic role of pocket gophers in the alpine zone of the Colorado Front Range. *Geografiska Annaler. Series A Physical Geography*, 60(3/4): 181–187.

Thorp, J. (1949). Effects of certain animals that live in soils. *The Scientific Monthly*, 68(3): 180–191.

Titus, J.H., Nowak, R.S. and Smith, S.D. (2002). Soil resource heterogeneity in the Mojave Desert. *Journal of Arid Environments*, 52(3): 269–292.

Tuba, Z., Csintalan, Z. and Proctor, M.C.F. (1996). Photosynthetic responses of a moss, *Tortula ruralis, ssp. ruralis*, and the lichens *Cladonia convoluta* and *C. furcata* to water deficit and short periods of desiccation, and their ecophysiological significance: a baseline study at present-day CO2 concentration. *New Phytologist*, 133(2): 353–361.

Underwood, J.G. and Van Pelt, W.E. (2008). A proposal to reestablish the black-tailed prairie dog (*Cynomys ludovicianus*) to southern Arizona. Nongame and Endangered Wildlife Program Draft Technical Report. Arizona Game and Fish Department, Phoenix, Arizona.

Van de Graaf, K.M. and Balda, R.P. (1973). Importance of green vegetation for reproduction in the kangaroo rat, Dipodomys merriami merriami. *Journal of Mammalogy*, 54(2): 509–512.

Verdolin, J.L., Lewis, K. and Slobodchikoff, C.N. (2008). Morphology of burrow systems: a comparison of gunnison's (*Cynomys gunnisoni*), white-tailed (*C. leucurus*), black-tailed (*C. ludovicianus*), and Utah (*C. parvidens*) prairie dogs. *The Southwestern Naturalist*, 53(2): 201–207.

Vleck, D. (1981). Burrow Structure and foraging costs in the fossorial rodent, *Thomomys bottae*. *Oecologia*, 49(3): 391–396.

Vorhies, C.T. and Taylor, W.P. (1922). Life history of the kangaroo rat, *Dipodomys spectabilis spectabilis* Merriam. *U.S. Department of Agriculture Bulletin*, 1091, 1–40.

Wagner, D.M. and Drickamer, L.C. (2004). Abiotic habitat correlates of gunnison's prairie dog in Arizona. *The Journal of Wildlife Management*, 68(1): 188–197.

Wallace, R.E. (1991). Ground-squirrel mounds and related patterned ground along the san andreas fault in central California. Report 91–149. US Department of the Interior. US Geological Survey.

Walsberg, G.E. (2000). Small mammals in hot deserts: some generalizations revisited. *BioScience*, 50(2): 109–120.

Ward, R.C. (1975). Principles of hydrology. London: McGraw-Hill.

Warner, T.T. (2004). The climates of the world deserts. In *Desert Meteorology* (pp. 63–166). Cambridge, UK: Cambridge University Press.

Westbrook, C.J., Veatch, W. and Morrison, A. (2013). Ecohydrology bearings - Invited commentary Is ecohydrology missing much of the zoo? *Ecohydrology*, 6(1): 1–7.

Whicker, A.D. and Detling, J.K. (1988). Ecological consequences of prairie dog disturbances. *BioScience*, 38(11): 778–785.

White, J.A. (2009). Summer burrows of Ord's kangaroo rats (*Dipodomys ordii*) in western Nebraska: food content and structure. *Western North American Naturalist*, 69(4): 469–474.

Whitford, W.G. and Kay, F.R. (1999). Biopedturbation by mammals in deserts: a review. *Journal of Arid Environments*, 41(2): 203–230.

Wilcomb, M.J.J. (1954, July 13). *A study of prairie dog burrow systems and the ecology of their arthropod inhabitants in Central Oklahoma* (Doctoral dissertation). University of Oklahoma, Norman, Oklahoma.

Wilkinson, M.T., Richards, P.J. and Humphreys, G.S. (2009). Breaking ground: Pedological, geological, and ecological implications of soil bioturbation. *Earth-Science Reviews*, 97, 257–272.

Williams, B.M. and Houseman, G.R. (2014). Experimental evidence that soil heterogeneity enhances plant diversity during community assembly. *Journal of Plant Ecology*, 7(5): 461–469.

Witwicki, D. (2013). Climate Monitoring in the Northern Colorado Plateau Network: annual report 2011. Natural Resource Technical Report NPS/NCPN/NRTR—2013/664. National Park Service, Fort Collins, Colorado.

Yensen, E. and Sherman, P.W. (2003). Ground squirrels. In Feldhamer, A.G., Thompson, B.C. and Chapman J.A. (Eds.), *Wild Mammals of North America: Biology, Management, and Conservation* (pp. 211–231). Batimore, Maryland: The Johns Hopkins University Press.

Young, P.J. (1990). Structure, location and availability of hibernacula of Columbian ground squirrels (Spermophilus columbianus). *American Midland Naturalist*, 123(2): 357–364.

Yurkewycz, R.P., Bishop, J.G., Crisafulli, C.M., Harrison, J.A. and Gill, R.A. (2014). Gopher mounds decrease nutrient cycling rates and increase adjacent vegetation in volcanic primary succession. *Oecologia*, 176(4): 1135–1150.

Zaitlin, B., Hayashi, M. and Clapperton, J. (2007). Distribution of northern pocket gopher burrows, and effects on earthworms and infiltration in a prairie landscape in Alberta, Canada. *Applied Soil Ecology*, 37, 88–94.

Water use and groundwater management

Chapter 15

Implications of spatially neutral groundwater management: Water use and sustainability in the Tucson basin

Violeta Cabello
Department of Human Geography, University of Seville, Spain

Nuria Hernández-Mora
Department of Human Geography, University of Seville, Spain

Aleix Serrat-Capdevila
UMI iGLOBES CNRS/Department of Hydrology and Atmospheric Sciences, University of Arizona, USA

Leandro Del Moral
Department of Human Geography, University of Seville, Spain

Edward F. Curley
Regional Wastewater Reclamation Department, Pima County, Arizona, USA

INTRODUCTION

Arizona has developed strong regulatory mechanisms to ensure long-term sustainable water use, and to integrate land and water use planning for the most populated areas (*Jacobs, 2009*). The sustainability objective in Arizona's water policy is based on the concept of "safe yield"; i.e., that the extraction of groundwater on a basin-wide and long-term basis is no more than is naturally and artificially recharged. This concept has been criticized by hydrologists, because it can be interpreted as implying that by achieving a balance between recharge and pumping there will be no detrimental impact on the aquifers and their dependent systems (*Zhou, 2009*). As a sustainability objective, the concept of safe yield may be considered as rather reductionist because it refers exclusively to the flows in and out of an aquifer, without taking into account other hydrogeological, socioeconomic and ecological criteria. Further, although limited, safe-yield as a management goal is nevertheless challenging to both implement and evaluate.

Until the arrival, in 1992, of Colorado River water through the Central Arizona Project (CAP) (see supra chapter 6), the city of Tucson and surrounding municipalities depended solely on groundwater for their water supply. As in other rapidly growing areas of Arizona, intensive groundwater pumping resulted in significant decreases in groundwater levels and land subsidence. Approval of the Groundwater Management Act (1980), and the resulting transformation of the institutional context for water management in Arizona, introduced changes in the way groundwater was managed and used in the Tucson basin. These included restrictions in water use patterns for municipal, industrial and agricultural users, through binding conservation programs. The arrival of CAP water brought a new water source to the region that helped to substitute for diminishing groundwater resources. A recharge and recovery program was created to manage the new "renewable resources"[1] that came with the CAP, thereby allowing the region to optimize water allocation by storing large volumes of Colorado River water underground, in overexploited aquifers.

The Tucson basin is now recognized as a reference for its conservation practices to curb demand and its innovative groundwater management system (*Jacobs & Holway, 2004; Megdal et al., 2014*). However, these practices are not exempt from critical assessment, since the techno-social fixes they present avoid facing the core challenge of uncontrolled urban growth head-on (*Hirt et al., 2008; Akhter et al., 2010*). To our knowledge, two elements of Tucson's water management system have not yet been evaluated: a) the impacts of water conservation programs on overall demand, and b) the spatial dynamics of the groundwater management system.

This chapter reviews the state of the art of current debates around sustainability objectives in Arizona water policy, with a focus on the Tucson basin area. The review was undertaken via a dialogue between water researchers and managers from Arizona and Spain, two different regions where the hydraulic paradigm has dominated water management practice (*Reisner, 1993; Sauri & Del Moral, 2001*). We analyze available data on water use and groundwater management, and compare it with other socioeconomic and environmental variables in order to provide insights into the limitations and challenges of current strategies to achieve safe yield. Specifically, we examine three relevant questions identified in collaboration with local stakeholders:

1. How has the water metabolism evolved since the approval of the GMA and the arrival of the CAP to the Tucson Basin?
2. Is water demand decreasing as a result of conservation programs?
3. How does the spatially neutral approach to groundwater management shape vulnerabilities in the socio-hydrological system?

This research uses a quantitative approach to the analysis of sustainability that builds on the concept of societal metabolism (*Giampietro et al., 2009, 2011, and 2014*) and is complemented by a thorough review of the academic literature and water planning reports, interviews with local experts, and participant observation of water planning meetings. The investigation was conducted in two phases, between February and July of 2013, and between November 2014 and March 2015. While a deeper understanding of the debate around sustainability in water governance in Arizona would require additional analysis of power relations (see *supra* chapter 7), the insights we gained can contribute to the discussion of ongoing and future water management challenges in the state.

The chapter is organized into five sections. First, we describe the institutional context for water management in Arizona. Then we introduce the conceptual framework

---

[1] The Arizona water community uses the term "renewable resources" to refer to the inflow of Colorado River water through the CAP. However, the consideration of Colorado water as renewable is questionable given the serious impacts that this interbasin transfer, coupled with all the other ones that the Colorado suffers, causes in the donor river basin, the severe drought-related variability of water availability, the uncertainty surrounding climate change predictions and the amount of energy required to pump Colorado water all the way to the Tucson basin.

Implications of spatially neutral groundwater management

and the methodology used. Section 4 discusses the results structured as i) A historical perspective on water use and planning; ii) A description of the evolution of the societal metabolism of water after the arrival of CAP; iii) A discussion of the interplay between conservation programs and water demand; and iv) A spatial analysis of groundwater dynamics. We conclude with a discussion of the effectiveness of current water management strategies to cope with long-term and spatially equitable[2] sustainability.

1 CHARACTERISTICS OF THE TUCSON BASIN

The Tucson basin is constituted by two wide alluvial valleys, bounded by mountain ranges, in which the city of Tucson (Pima County) is located. The basin overlies the interconnected aquifers of the Avra Valley and the Santa Cruz River (*Figure 1*), and this delimitation is used by for water planning by the Arizona Department of Water

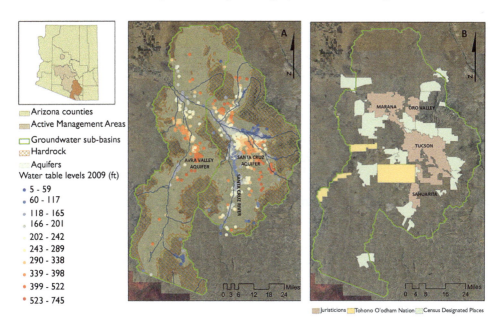

Figure 1 A – Tucson basin location and groundwater levels. B – Urban areas.

---

2 Equity implies a social or political consensus about the 'fairness' or 'justice' of the distribution of costs and benefits of a policy or program. Yet achieving a consensus concerning the fairness of a particular distribution is almost impossible. Thus, equity is a complex and value-laden concept (Truelove, 1992). However, the notion of 'spatial equity' enjoys a long tradition in spatial planning practice. In a physical sense, spatial equity can be understood as the equitable development of land use. In a socio-economic sense it can refer to the equitable flow of goods and services from one spatial arena to another. In both senses, spatial equity is a parameter for sustainable development and can be defined as both a process and an outcome. As process, it involves the redistribution of the overall resources and development opportunities and/or the optimization of locally existing resources and development opportunities of an area. As an outcome, it envisions a region or area where such redistribution or optimization is achieved and sustained (*Buhangin, 2013; Kunzmann, 1998*).

Resources (ADWR), which established the Tucson basin as a management unit (the Tucson Active Management Area, or TAMA), via the 1980 Groundwater Management Act. The Santa Cruz River used to flow in a Southeastern-Northwestern direction, as did the groundwater flow of the underlying aquifer, until aquifer overdraft caused the water table to drop and the river to dry up during the second half of the 20th century. Most of the runoff and aquifer recharge originates from higher precipitation rates along the mountain front during both winter rainfall and monsoon summer storms. Ephemeral channel recharge from storms in the basin can also be significant. After Phoenix, the TAMA is the second most populated region in Arizona, with a total population of one million people distributed over four main urban areas (City of Tucson, and the towns of Marana, Oro Valley and Sahuarita), other urban sprawl areas (Census Designated Places) and parts of the Tohono O'odham Nation.

2 INSTITUTIONAL CONTEXT FOR WATER MANAGEMENT IN ARIZONA

The evolution of water law and management in Arizona has been characterized by an ongoing effort to augment water supplies to support unconstrained economic and population growth (*Waterstone, 1992; Akhter et al., 2010*). The institutional context for water management consists of a complex system of regulations, norms, agencies and public and private operators that have evolved over time in response to changing socioeconomic, political and technological realities.

Groundwater use in Arizona was largely unregulated until the approval (in 1980) of the Groundwater Management Act (GMA) (*Gastelum, 2012*), while surface water law is governed by the prior appropriations doctrine. Before 1980, groundwater abstractions were only limited by the reasonable use doctrine (*Jacobs, 2009*). Starting in the 1940s, strong socioeconomic and population growth resulted in significant aquifer overdraft and land subsidence. By the 1970s it was clear that something had to be done to regulate groundwater pumping. In 1976 the Arizona legislature created a groundwater commission to write a groundwater law, but political resistance from agricultural users (who held a majority of groundwater rights) prevented any proposal from advancing. Negotiations finally succeeded when the Federal Government conditioned the approval of funding for the construction of the Central Arizona Project (CAP) to the passing of groundwater management rules in Arizona (*Akhter et al., 2010*).

The GMA designated four Active Management Areas (AMAs) in parts of the state where groundwater pumping was particularly intense around major urban and agricultural areas (see Map 4: 20). A groundwater management goal was established in each AMA, to be achieved by 2025 through the implementation of 5 consecutive Management Plans (MPs). The management goal for the Phoenix, Tucson and Prescott AMAs is to achieve safe yield. The goal for the Pinal AMA is to maintain the agricultural-based economy for as long as possible. In 1995 a portion of the Tucson AMA was separated out and became the Santa Cruz AMA. Its management goal is to maintain safe yield and prevent local water tables from experiencing long term declines.

Within the AMAs, existing groundwater uses prior to 1980 received a *"grandfathered right"*, and a moratorium on new irrigated agricultural land was imposed (*Megdal et al., 2014*). Management plans (MP) for each AMA established mandatory conservation goals for groundwater users that apply to most non-exempt wells (wells that pump in excess of 35 gallons/minute or 70.000 m^3/year) in the agricultural, industrial and municipal sectors (*Jacobs, 2009*). The GMA established clear guidelines for the first three MPs but was vague on the requirements for the 4th and 5th, given the uncertainties associated with such a long-term planning horizon. Finally, the GMA created the Arizona Department of Water Resources (ADWR), centralizing all quantity-related water management responsibilities.

The three first MPs (1985–1990, 1990–2000, and 2000–2010) followed specific guidelines established in the GMA. As of October 2015 (when this paper was completed) the IV MP had not yet been approved and the III MP's rules continue to apply. MPs are primarily regulatory documents establishing conservation programs for the different sectors (municipal, agricultural and industrial). They are not true management plans in the sense of roadmaps towards achieving objectives (*Megdal et al., 2008: 35*). Management per se is done by providers in a decentralized governance regime, without regional (basin scale) common planning over resources allocation.

The CAP is the primary source of renewable water supplies in central Arizona. Every year it delivers 1.6 MAF (1900 Mm^3) of Colorado River water to portions of the Phoenix, Pinal and Tucson AMAs (Prescott and Santa Cruz AMAs do not have access to CAP water), representing 57% of Arizona's 2.8 MAF entitlement of Colorado River water. The Central Arizona Water Conservation District (CAWCD) was created to manage and operate the CAP and generate the resources to repay the federal government for the investment. To help ensure long-term water supply, given that Arizona's CAP water entitlement exceeded instate demand, a groundwater recharge and storage system was devised to utilize Arizona's surplus water and firm its supply from Colorado River water. Those entities that recharge water get groundwater recovery credits for the future. There are two mechanisms for credit generation:

- Underground Storage Facilities (USFs): These are areas where CAP or reclaimed water is physically recharged, either through constructed injection wells or recharge basins, or other managed recharge mechanisms, by a diversity of private and public operators. This water can then be recovered (pumped) in the form known as CAP/reclaim recovered water.
- Groundwater Saving Facilities (GSFs): Also called in-lieu or indirect recharge, these are locations where CAP water or effluent is primarily used by irrigation districts instead of their irrigation groundwater rights. The surface water provider gets a groundwater credit for the amount of water that would have otherwise been pumped.

The program distinguishes between water stored for recovery in the same calendar year (recovered water or short-term credits) or in a later year (long-term storage credits). In the latter case, 5% of each acre-foot of CAP water recharged or not extracted is considered to be the *"cut to the aquifer"*, devoted to overdraft recovery. In the case of reclaimed water the cut to the aquifer is 50% if it is recharged via a managed facility, while reclaimed recharge from constructed facilities has no cuts.

Given the expectation that the municipal water sector would continue to grow, the Assured Water Supply (AWS) program was created to link water and land use planning (*Jacobs, 2009*). The draft rules set by the ADWR in 1988, that restrict allowable groundwater declines, encountered strong opposition from the development community, agricultural sector and cities without CAP access (*CAGRD, 2014: 17*). The outcome was the AWS program, a new rules package (approved in 1995) that requires all new urban developments to provide proof of physical, legal, and continuous access to a 100-year supply of water.

The Central Arizona Groundwater Replenishment District (CAGRD) was created in 1993 to facilitate municipal water users meeting the AWS rules. It encompasses the Phoenix, Tucson and Pinal AMAs. Membership in CAGRD allows landowners and water providers without access to CAP water or other renewable supply to use mined groundwater to prove AWS. Members pay the CAGRD to replenish any water pumped in excess of AWS rules. The CAGRD thus serves a double function of firming larger amounts of CAP water while at the same time facilitating development and growth in the AMA regions by ensuring 100 years of water supply to those municipal users outside CAP service areas. The CAGRD has priority over the recharge capacity of CAWCD sites (*CAGRD, 2014: 11*).

A final but important piece of the institutional puzzle for water management at the state level is the Arizona Water Banking Authority (AWBA), created in 1996 with the double purpose of allowing intrastate and interstate water banking and of facilitating the firming of Arizona's full Colorado water entitlement. Funding for the operation of the AWBA comes from a property tax on all real-estate owners in the 3 CAP counties (Maricopa, Pinal and Pima), and from a fee on groundwater pumping and state appropriations (*Megdal et al., 2014*). Until December 2013 AWBA had spent $207.9 million and stored 3897 MAF (4806.9 Mm^3) in long-term storage credits, the majority in Phoenix and Pinal AMAs (*AWBA, 2013*). AWBA does not hold rights and it does not operate a water market. It also does not own or operate storage facilities and is not responsible for recovering the water it stores—the CAP recovers the water in times of shortage (*Jacobs, 2009*). The target of the AWBA is to store up to 3.6 MAF (4493 Mm^3) to ensure long-term municipal uses in times of shortage (*AWBA, 2013*).

The ADWR's functions are mainly related to conservation programs, data collection, water accounting and information generation and technical support to regional water management processes within the AMAs (*ADWR, 2015b*). The GMA established Groundwater Users Advisory Councils (GUAC) in each of the AMAs to act as intermediaries between the multiple parties involved in the water management networks and the ADWR and AWBA. The Tucson AMA is an acknowledged example of active regional cooperation. Besides the GUAC, several initiatives have been undertaken in the last 15 years analyzing and promoting regional water policies. The Institutional and Policy Advisory Group (IPAG) was specifically formed to develop the recharge plan for the TAMA in 1995.[3] Recently, a new working group called the Safe Yield Task Force was created to coordinate efforts towards the achievement of the AMA's management goal.

3 http://www.azwater.gov/azdwr/WaterManagement/AMAs/TucsonAMA/TAMA_GUAC.htm.

3 METHODS

The objective of this chapter is to delve into the debates about sustainability of water management in the TAMA, focusing on three specific issues: 1) the changes in the water metabolism driven by the GMA and the arrival of CAP water to the TAMA; 2) the effects of conservation programs on water use; and 3) the spatial dynamics of groundwater management. For this purpose, the analysis is based on the theoretical and methodological framework provided by the Multi-Scale Analysis of Societal and Ecosystem Metabolism for water use analysis (*Giampietro et al., 2009; Madrid et al., 2013*). Time series data regarding the TAMA water budget are analyzed in relation to other socioeconomic variables and spatial information on groundwater management. These quantitative approaches are complemented by a review of the literature and planning reports, interviews, and participatory observation of water management meetings.

3.1 Societal metabolism

The concept of societal metabolism refers to the processes of appropriation, transformation and disposal of energy and materials to sustain socio-ecological systems (*Martinez-Alier & Schlüpmann, 1987; Giampietro et al., 2011*). These are understood as complex hierarchical systems operating at multiple levels of organization and different spatial and temporal scales (*Allen, 2008*). The functioning of such a system is investigated at three analytical levels: the whole social system extracting resources and disposing wastes (level n), the different sectors of the system among which resources are distributed (lower levels n−x), and the environmental context that provides services and is impacted by these activities (upper levels n+x). While ecosystem processes and integrity pose the external constraints for feasible societal metabolic patterns, the internal constraints are imposed by institutional rules and cultural values. These constraints show up as non-linear interactions between and within levels. When specifically focused on water use, the approach is known as water metabolism (*Madrid et al., 2013; Madrid & Giampietro, 2015*) and addresses the interplay between the water cycle (n+2), impacts on ecosystem and their water dependency (n+1) and society (n).

The methodological approach used is the Multiscale Integrated Analysis of Societal and Ecosystems Metabolism (MuSIASEM), an environmental accounting scheme applied for the water-energy-land-food nexus assessment (*Giampietro et al., 2014*). It builds on the flow-fund model of *Georgescu-Roegen (1971)* to generate multilevel matrices that contain and connect different types of variables. Fund variables are those that remain the same or that we want to conserve during the analytical timeframe; they describe the structure and size of the system. Flow variables are the resources used, or products generated, to maintain structural fund elements. Typical social fund variables are land used, human activity and infrastructures. Ecological funds are biodiversity, soils or hydrologic patterns.

While most environmental accounting schemes consider natural resources to be stocks, there is a fundamental difference between the treatment of funds and stocks in MuSIASEM, which differentiates between renewable and non-renewable resources (*Giampietro & Lomas, 2014*). A flow of water, energy or wood can come from a fund if it is extracted under renewability rates (like sustainable managed forestry or sustainable aquifer yield) or from a stock if it is depleting non-renewable resources at human scales like fossil fuels or aquifers reserve. Flows and funds are quantified in absolute terms (extensive variables) on a multi-level basis aggregating from lower levels (households, specific economic activities) to the whole social system. The combination of flows and funds variables generates indicators (flow/fund, fund/fund intensity ratios) that allow a comparison of metabolic patterns of resource use. The approach to the interphase of socio-ecological systems is twofold: on the one side quantitative, through the analysis of environmental impacts of resource extraction and waste disposal, and on the other side qualitative, through the analysis of the institutional rules and policies that shape these physical interactions (*Cabello et al., 2015*).

3.2 Application to the Tucson basin

The methodology was deployed in four steps. We first analyzed the evolution of water flows in the TAMA water budget, using a 25 year long data series for the period 1985 to 2009–10, disaggregated per source and sector for the whole basin. The series and combined water sources per sector were plotted in an interactive visualization type Icicle tree[4] using the Quadrigram software (*www.quadrigram.com*). *Table 1* shows the variables used and *Table 2* lists the data sources; all the graphs and tables presented in the results section were produced using data from these sources. We maintain the same nomenclature for water flows as for the water budget.

Next, to address structural changes that occurred since recharged CAP water began to be recovered, we analyzed the evolution of societal metabolism of water between 2000/01 and 2010/11. Our analysis included societal funds, land use and human activity, and water flows per end use sector. Land use and cover categories were aggregated from those of the 2001 and 2011 National Land Cover Databases. Human activity was calculated from the American Census demographic, economic and employment data for 2000 and 2010. Note that the methodology followed in both censuses is different, in that the former is an extensive one year inventory of the entire population while the latter contains the average variables of surveys to population samples during different years. Data for 2010 are averages of 5 years. Water uses per sector were averaged for the previous decade (1990–99 and 2000–09) in order to compare tendencies.

In the third stage, we analyzed the evolution of water conservation targets for the municipal and agricultural sectors. The different components of municipal demand were included in the water budget alongside the population served by these subcomponents (large municipal residential and none residential, small municipal and exempt wells). Gallons per capita per day were calculated by simple division of those variables. Agricultural demand was contrasted with precipitation and crop prices data. Precipitation time series for the weather station in the city of Tucson were obtained from the National Weather Service Forecast Office. Data for evolution of crop patterns and prices were obtained from the National Agricultural Statistics Service (available starting in 1996).

---

4 https://philogb.github.io/jit/static/v20/Jit/Examples/Icicle/example2.html.

Table 1 Water metabolism variables for the Tucson basin.

	Extensive variables	Unit	Description
Flows	Water sources	AF/mm³	
	CAP direct		Water from CAP that is directly used without previous recharge
	Groundwater in-lieu		Water from CAP that is used instead of pumping groundwater
	CAP recovered		Water pumped from aquifers in exchange of previously recharged CAP water
	Reclaimed		Wastewater effluent directly reused after treatment
	Reclaimed recovered		Water pumped from aquifers in exchange of previously recharged wastewater effluent
	Groundwater		Water pumped from aquifer
	Overdraft		Difference between total water pumped from aquifers and natural + artificial recharge. Calculated in the water budget on a basin wide basis
	Water use		Sum of total gross water use per each of the sectors
	Municipal		Water supplied by municipal providers for residential and non-residential use. It is composed by large provider's residential, large non-residential (Other urban services), lost and unaccounted, small providers, exempt wells and deliveries to individual. Exempt wells are estimated as 1 AF of annual demand per every four wells
	Mining		Water withdraw by mines
	Other economic sectors		Water used by economic sectors outside the municipal supply network: dairy and feedlot; sand and gravel extraction; electric power generation; golf and turf facilities; other
	Agriculture		Water used by agricultural sector
	Indian nations		Water used by Tohono D'Oham nation and Pascua Yaqui tribes
Funds	Human activity	Hours	Population in a given year per 365 days per 24 hours
	Households		Hours of non-paid activities, calculated as the difference between paid work hours and total human activity. The required data to disaggregate this sector are the Time Use Surveys which are only available in the United States at the national level but not at the state level.
	Paid Work		Hours employed in paid work activities. Calculated as the sum of employment in each sector per average

(Continued)

Table 1 (Continued)

	Extensive variables	Unit	Description
	Land uses and covers	Miles/acres/ hectares	
	Forest		Sum of deciduous and evergreen forest surface categories of the National Land Cover Databased (NLCD)
	Shrubs		Shrub category of the NLCD
	Water bodies		Sum of water bodies, woody wetlands and herbaceous wetlands of the NLCD
	Barren land		Barren land category of the NLCD – mines area
	Cattle grassland		Sum of grassland and pastures categories of the NLCD
	Mining		Digitalized over orthophoto 2014
	Urban		Sum of high, medium and low density and open space categories of the NLCD
	Crops		Crop category of the NLCD
	Intensive variables		
Fund/fund	Employment	%	Hours in each economic sector out of total working hours in a year
	Dependency ratio	%	Hours of unpaid activities (households) out of total hours in a year
	Land occupation ratio	%	Land employed in productive human activities out of total land minus hard rock (not available land)
	Housing units density	Housing number/mile^2	Number of houses per land unit
Flow/fund	Income per capita	$/capita	Gross income per capita in a year
	Gallons per capita day	Gallons/cap*day	Municipal daily water demand divided by total population served
	Water use density	Acrefeet/acre	Water use per acre of land used
	Water use intensity	Gallon/hour	Water use per hour of total human activity
	Crop prices	$/lb	Annual price of agricultural commodities received by farmers

Finally, we conducted a spatial analysis of groundwater management. The analysis considered available GIS data for groundwater recharge and recovery sites, location of groundwater users and the changes in aquifer levels between 2000 and 2010. The latter were interpolated from point measurements via Inverse Distance Weighting using ArcGIS 10.1. Long-term groundwater storage credit data for each recharge area was only available for the AWBA credits. The long-term storage credits held by other institutions (about 50% of all long term credits) were inferred by combining the ADWR total accounting per owner updated in February 2015 (*ADWR, 2015a*), the annual status report of the TAMA recharge plan (*ADWR, 2007*) and data from CAP recharge sites (*CAP, 2015*). Being based on a series of assumptions, the estimates cannot be considered to be fully accurate, but can be deemed sufficiently good for the purpose of establishing a spatial reference regarding where the water is being stored.

Table 2 Data sources.

Data Type	Sources	Links (Accessed February 2015)
Rainfall	National Weather Service Forecast Office	http://www.wrh.noaa.gov/twc/climate/reports.php
Shallow groundwater areas	Pima Association of Goverments	http://gismaps.pagnet.org/subbasins/#/MapUser
Water table levels	Pima Association of Goverments	http://gismaps.pagnet.org/subbasins/#/MapUser
Wells inventory	Arizona Water Resources Department	https://gisweb.azwater.gov/waterresourcedata/WellRegistry.aspx
Artificial recharge	Arizona Water Resources Department	http://gisdata.azwater.opendata.arcgis.com/
Long-Term Storage credits	Arizona Water Banking Authority	http://www.azwaterbank.gov/Ledger/defaultIntrastate.aspx
	Arizona Water Resources Department	http://www.azwater.gov/azdwr/WaterManagement/Recharge/default.htm
	Central Arizona Project	http://www.cap-az.com/index.php/departments/recharge-program
Water accounting areas	Pima Association of Goverments	http://gismaps.pagnet.org/subbasins/#/MapUser
Water budget	Arizona Water Resources Department	http://www.azwater.gov/AzDWR/Watermanagement/AMAs/TucsonAMA/TAMAOverview.htm#waterbudget
Demography, housing, income & employment	American Census FactFinder	http://factfinder.census.gov/faces/nav/jsf/pages/searchresults.xhtml?refresh=t#
Land covers	Multi-Resolution Land Characteristics Consortium	http://www.mrlc.gov/
Crops and prices	National Agricultural Statistics Service	http://www.nass.usda.gov/Statistics_by_Subject/index.php?sector=CROPS

3.3 Collaborative research and participant observation

Our interest in the research questions addressed by this chapter arose from a series of interactions with local stakeholders in regards to water issues in the Tucson basin. Our work is situated in a constructivist context to the perspective known as post-normal science (*Ravetz and Funtowicz, 1993*). We consider that sustainability science must pay especial attention to the question *who reframes scientific questions?* (*Filardi, 2015*). For that reason, we proceeded to design this work in an iterative manner. In the first phase (February-July 2013) we conducted a preliminary literature review, and an interview with a local water manager allowed us to frame a draft set of scientific questions that were presented, reframed and prioritized in a participatory workshop in April 2013 with participation of University of Arizona experts and local stakeholders. Key issues identified were:

- The effect that changes in the socioeconomic structure have over water demand.
- The effectiveness of TAMA Management Plans for achieving safe yield by 2025.

- The impact of the groundwater credit system on the present and future dynamics of the water budget in the Tucson Basin.
- The impact of groundwater dynamics on biodiversity conservation.

The bulk of the research was then conducted between November 2014 and March 2015, during which time we attended two regional water management meetings as participant observers – the Safe Yield Task Force meeting on January 23rd and the Groundwater Users Advisory Committee on February 28th, 2015 – where discussions were held regarding how regional planning is moving forward to face identified management challenges. Preliminary results were also discussed and validated with local stakeholders.

4 RESULTS

4.1 Evolution of water use

In this section we explore the evolution of the TAMA as a socio-hydrological system since the approval of the GMA, linking changes in the institutional context to those in water use. The information presented here is based on a thorough review of water planning reports (*ADWR 1999, 2008, and 2010a; AWBA, 2013, 2014; Megdal et al., 2008; and TAMA, 1998*) in combination with data from the last update of the TAMA water budget until 2010. The data presented in *Figure 2*, using the Icicle visualization, illustrate the evolution of the different sources of water used in the whole Tucson basin (big upper square) and per sector (four small lower squares) in 1990, 2000 and 2009 (different colors are used for each water source). In addition, *Figures 3* and *4* show the temporal evolution of the data.

1980–1990: Responding to challenges. While the CAP was being constructed, the first TAMA MP boosted water conservation programs by setting conservation goals for each sector. A target of 140 gallons per capita day (GPCD) (530 liters per capita per day) was set for the municipal sector. The Base Conservation Program (BCP) approved for the agricultural sector established groundwater allotments based on irrigation efficiency targets,[5] water duties[6] and water duty acres for the reference period of 1975 to 1979. Specific programs were developed for each type of industrial use permit. Mandatory water use reporting requirements were set and water accounting started in 1985. As illustrated in *Figure 2a*, all sectors relied almost exclusively on groundwater during this period, with the exception of some reclaimed water used by the municipal and agricultural sectors. Indian nations represented a small share of total water demand (1%) while mining was already relevant (*Figure 4*). The municipal sector had already become the biggest water consumer, steadily growing from 41 to 48% of total water demand during this period, while agriculture fell from 42 to 32% of overall water demand as a result of the gradual reduction in irrigated acres (see *Figure 3*).

---

5 Efficiency defined as final water uptake per water delivered.
6 Calculated for each farm unit as irrigation requirements divided by total acres planted from 1975 to 1979 and multiplied by irrigation efficiency target.

Implications of spatially neutral groundwater management 303

Figure 2 Sources of water used for the TAMA (upper half of the figure) and per sector (lower half) in 1990 (A), 2000 (B) and 2009 (C).

1990–2000: Adapting. CAP water arrived to Tucson in 1992 (*Figure 3*). One of the main objectives of the 2nd MP was to overcome legal, institutional and structural barriers for utilization of new supplies from CAP and reclaimed water (*Megdal et al., 2008: 90–91*). During this period, most of the laws, programs and institutions in place

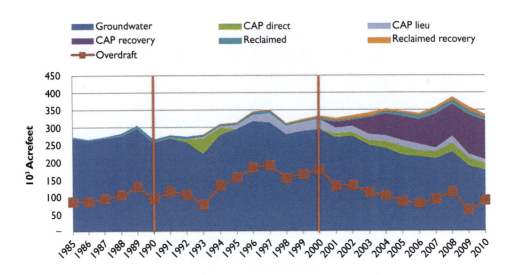

Figure 3 Evolution of water use per source and groundwater overdraft.

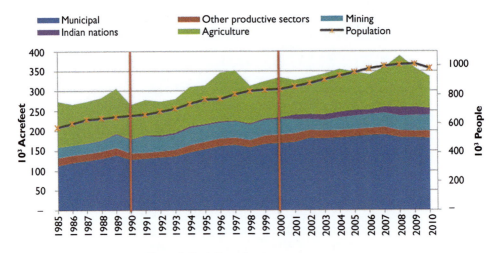

Figure 4 Evolution of water use per sector.

to firm CAP water (for instance AWBA or CAGRD) were created as described in section 2.2. In the TAMA, the regional recharge plan was enacted as a new device for achievement of the safe yield goal by storing excess CAP water underground (*IPAG, 1998*). While the second MP renewed conservation programs, it also introduced flexibility measures in both the agricultural sector – to facilitate adaptation to the evolution of market for agricultural products —, and in the municipal sector for small providers who had encountered difficulties achieving the 140 GPCD target. For agriculture, a highly controversial efficiency target of 85% was set during this period. In addition, farmers who did not use their entire groundwater allotment in one year were allowed to "bank" this water as "flexibility credits" for future recovery (*Fleck, 2013*).

The city of Tucson started using CAP water for municipal supply in 1993. It was treated to drinking standards and delivered through a water distribution system that had only conveyed groundwater in the past. Due to the different chemical composition and pH of the CAP water, it dissolved and re-mobilized mineral concretions that had accumulated inside the pipes over the years, resulting in unappealing brown water coming out of the taps. The consumer protests that ensued led to abandonment CAP water for direct municipal use after less than 2 years. Tucson reverted to groundwater use while alternative solutions were being developed to enable indirect use of the CAP water for the city's water supply.

Groundwater use by the mining sector increased significantly in 1991 to 8449 AF (10 Mm^3), remaining constant for the rest of the decade. According to the TAMA water budget, the groundwater in-lieu program started in 1992, rerouting direct CAP use to agricultural production (albeit not in a significant share until 1998), in exchange for the accumulation of long-term storage credits (LTSC). Municipal providers subsidized the cost of part of this CAP water to farmers accruing the generated long-term credits in exchange for municipal groundwater pumping for residential water supply. The result of all these parallel processes was that the annual overdraft of groundwater diminished in 1993 but began increasing again a year later to peak at 189,916 AF (154 Mm^3) in 1997 (*Figure 3*).

2000–2010: Complexifying. The 3rd MP inaugurated the decade of groundwater storage and recovery. Between 2001 and 2010 seven different sources of water were used in the Tucson AMA (see *Figures 2–4* and *Table 1* for explanation): groundwater, direct use of CAP, CAP in lieu, CAP recovered, reclaimed, reclaimed recovered as well as small quantities of surface water or low quality groundwater. While all sectors diversified their sources of water, the greatest change observed throughout this period was in the municipal sector, which by 2009 was using 60% of recovered CAP water, along with water from five different other sources. The recharge infrastructure and institutional framework created in the previous decade enabled the increasing municipal demands to be met, while simultaneously replacing direct groundwater use with recovered CAP water, so that the annual groundwater overdraft started to decrease significantly (*Figure 3*). Another noteworthy change was the reallocation of CAP water to the Indian nations and tribes following the Arizona Water Settlements Act of 2004. As observed in *Figure 4*, the agricultural sector drives overall variability in demand and, in turn, the instability of annual groundwater withdrawals. In addition, conservation programs were substantially softened during the 3rd MP, substituting conservation targets with the Best Management Practices program (BMP) that tailors the improvements towards conservation to each end-user, instead of setting a common goal.

4.2 Evolution of societal metabolism

In this section, with the aim of widening the discussion from water flows to other relevant dimensions of sustainability, we compare two snapshots (for 2000 and 2010) of the societal metabolism of water in the Tucson basin. *Table 3* shows societal funds and moving average water flows for the two decades, alongside some metabolic indicators (intensive variables). Indian nation demand has been disaggregated and added to final subsectors (municipal, agriculture, and other economic sectors).

Table 3 Societal metabolism evolution during the 3rd MP.

		Land use (miles²) 2000	2010	Human activity (10⁶ hr) 2000	2010	Water use (10³ AFY) 2000	2009
n+2	Tucson basin	3871					
n+1	Forest	162	145				
	Shrubs	3235	3216				
	Water bodies	7	10				
	Barren land	17	16				
n	Land occupation	451	486	Total human activity	6810	7990	Gross water use	306	346
n−1				Paid Work	501	657	Economic sectors	197	209
n−2	Crops	42	43	Agriculture	1.4	2.3	Irrigation	97	110
	Grassland	52	53				Dairy & feedlot	0.07	0.1
	Mining	NA	50	Mining	2.5	4.4	Mining	39	34
				Building	38.7	40	Sand & gravel	4.1	3.9
				Manufacturing & Retail	140	163	Electric power	2.1	3.5
	Urban & developed	307	340	Real State & financial	29	35	Golf & turf facilities	7.4	8.4
				Other urban services	254	362	Other urban services	39	43.5
				Government & military	35	50	Other	7.2	5.3
n−1				Households	6308	7333	Residential	109	136
n	Land occupation ratio (%)	0.19	0.21	Dependency ratio (%)	93%	91%	Water use density AF/acre)	1.06	1.11
	Housing units density (houses/mile²)	1.0	1.2	Income ($/cap)	19,959	25,454	Water use intensity (Gallon/hour)	14.67	14.11

During this period, the land occupation ratio increased by two percent, driven mainly by the urbanization of shrubland areas, with an average annual growth ratio of 3.3%. In addition, the housing density rose from 1 to 1.2 houses per square mile. A significant fact is that the small surface area devoted to agriculture surpassed that allocated to large-scale mines. Conifer forested areas decreased by 11.7%, mostly in the Northwest Catalina peaks. A positive environmental change was the increase in surface area of water bodies by 40%, especially wetlands, partially because of the groundwater recharge sites but also due to riparian restoration projects.

In regards to human activity, the ratio of total working hours to total human activity increased despite increased unemployment in many urban areas, especially for those with lower incomes such as South Tucson, Summit, Three Points and Drexel Heights. This was compensated for by jobs generated in new urban areas, resulting in an overall employment rise of 13%. The economic model of Arizona has been based on the services sector coupled to urban growth (*Jacobs, 2009*). Indeed, the services sector grew more in terms of employment generation, particularly in education, health, professional science, recreation and food services. This unveils the role of the University of Arizona as an important economic driver for the region. In addition, Arizona is famous as being a destination for winter seasonal retirees who help to boost the services economy. The demographic evolution shows two clear trends: a process of ageing and a permanent domination of the group aged between 18 and 25. On the other hand, the building and real estate sectors lost importance in regards to fraction of the total economy, although both grew in absolute terms. Agriculture and mining are smaller, but yet increasing sectors. The overall income per capita increased by 27%.

Most water uses are positively correlated with the evolution of the employment pattern. For instance the sand and gravel water use decreased with the declining weight of the building sector in the overall economy. Main water use increases were observed in residential and urban economic activities (non-residential municipal), in parallel to the growth of the services sector and the expansion of urban areas. Mining is the only activity that grew in employment without mirroring increments in water flows, thus becoming more efficient per hour of human activity. On the other hand, agriculture augmented its average consumption by 13% during this decade. Overall water efficiency improved per hour but decreased per acre (from 2032 m^3/ha in 2000 to 3432 m^3/ha in 2010) linked to the process of densification of urban areas. From a sustainability perspective, it is important to point out that the TAMA water management system depends on two external resources:

i. Imports of practically 100% of food requirements since agricultural production is mainly devoted to cotton and cattle-feeding products.
ii. Low-cost energy from the Colorado dams, and the availability of the Navajo Generating Station for pumping CAP water and is lifting it 2900 feet from the Colorado to South Tucson city.

Regarding the latter, the CAP is the major single energy consumer in Arizona, with an annual consumption of 2.8 million megawatt-hours (*CAP, 2010*). Ninety percent of this electricity is supplied by the Navajo Generating Station coal-fired power plant in Page, which also supplies energy to the Tucson Electric Power Company. According to *Eden et al. (2011)*, the estimated energy intensity of CAP water when it reaches Tucson is 3,140 KWh/AF (2.54 KWh/m^3), which is four times larger than the average for groundwater pumping. Interestingly, the current (2014) rate for CAP water is only 140 $/AF (0.11 $/m^3), thanks to good energy efficiency management and the revenues obtained from sales of surplus NGS energy (*Eden et al., 2011*). As shown in *Table 3*, water used for electric power generation within the Tucson basin is a small but increasing share of the overall budget. Increasing regulations over emissions and shortage predictions in the Colorado River basin are pinpointed as vulnerabilities of the system to an increase in energy prices (*Cullom, 2014*).

4.3 Is water conservation curbing demand?

As described in section 4.1, the use of water conservation programs was a core management device during the first three MPs, because it was specifically required by the GMA. Nevertheless, MP goals and requirements have evolved towards increasing flexibility and adaptability for each individual end-user, to the point that their effectiveness is currently being questioned (*Megdal et al., 2008; Fleck 2013*). The general accepted view is that demand is decreasing because of a reduction in the GPCD in the municipal sector. In what follows we examine available data from the TAMA water budget. The data are given for entire sectors, and are only disaggregated for municipal demand into the categories shown in *Figure 5*. Data for agricultural uses only indicates overall demand and irrigable acres, but does not identify actually irrigated land. The problem with this data format is that it does not allow us to distinguish the effects of conservation programs on demand evolution from other drivers like climate, land use or market changes (*Megdal et al., 2008*).

As shown in *Figure 5*, 58% percent of municipal demand is residential, supplied by large water providers within what are called service areas. This demand grew continuously until 2002, whereupon it stabilized. From 2007 to 2009, overall large provider residential demand decreased by 1223 AF (1 Mm³), and the GPCD also decreased to 97 GPCD (370 lpcd) in 2009 (down from 122 GPCD in 1989). On the other hand, large-provider non-residential deliveries increased during the last decade, and lost and unaccounted municipal water uses remained stable. Small providers and exempt wells[7] are a very small share of the total municipal demand, but have very high GPCD (181 and 645 GPCD per capita in 2009 respectively). The significant decrease in overall demand between 2007 and 2009 comes from the removal of one category from the overall accounting: delivery to individual users that are described

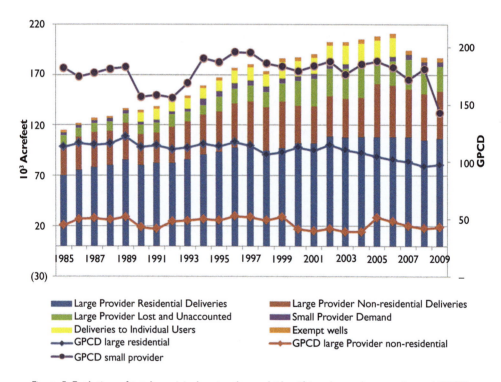

Figure 5 Evolution of total municipal water demand, identifying demand categories and GPCD.

7 Estimated as 1 AF of annual demand per every four wells.

as non-irrigation users with conservation requirements, including turf and cooling facilities. Between 2000 and 2009, the population in the TAMA region increased by 173,864 people, but decreased in 2010 (for the first time on record). The increase did not mirror increases in large-scale domestic demand. Updated data presented by the ADWR at the GUAC meeting[8] of February 2015 confirmed the decreasing tendency in domestic demand, both in absolute and relative terms.

The agricultural sector is a different and very complex reality. The GMA limited the possibility of increasing irrigable acres. Since 1995 these have remained relatively stable at around 36,200 acres (14,500 has, 1% of the total TAMA area), when 6210 acres of irrigation grandfather rights were bought by Tucson water and transformed into non-irrigation rights (*ADWR, 2015b*). There is no available data on actual irrigated acres per year per irrigation district, nor of the evolution of irrigation systems that could allow an assessment of the effects of conservation programs on agricultural demand. Average agricultural efficiency has increased from 50% to 80–90% as a result of the BMP program (*ADWR, 2015b*). Nonetheless, the literature is skeptical in regards to these results (*Wilson & Needham, 2006; Bautista et al., 2010*). A very generous water allotment from the beginning and the introduction of flexibility accounts are pointed out as primary causes for their ineffectiveness. According to these authors, conservation programs for the agricultural sector are so flexible that most farmers didn't even change to the supposedly more flexible BMP program but, rather, remained in the initial Base Conservation Program.

Wilson and Needham (2006) and *Fleck (2013)* show rather than the conservation programs of the GMA, it is commodity prices (especially for cotton and alfalfa, which are water intensive crops) and rain that are the main explanatory factors driving agricultural water demand variability in central Arizona. *Figures* 6 and 7 show the evolution of agricultural water use, precipitation and the prices of the three main crops planted in the Tucson basin (cotton, hay and wheat). Agricultural demand is highly variable on a year-to-year basis, but fluctuates around a rather stable average. Until 1998, demand had a negative correlation with precipitation (Pearson −0.63) but since then, this relation is

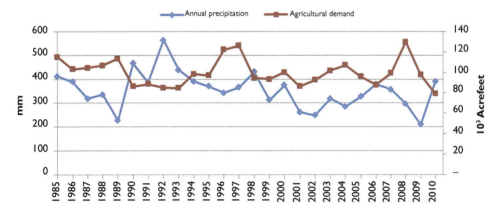

Figure 6 Agricultural demand and precipitation.

---

8 http://www.azwater.gov/azdwr/WaterManagement/AMAs/TucsonAMA/documents/FinalAgenda-TucsonAMAGUAC2.26.15.pdf.

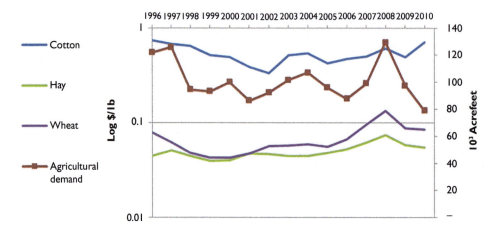

Figure 7 Agricultural demand and crop prices.

much less obvious. The 1996 Federal Agricultural and Improvement Reform Act decoupled crop prices and government subsidies from production, and increased planting flexibility (*Frisvold, 2007*). Separating out the composite effect of this legislation from the evolution of crop prices and precipitation would require an econometric model that is outside the scope of this work. Nevertheless, *Figures* 6 and 7 show that from 1996 onwards, the peaks in prices (especially for cotton) mirror peaks in water demand even when precipitation is not below the mean (Pearson 0.45 for cotton price, 0.3 for wheat, 0.44 for hay and −0.2 for precipitation). In 2008 peak water demand for the decade coincided with both lower precipitation and peak prices for all crops.

From the analysis in the previous sections we can conclude that:

i. Overall water demand trend in the Tucson basin has continued to increase over the past 25 years although the pace of increase has slowed by one third during the last decade (with respect to 1990–2000);
ii. Large municipal providers are making progress both in terms of cutting domestic demand as well as reducing groundwater overdraft;
iii. For the other water use sectors analyzed, conservation has not been very effective as a demand reduction strategy; and
iv. Agriculture, being highly affected by crop prices and precipitation, drives annual variability of overall Tucson basin demand and groundwater use.

The capacity to continue curbing demand in the future by increasing conservation is considered small (*Megdal, 2015; ADWR, 2015b*). Instead, the ADWR plans to turn the core management strategy for the forthcoming 4th MP to supporting regional cooperation towards achieving safe yield during the next 10 years (*ADWR, 2015b*).

4.4 A spatial assessment of groundwater management

Undoubtedly, the main management strategy for achieving the TAMA goal of safe yield is the substitution of groundwater overdraft by other resources. Taken together, the total volume of CAP water and wastewater is three times the groundwater available through natural recharge. From 1993 to 2009, an average of 53% of total artificial recharge was recovered annually for municipal and industrial users, 1.6% was lost through evaporation in recharge sites, 7.4% remained as cut to the aquifer, and the rest was stored as LTSC. The continuous increase of recharge capacity coupled with the renaming of most municipal groundwater withdrawals as recovered water, resulted in a technical achievement of safe yield on a basin-wide scale (SYTF, 2015). However, the spatial distribution of this achievement is not homogenous.

As depicted in *Figure 8A*, there are 12 USF sites in the Tucson AMA – 7 recharging reclaimed water and 5 recharging CAP water – plus 6 GSF located in agricultural sites. Most of the recharge occurs in the Avra Valley and Pima mine road CAWCD sites, and uses CAP water. Most of the recharge of effluent takes place north of Tucson city. Groundwater recovery is mostly done by Tucson Water in the area of influence of the Avra Valley (CAP) and Sweetwater (effluent) recharge sites and delivered to the city (*ADWR, 2010a: 52*). However, 90% of recovery and withdrawal wells are scattered throughout the municipal service area, with an important concentration in the large Mission and Sierrita Mine sites (located in southeastern Pima County), which are spatially disconnected from recharge areas (see *Figure 8A and B*).

Arizona statutes require that groundwater recovery for municipal providers be located either within a 1 mile of a USF site or in areas where groundwater decline is less than 4 ft/year (1.22 m/year). This limitation does not apply to those municipal users that join the CAGRD to meet the AWS requirements and can withdraw groundwater anywhere within their service or member land (ML) areas. This was seen by municipal providers to be a major equity problem in the region (*Megdal et al., 2008: 24*). Indeed, many of these providers have transferred their LTSCs to the CAGRD to enjoy the same advantages (*ADWR, 2010a: 55*). As observed in *Figure 7B*, the CAGRD service area embraces all municipal providers while new member lands have three hotspots in northwest Catalina Mountains, eastern Vail and south Green Valley, all primary development areas within the TAMA. In 2009, 50% of groundwater (not recovered) pumping for municipal use was allocated to new developments, 37% as groundwater allowed under the AWS rules and 13% as excess groundwater that has to be replenished by the CAGRD.

The last piece of this complex puzzle is the Long-Term Storage Credit system. The most recent update of credits accrued in 2014 showed a total of 1.4 M AF (1129 mm³, nearly four times total water demand in 2010), an increase of 80% since 2009 (see *Table 4*). During the last five years, the AWBA has been especially focused on recharge

Table 4 Water resources (AFY).

1908–2010	Funds	Precipitation (mm)	209–670
		Average	379
		Average natural recharge	81,964
2009	Flows	CAP inflow	197,289
		Reclamation	50,904
		Artificial recharge (CAP + reclaimed)	202,201
	Stocks	Recovery	124,118
		Long-Term Credits	798,844
		USF-CAP	630,545
		USF-Effluent	89,583
		GSF	78,716

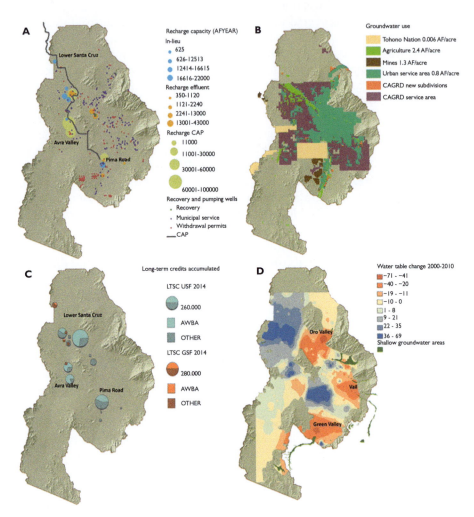

Figure 8 A - Recharge sites and capacity; B - location of water users; C - accrued LTSCs per site; D - evolution of groundwater levels between 2000 and 2010 (feet) and shallow groundwater areas.

within the Tucson basin, accounting for 50% of the total LTSC. Other major owners are Tucson Water (15.6%), CAGRD (8.6%), Tohono O'odham Nation (6.2%), the Bureau of Reclamation (5%) and the Rosemont mine company Augusta Corporation (3%) (*ADWR, 2015a*). In addition, there are 18 other entities owning less than 2% of the credits including small municipal providers (Marana, Oro Valley, Vail, Metrowater) and one irrigation district. As shown by *Figures 8C* and *D*, accumulation of credits has been responsible for the recovery of aquifer levels in Avra valley and along Pima mine road. The rate of annual recovery of LTSC is around 1%. These credits can be recovered from anywhere within an AMA as long as consistency with management plan goals is maintained, and the recovery is inside or within three miles of the service area of a municipal provider or irrigation district. The credits owned by

AWBA have the purpose of assisting municipal and industrial uses in case of shortage, meeting Indian water rights and fulfilling management goals. They have a specific recovery plan (*AWBA, 2014*).

There is no available spatial data online that provides an exact accounting of recovery and pumping. Nevertheless, water table levels are monitored and their evolution from 2000–2010 is displayed in Fig. 8 D.[9] It can be seen that the areas where groundwater credits are being accrued are those undergoing water table rises of up to 60 feet (18 m). Groundwater levels in the central part of the city of Tucson have also been rising, since the recovery in Avra Valley enabled Tucson Water to turn off its central well (that was driving the major cone of depression and land subsidence in the TAMA). On the other hand, few areas of water table decline remain. Peak declines of up to 71 feet (21.6 m) are observed in north-east Oro Valley area where the major use sector is urban. The second relevant drawdown area is the southern Green Valley where some of the largest mines coincide with new developments and a large irrigated area, all of which rely mainly on groundwater. In addition, the eastern area of Vail has experienced similar average decreases of 44 feet (13 m) in the last ten years. As can be seen in *Figure 8D*, the mountain ranges around the Santa Cruz valley are home to the largest riparian ecosystems in what are known as shallow groundwater areas (*SGWA, PAG, 2012*). These are sustained by natural recharge over high bedrock, but many connect to areas of the aquifer with declining levels. Within the Tucson basin there are 20,537 acres of SGWA connected to wider systems (*Figure 8 D*), 46% of which overlap with areas of the aquifer having declining levels. It is noteworthy that there have been very few areas showing declines over 40 feet during the ten years monitored and in which recovery was forbidden.

In 2013, the ADWR launched a public consultation regarding a proposal named Enhanced Aquifer Management (*ADWR, 2013*) that aimed to encourage groundwater recovery nearby recharge sites. It consisted on a calibration of percentage cuts to the aquifers depending on the distance to the recharge site: 0% within 1 mile buffer, 10% after 1 mile but within the AMA, 20% outside of the AMA. All comments to the proposal were negative arguing that any disincentive to use CAP water would turn users towards groundwater again, resulting in increased water costs to customers or negatively affecting the emerging LTSC market (*Brooks, 2013; Tucson Water, 2013*). Alternative proposals included limiting pumping in areas with declining groundwater levels, limiting the allowable declining rate, or setting a tax based on observation of impacts in declining areas (*Brooks, 2013*). The final outcome of the discussion was twofold: 1) a requirement to improve information, and 2) a proposal to construct more pipes to allow CAP water to be delivered to more areas within the TAMA. On one hand, the Safe Yield Task Force recently proposed subdividing the Tucson basin into seven water accounting areas (WAAs) as a tool to improve water planning (*ADWR, 2015b*). On the other hand, water providers are also working on cooperative Wheeling Programs with the aim of building the infrastructure required to deliver CAP water to all urban service areas experiencing declining water tables.[10]

---

9 The figure shows interpolated data for monitored wells between September 2009 and March 2010. For a detailed visualization of wells location and levels visit the interactive map of Pima Association of Government http://gismaps.pagnet.org/subbasins/#/MapUser.

10 http://www.azwater.gov/azdwr/WaterManagement/AMAs/documents/SAWUA_TW_EAMPresentation06042014.pdf.

5 DISCUSSION: GROWTH, SUSTAINABILITY AND SPATIALLY NEUTRAL GROUNDWATER MANAGEMENT

In this chapter, we have examined the evolution of water metabolism with particular focus on the changes induced by the arrival of CAP water to the TAMA, and with the aim of contributing to the debate regarding water management strategies to achieve sustainability objectives in the Tucson basin. The goal of safe yield imposed by the Groundwater Management Act has been pursued by a combination of i) reducing demand for existing uses through conservation practices (i.e. improving efficiency), ii) limiting the expansion of new demands and iii) bringing new resources to the region to substitute for the use of groundwater. Dissecting the effect of each of these strategies is a difficult task, since multiple interconnected layers of regulations have been overlaid during the past 30 years without a discrete assessment being carried out. Here, we have analyzed the available data and pinpointed limitations in information.

We have shown that construction of the CAP was a tipping point in the water metabolism of the area, in the sense that it brought about a drastic reconfiguration and diversification of water sources for the different sectors, while fueling the economy. This was enabled by increasing infrastructural and institutional complexity to make full use of what are deemed renewable resources from the Colorado River. Infrastructural complexity was deployed through a system of new facilities for recharge and storage, and by constructing new wells and pipelines to transport recovered water to the denser urbanized Tucson area. Institutional complexity was achieved through a series of new laws, programs, institutions and cooperative agreements that multiplied the decision-making nodes of a decentralized governance network.

Regarding the control of water demand, we have shown that, despite population growth, large municipal providers have managed to stabilize urban demand by reducing demand per capita. Therefore, if not reducing overall demand, at least the sector is now balancing savings against new demand. Other municipal components do not seem to be making significant progress and the apparent slight reductions in total municipal demand are mainly due to a change in accounting rules. Further, conservation programs for agriculture seem not to be having the foreseen impact. On an annual basis, irrigation demand varies about a rather stable average, driving peaks in both the total Tucson basin demand and groundwater pumping on dry years and/or periods of high commodity prices. Since 2000, the Indian Nations have become significant players in the overall budget. Total water demand in the Tucson basin has grown continuously, although a slowdown in the pace of growth was observed from 2000 to 2010, in comparison with the previous decade. CAP water has partially replaced groundwater withdrawals, therefore contributing to overdraft reduction.

With respect to growth limiting measures, the binding non-expansion rule for agriculture has been effective in controlling demand. Mines and other economic sectors have no limits imposed on their permits. The data indicate that mines have become more efficient in water use, but that their local impacts on water table levels are still very significant. Water uses are in general coupled to the evolution of the economic sectors with a clear predominance of urban services. The Achilles heel of Arizona water problems is that of limiting growth in the urban sector, since the dominant economic model is tied to urban expansion (*Akhter et al., 2010*). All attempts to set constraints regarding groundwater overdraft that might affect development have been systematically thwarted. From 2000 to 2010 the development sector lost weight in the economy, but this is perceived as associated with the volatility of the housing market after 2008. According to the CAGRD Operation Plan 2014, the annual rate of membership drastically dropped since 2009, and so did their replenishment obligations. Most land lots have not been built upon and current projections show construction increasing over the next 10 years and peaking in 2021. Coupled with this, municipal water demand is projected to grow until 2045 (*CAGRD, 2014: 49–51*) by nearly 29.000 AF (35 Mm^3) in the Tucson AMA. It is however the lowest of the projections for the three CAGRD AMAs.

The lack of spatial disaggregation of the water budget makes it difficult to assess the extent to which improvements in efficiency in some urban areas are enabling growth in others. What seems clear is that there is a disconnection between recharge and recovery in some areas and that local impacts on the water table are still important. The technical achievement of safe yield at a basin level is uneven and there are wide areas in which overdraft continues to occur, especially in new development locations. Larger biodiversity hotspots are dependent on shallow groundwater, and some of them are partially located over areas with declining aquifer levels.

The new category of *recovered water* enables continued mining of groundwater without being properly accounted for in the overdraft. A proper accounting should reflect which part of the recovered water is actually CAP, which is reclaimed water (for instance the water that Tucson Water transports from Avra Valley to the city), and which is not (all the water recovered outside the area of impact of the recharge site), and should split the accounting of safe yield into different sub-regions according to that. The WAAs project is a good step in this direction. The regional network for water governance is aware of the impacts of the ill-defined spatial management strategy and is negotiating solutions. While it was initially proposed to constrain recovery near recharge, it seems instead that the final bet is for bringing recharge close to recovery through an expansion of the CAP infrastructure to reach more areas within the TAMA. Some have argued this is a straightforward solution to the current depletion problems (*Tucson Water, 2013*), but at the same time this view may not properly account for the expected shortage of Colorado water acknowledged by CAP managers. The AWBA recovery scenarios until 2024 for M&I and Indian uses in the TAMA can be largely met with 66% of its actual storage (*AWBA, 2014: 46*). The main recovery mechanism that has been proposed is the exchange of short-term annual credits of municipal providers for LTSCs accumulated near recharge sites (*AWBA, 2014: 55*). Agriculture has low priority access to CAP water and thus it is the most vulnerable sector to potential Colorado water shortages. Nevertheless, it has grandfathered rights that could again increase the pressure on groundwater use. The AWBA recovery plan does not mention safe yield at all, and so far there is no assessment of how recovery of the different credits by other different owners would impact the management goal.

In conclusion, the problem of how to reconcile the positive and negative impacts of urban growth remains the eternally unresolved debate in the Tucson basin and in the American Southwest. Questions regarding potential physical, socioeconomic or environmental limits to growth are not even "on the table" in Arizona. Water scarcity imposes a key limiting factor on the current urban growth-based economic model. However, an increasingly sophisticated governance regime has been devised to try to overcome this limitation. Safe yield is a laudable management goal that has triggered important changes in the water metabolism. Yet, the discourse regarding CAP as a renewable resource, and the use of creative accounting devices veil an unequal distribution of impacts and vulnerabilities derived from the spatially neutral approach to groundwater management. How this spatial inequity will be resolved is likely to characterize the sustainability debate over the next ten years, when the GMA is due to be assessed. Achievement of safe yield might be possible in most areas if new pipes are constructed to deliver CAP water to those locations, as long as no severe shortage in the Colorado River occurs. Whether this is a resilient or a ceteris paribus strategy that increases vulnerability will be seen over the course of the next decade. Any prior hypothesis would require a much more detailed analysis of disaggregated spatial data of water uses and sources that is not available at the moment.

ACKNOWLEDGEMENTS

The authors would like to thank the SWAN project for funding this work, its coordinator Dr. Frank Poupeau and the team leaders for pursuing the development of this book and supporting the students group with diligence. We would also like to acknowledge the generous time and thoughts of Linda Stitzer, Claire Zucker and Rita Bodner to help reframe the questions we look at in this paper, and those of the interviewees that helped us to understand the great complexity of the Tucson basin water system. Claire Zucker, Mead Mier and John Pope from the Sustainable Environment Program of the Pima Association of Governments have greatly contributed to this research by facilitating required spatial data, the transparency of this institution is exemplary. In addition, Sharon Megdal made excellent comments to this paper that substantially increased its accuracy and readability. Finally we would like to thank all the students and researchers in the SWAN project for the fantastic intellectual and cultural exchange, their patience and openness towards collective learning of what interdisciplinary and transdisciplinary mean.

REFERENCES

Akhter, M., Ormerod, K.J. and Scott, C.A. (2010). Lost in translation: resilience, social agency, and water planning in tucson, arizona. *Critical planning*, 17: 46–65.

Allen, T.F.H. (2008). Hierarchy theory in ecology, in Sven Erik Jørgensen and Brian D. Fath (eds.) *Encyclopedia of Ecology*. Oxford: Academic Press: 1852–1857.

ADWR—Arizona Department of Water Resources (1999). *Third Management Plan for the Tucson Active Management Area, 2000–2010*. Accessed February 2015: http://www.azwater.gov.

ADWR—Arizona Department of Water Resources (2007). *Annual status report: underground water storage, savings and replenishment (recharge) program*. Accessed February 2015: http://www.azwater.gov/azdwr/WaterManagement/Recharge/default.htm.

ADWR—Arizona Department of Water Resources (2008). *Regional recharge plan. Tucson Active Management Area Institutional and Policy Advisory Group*. Accessed February 2015: http://www.azwater.gov/azdwr/WaterManagement/AMAs/TucsonAMA/documents/RRP_Ch1_and_Ack.pdf.

ADWR—Arizona Department of Water Resources (2010a). *Draft Demand and Supply Assessment 1985–2010. Tucson Active Management Area*. Accessed February 2015: http://www.azwater.gov/AzDWR/WaterManagement/Assessments/default.htm.

ADWR—Arizona Department of Water Resources (2010b). *Enhance aquifer management: alternative cut to the aquifer*. Accessed February 2015: http://www.azwater.gov/azdwr/WaterManagement/AMAs/EnhancedAquiferManagementStakeholderGroup.htm.

ADWR—Arizona Department of Water Resources (2015a). *Long Term Storage Account (LTSA) Summary*. Updated on February 23, 2015. Accessed February 2015: http://www.azwater.gov/azdwr/WaterManagement/Recharge/default.htm.

AWBA—Arizona Water Banking Authority (2013). *Annual plan of operation*. Accessed October 2015: http://www.azwaterbank.gov/Plans_and_Reports_Documents/documents/Final2013AnnualReport70114withletter.pdf.

AWBA—Arizona Water Banking Authority (2014). *Recovery of water stored by the AWBA. A Join plan of AWBA, ADWR and CAP. Draft*. Accessed February 2015: http://www.azwaterbank.gov/Plans_and_Reports_Documents/Annual_Plan_of_Operation.htm.

Bautista, E., Waller, P. and Roanhorse, A. (2010). *Evaluation of the Best Management Practices Agricultural Water Conservation Program*. Phoenix, AZ: Arizona Department of Water Resources.

Brooks, C. (2103). *ADWR Enhanced Aquifer Management Proposal* – Evaluation comment. Accessed February 2015: http://www.azwater.gov/azdwr/WaterManagement/AMAs/EnhancedAquiferManagementStakeholderGroup.htm.

Buhangin, J. (2013). *Spatial Equity: A Parameter for Sustainable Development In Indigenous Regions*. WIT Press: Southampton.

Cabello, V., Willaarts, B., Aguilar, M. and Del Moral, L. (2015). River basins as socio-ecological systems: linking levels of societal and ecosystem metabolism in a Mediterranean watershed. *Ecology and Society, 20*(3) art. 20.

CAGRD (Central Arizona Groundwater Replenishment District) (2014). Plan of Operation 2015. Draft. Available at: http://www.cagrd.com/index.php/operations/plan-of-operation.

CAP—Central Arizona Project (2010). The Navajo Generation Station Whitepaper. Quoted in: Background on Navajo Generating Station Impacts to Arizona in Anticipation of Proposed EPA Coal Plant Emissions. Accessed August 2015: https://energypolicy.asu.edu/wp-content/uploads/2012/03/NGS-brief-sheet.pdf.

CAP—Central Arizona Project (2015). Accounting of deliveries and LTSCs from Central Arizona Water Conservation District sites. Accessed February 2015: http://www.cap-az.com/departments/recharge-program.

Cory, D.C., Evans, M.E., Leones, J.P. and Wade, J.C. (1992). The role of agricultural groundwater conservation in achieving zero overdraft in Arizona. *Water Resources Bulletin* 28(5): 889–901.

Cullom, C. (2014). *Colorado River update: risks and vulnerabilities*. Presentation in Sustainable Water Action project visit to CAP facilities.

Eden, S., Scott, C.A., Lamberton, M.L. and Megdal, S.B. (2011). Water-energy interdependencies and the Central Arizona project. In D.S. Kenney and R. Wilkinson (Eds.) *The water-energy nexus in the American west*. Williston: Edward Elgar Publishing Inc., 109–122.

Filardi, C.A. (2015). *Place in the World—Science, Society, and Reframing the Questions We Ask*. 1st International Conference on Citizen Science. Opening plenary. San Jose, California. Accessed March 2015: http://citizenscienceassociation.org/2014/11/07/keynote-speaker-announced-for-citizen-science-2015/.

Fleck, B.E. (2013). *Factors affecting agricultural water use and sourcing in irrigation districts of central Arizona*. Msc thesis submitted to the faculty of the Department of Agricultural and Resource Economics. Accessed February 2015.

Frisvold, G.B., Wilson, P.N. and Needham, R. (2007). Implications of federal farm policy and state regulation on agricultural water use. In B.G. Colby, & K.L. Jacobs (Eds.) *Arizona Water Policy: Management Innovations in an Arid, Urbanizing Region*. Washington: Resources for the Future, 137–156.

Gastelum, J.R. (2012). Analysis of Arizona's water resources system. *International Journal of Water Resources Development,* 28(4): 615–628.

Georgescu-Roegen, N. (1971). *The Entropy Law and the Economic Process*. Cambridge: Harvard University Press.

Giampietro, M., Mayumi, K. and Ramos-Martin, J. (2009). Multi-scale integrated analysis of societal and ecosystem metabolism (MuSIASEM): Theoretical Concepts and Basic Rationale. *Energy*, 34(3): 313–322.

Giampietro, M., Mayumi, K. and Sorman, A. (2011). *The Metabolic Pattern of Societies: Where Economists Fall Short*. New York: Routledge.

Giampietro, M., Aspinall, R.J., Ramos-Martin, J. and Bukkens, S.G.F. (2014). *Resource Accounting for Sustainability Assessment: The Nexus between Energy, Food, Water and Land Use*. Routledge Explorations in Sustainability and Governance. New York: Routledge.

Giampietro, M. and Lomas, P. (2014). The Interface between Societal and Ecosystem Metabolism. In M. Giampietro, R.-M. Aspinall, S. Bukkens (Eds.) *Resource Accounting for Sustainability: The Nexus between Energy, Food, Water and Land*. New York: Routledge: 33–48.

Hirt, P., Gustafson, A. and Larson, K. (2008). The mirage in the Valley of the Sun. *Environmental History*, 13: 482–514.

IPAG—Tucson AMA Institutional and Policy Advisory Group (1998). Regional recharge plan. Available at: http://www.azwater.gov/azdwr/WaterManagement/Recharge/default.htm.

Jacobs, K. (2009). *Groundwater management issues and innovations in Arizona*. Unpublished work, available at: http://admin.cita-aragon.es/pub/documentos/documentos_JACOBS_2-4_26d10f50.pdf.

Jacobs, K. and Holway, J. (2004). Managing for sustainability in arid climates: Lessons learned from 20 years of groundwater management in Arizona, USA. *Hydrogeology Journal*, 12: 52–65.

Kunzmann, K.R. (1998). Planning for spatial equity in Europe. *International Planning Studies*, 31, 101–120.

Madrid, C. and Giampietro, M. (2015). The water metabolism of socio-ecological systems: reflections and a conceptual framework. *Journal of Industrial Ecology*, 19(4) (September issue, complete on proofs).

Madrid, C., Cabello, V. and Giampietro, M. (2013). Water-use sustainability in socioecological systems: A multiscale integrated approach. *Bioscience*, 63(1): 14–24.

Martinez-Alier J. and Schlüpmann, K. (1987). *Ecological economics: energy, environment, and society*. Oxford: Basil Blackwell.

Megdal, S.B., Smith, Z.A. and Lien, A.M. (2008). *Evolution and Evaluation of the Active Management Area Management Plans*. Report of the Water Resources Research Center.

Megdal, S.B., Dillon, P. and Seasholes, K. (2014). Water banks: Using managed aquifer recharge to meet water policy objectives. *Water*, 6: 1500–1514.

NWS-NOAA, National Weather Service Forecast Office, Tucson, Arizona. Accessed 4th February 2015: http://www.wrh.noaa.gov/twc/climate/tus.php. NOT IN THE TEXT

PAG—Pima Association of Governments (2012). *Shallow groundwater areas in Eastern Pima County*. Accessed February 2015: https://www.pagnet.org/documents/water/SGWAReport2012.pdf.

Ravetz, J.R. and Funtowicz, S. (1993). Science for the Post-Normal Age. *Futures*, 25(7): 739–755.

Reisner, M. (1993). *Cadillac Desert: The American West and Its Disappearing Water*. New York: Penguin Books.

Sauri, D. and Del Moral, L. (2001). Recent developments in Spanish water policy. Alternatives and conflicts at the end of the hydraulic age. *Geoforum*, 32: 351–361.

Tucson Water (2013). *ADWR Enhance Aquifer Management—Evaluation comment*. Accessed February 2015: http://www.azwater.gov/azdwr/WaterManagement/AMAs/EnhancedAquiferManagementStakeholderGroup.htm.

Truelove, M. (1992). Measurement of spatial equity. *Environment and Planning C: Government and Policy*, 11: 19–34.

Waterstones, M. (1992). Of dogs and tails: Water policy and social policy in Arizona. *Water Resources Bulletin. America Water Resources Association*, 28:3, 479–486.

Wilson, N. and Needham, R. (2006). *Groundwater Conservation Policy in Agriculture.* Contributed Paper, 6th Conference of the International Association of Agricultural Economists Gold Coast Convention & Exhibition Centre. Queensland, Australia.

Zhou, Y. (2009). A Critical Review of Groundwater Budget Myth, Safe Yield and Sustainability. *Journal of Hydrology, 370*(1–4): 207–213.

OTHER SOURCES

ADWR—Arizona Department of Water Resources (2015b). Interview with the author.

Megdal, Sharon. (2015). Interview with the author.

SYTF—Safe Yield Task Force. 2015. January meeting. Tucson Metropolitan Water District. Tucson, Arizona.

Chapter 16

Groundwater dynamics: How is Tucson affected by meteorological drought?

Natalia Limones
Department of Human Geography, University of Seville, Spain

INTRODUCTION: CONTEXT RATIONALE AND OBJECTIVES

Rainfall does not provide a constant source of water; it is characterized by remarkable inter-annual and seasonal variability. Drought is a temporary anomaly, more or less prolonged, during which rainfall is below normal, resulting in a significant reduction of water resources (*National Drought Mitigation Center, 2015*). Drought is a natural risk. It is particularly common and extreme in the semi-arid regions such as in Tucson, where aridity, the systematic lack of moisture and water, make it a natural feature that defines the local climate. Climate change is expected to increase the risk of drought because rising temperatures incrementally increase the evaporative potential of the air, which combined with erratic rainfall fosters less water availability and longer dry periods that may be very intense (*Limones, 2013*). Drought usually leads to water scarcity, but not necessarily, because it is possible to mitigate the negative environmental and social effects with proper planning and management tools (*WWF, 2012*). In order to understand the interaction of droughts and water shortages, it is necessary to analyze the variables that influence the social response to different aspects of the water cycle. Additionally it is necessary to take into account the whole hydro-social system to further develop potential prediction methods of physical and social system response.

The objective of this study is to start understanding hydrological drought in Tucson by analyzing the fluctuations of the water table. This main objective is broken down into the following specific goals to be accomplished:

- Analyze the existing connections and causalities between meteorological drought and drought reflected in water tables. A perfectly linear relationship between the two is not expected, instead it is expected that there will be different numbers of dry sequences, different durations, intensities, lags, and accumulation of effects.
- Link the results to management circumstances and other human influence affecting the basin to forecast a pattern of hydrological response to meteorological drought for each situation.

This study will be an initial evaluation. It is not intended to be exhaustive; rather it aims to perform a first exploration of the connection between the two phenomena in the study area as a starting point for future research. The main justifications of this analysis may be summarized by the following facts. First, the relative lack of studies on how meteorological drought drives anomalies in groundwater tables compared to those on droughts in precipitation or streamflows. It is paradoxical, because understanding the functioning of hydrological drought in a catchment is an important step in the determination of water availability and potential supply. Hydrological drought is more relevant to water shortages and the status of the resources than meteorological drought. Second, there is limited literature available that explores the relationship between meteorological drought and the effects on Tucson's groundwater. Third, since 2000 there has been a shift towards the use of Central Arizona Project water supplies rather than groundwater in the Tucson area, according to the Tucson Active Management Area (TAMA) water budget (*Arizona Department of Water Resources, 2009*). Because groundwater is supposed to be less overexploited, there may be less noise in the relationship between meteorological anomalies and their impacts on water table changes.

Two general hypotheses lead this research:

- According to the functioning of the hydrological cycle, drought in groundwater tables should reflect the pattern of meteorological drought, but with notable variations attributable to the conditions imposed by the physical characteristics of the catchments and the aquifer, especially noticeable in the main parameters of drought: number of dry sequences, duration, intensity, etc. and with some lag, reflecting the specific recharge and groundwater flow processes. In the basin and range landscape of Southeastern Arizona, the main recharge mechanisms are mountain front recharge and ephemeral streambed recharge.
- Knowing these variations and relationships between droughts in both variables, meteorological drought could provide some predictive value with respect to groundwater levels, an advantage that would be very useful for resource planning in deficit situations.

1 ANALIZING DROUGHT IN THE TUCSON AREA

1.1 Types of drought

First, clarifying some concepts and ideas will facilitate the purposes of this paper. According to the website of the National Drought Mitigation Center, the most popular categorization of drought is made in terms of the variables that measure the physical water deficit: meteorological, agricultural and hydrological drought. Each one of them is defined on the basis of the deficit in comparison to the normal amount of water expected in rainfall (meteorological drought), in soil and vegetation (agricultural or edaphic drought) or in water bodies (hydrological drought in groundwater, streams, lakes, etc.). In this paper we will use the groundwater "version" of hydrological drought, also referred to as hydrogeological drought.

In the literature there is a consensus about associating hydrological drought with the effects of low precipitation periods on the levels of aquifers, lakes and rivers (*García Prats, A., 2006; UNESCO/OMS, 1992*). Although both phenomena can overlap, their onsets (and ends) are correlative and deferred in time and the sequence of them occurs only if the rainfall deficit lasts long enough and/or is sufficiently intense. According to some authors (*Bazza, M., 2014; Garrido, N. et al. 2006*), the distinction between meteorological and hydrological drought is based on the fact that the former is explained as the deviation of the climatic values with respect to normal values, and it is the natural trigger. On the other hand, hydrological drought depends on catchment management and this dependency brings in human influence to modulate the natural relationship between rainfall and groundwater. However, it may be used as a measure of the rainfall drought impacts on runoff and the aquifer. In fact, *Marcos Valiente (2001)* argues that, because hydrological drought is delayed with respect to rainfall drought, logically it is not a good indicator of the onset of the deficit but is tremendously important to show the intensity and impacts of the meteorological drought.

1.2 Hydrogeological drought studies in Tucson

As mentioned above, there are knowledge gaps about the relationship between rainfall drought and the effects on the level of Tucson groundwater. Nevertheless, many institutions and reports recognize the importance of the depletion of groundwater resources in times of drought, but they tend to link it to socio-economic activities. In contrast, *Glennon and Maddock (1994)* explain that the exhaustion of aquifers has contributed to the disappearance of the perennial desert rivers and their riparian habitats. The State of Arizona Hazard Mitigation Plan (*Arizona Department of Emergency and Military Affairs, 2013*) explains the particular impacts of groundwater drought in terms of increased wildfire activity, the loss of vegetation, livestock and wildlife and also emphasizes that there are secondary effects such as ground subsidence and erosion, dust storms and flooding due to the loss of forest. However, they assume that only human activity is causing the problem, when it could be a combination of several drivers. The great majority of the literature stresses that continued drought can affect groundwater levels if well owners continue to pump water when there is a lack of rainfall, but it is not that common to find reflections on the consequences of the natural phenomenon.

Despite this focus on the impact of human activity on groundwater levels, there are some interesting initiatives to unravel the natural linkage at different scales and from different points of view. For example, *Morehouse, Carter and Tschakert (2002)* analyzed the variability of the proportion of rainfall that became recharge in the area depending upon antecedent moisture conditions and rainfall intensity. In the field of practitioners, the USGS, through its website on Drought (http://water.usgs.gov/ogw/drought/), emphasizes the importance of the interconnection of groundwater and drought: *"Droughts can significantly impact the Nation's groundwater resources while the drought is occurring and for some time afterward (…). Climate can be a key, but underemphasized, factor in ensuring the sustainability and proper management of groundwater resources"*. The USGS website points out the difficulty of determining drought impacts on the water table, stressing that the water levels reflect the shortage of rainfall after a prolonged time, very often after some years. In this regard, the USGS Climate Response Network is a set of Groundwater Level Network wells that are used to observe the effects of meteorological droughts on groundwater levels as a first approach, but the problem is that there are still few wells represented and it doesn't include any wells from southern Arizona.

Both the USGS Arizona Active Water Level Network and the AWRD Annual Drought Preparedness Reports include more measurement points but, in the end, only the water table levels are displayed and the fluctuations are explained without considering the seasonal effect or calculating any index that could facilitate the proper interpretation or comparison of other water cycle variables. However, other types of drought such as the meteorological one or the hydrological drought measured in streamflows are subject to more analysis in these reports and websites.

Importantly, the Groundwater, Climate and Stakeholder Engagement (GCASE) Project of the Water Resources Research Center of the University of Arizona addresses the relationship between meteorological and hydrological drought in the same study area used in the research for this paper (*University of Arizona, 2014b*). The ultimate goal of the GCASE Project is to ensure a safe yield condition to avoid significant declines in groundwater storage. However, the Project is intended to evaluate the response of the aquifers to future climate scenarios rather than designed to focus on the historical influence of drought. All the efforts aimed at quantifying the nexus between droughts and the hydrological cycle, could help to validate the modelling performed for this project.

On the other hand, the United States Section of the International Boundary and Water Commission organized a Public Meeting in Tucson in September 2014 called "Regional Groundwater Issues and Drought Planning", addressing groundwater management under drought conditions as a central topic and presenting some of the results from the GCASE Project (*University of Arizona, 2014a*). Clearly, the recent southwestern drought is motivating conversation and raising awareness in all the sectors about the nexus between drought and groundwater.

1.3 Study area and time span

The Tucson Active Management Area consists of three groundwater sub-basins: the Avra Valley sub-basin, the Upper Santa Cruz sub-basin and the Altar Valley sub-basin. The selected study area is the Upper Santa Cruz aquifer because it encompasses the whole Tucson metropolitan area, the most populated area of the TAMA (*Barker, 2009*) and the results obtained from it should be a good benchmark of how hydrological drought behaves and affects the region. Specifically, the data used comes from three water table monitoring wells located in the City of Tucson, particularly in the southern part (See *Figure 1*).

This study considers the interval from 1980 until 2012, which in principle should be a sufficiently long time period to yield statistically significant conclusions. That time span is conditioned by the data availability, as explained further in the following section.

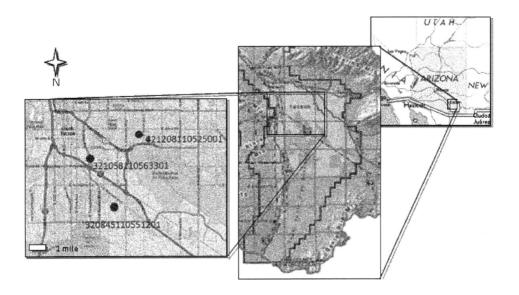

Figure 1 Map of the Upper Santa Cruz aquifer borders (central image) with the details of the three points of water table measurement used in this study (image on the left). Source: Adapted from *Bota & Mason, 2006* (central) and USGS Waterdata web (left).

2 METHODOLOGY AND DATA

2.1 Methodology

In the study, there are three different stages of analysis:

- First, a climatic and piezometric characterization of the study area was conducted.
- Second, an analysis of the relationship between rainfall and water table depth was performed on a monthly basis. The patterns recorded in rainfall and water tables were compared in order to assess whether these variables follow a pattern of interconnection and, if so, were evaluated to determine possible explanations. Apart from the correlations between these variables, it is important to analyze precisely the amounts of precipitation and water levels in different circumstances, such as in peaks or minimum values. This prevents interesting details that can clarify a pattern of disparity or correspondence from being masked by all-encompassing calculations. This intermediate analysis is important because it tests whether there is a strictly linear connection between the two variables. If such a connection exists then the information on drought obtained from the water tables is simply redundant. On the other hand, if such a linear connection does not exist, then drought analysis will be improved by the addition of water table data to the rainfall data.
- Last, the *Precipitation Drought Index* (PDI) is applied to both rainfall and water tables to identify and evaluate the different dry sequences of the two series

(the moments when the index is below 0.5). The same principles used in the previous step will be used to compare the behavior between the two variables.

2.2 Data sources

Rainfall data was extracted from the Climate Research Unit TS3.21 database, using a 0.5° latitude and longitude spatial resolution grid product that can be obtained through the British Atmospheric Data Center website, and which has a monthly time span from January 1901 until December 2012, which is sufficient for this investigation. The time series of the cell containing Tucson was extracted from the database and considered to be significant for the study area. The Upper Santa Cruz water tables values were obtained from the USGS National Water Information System database, provided directly through the website: http://waterdata.usgs.gov/az/nwis/uv/?referred_module=gw.

To characterize the regimes, to study the evolution of the values, and to apply the drought index it is important that the series have enough continuity and historical length. Consequently, only three geographical points (*Figure 1*) could be used. The three points selected have monthly records from 1980 to the present although they have some discontinuities. However, in this case, this is not a material disadvantage *per se*, because the metropolitan location of the three available points makes them very relevant for the interpretation of the results in the study area. The database registers the points with complete codes (*Figure 1*), but throughout the study they are abbreviated to the last four digits, being 5001 (NE point), 3301 (NW point) and 1201 (S point).

Water table levels at these points are always below 100 feet (30 meters), so it is considered deep groundwater. Despite their proximity, the depth differences are significant, as it is expected for an aquifer whose wells show average water depths ranging from 4 to 600 feet (1.2 to 180 meters) (*Arizona Department of Water Resources, 2015b*).

Deep groundwater renewal times are expected to be long as its connection with precipitation is not immediate. Moreover, deep groundwater has little interaction with surface runoff and, as a result, its storage is more stable. Consequently, these types of groundwater bodies are good references from which to measure the impacts of drought because their only outputs are the abstractions and groundwater discharge which is too slow to be significant on short time scales.

2.3 The PDI: Precipitation Drought Index

For drought analysis, the Precipitation Drought Index (PDI) (*Pita-López et al., In press*) has been selected because of its computational simplicity and because it is considered a suitable drought index for semi-arid climates (*Limones, 2013; Pita-López, 2000*), where the variables of the hydrological cycle and their anomalies are very difficult to adjust to a normal curve (*Lloyd-Hughes and Saunders, 2002*). PDI is a drought index based on the calculation of monthly rainfall anomalies accumulated and standardized, similar to the widely accepted Standardized Precipitation Index (SPI) (*McKee, T.B. et al., 1995*).

Normally, drought indices are computed by establishing a cumulative precipitation anomaly over normal rainfall values, which is standardized in a subsequent step with the aim of making the values comparable between different study areas, or even applied to different variables in the hydrological cycle, as intended in this study. The application of a drought index requires:

a   choosing a time scale,
b   establishing a benchmark to express the value considered normal,
c   calculating the anomaly of the period compared to that normal value,
d   suggesting a procedure to accumulate successive anomalies,
e   standardizing the values of accumulated anomalies, and
f   establishing drought thresholds.

PDI uses monthly rainfall series as input, which is an appropriate level of aggregation for measuring drought in semi-arid climates (*Pita-López, 1995*). Instead of the average, the selected reference value expressing the normal monthly rainfall is the median because it is less affected by potential histogram skewness. The index is calculated in three steps, summarized as follows:

1   For the evaluation of water shortage experienced in a month compared to the normal value for the same month of the year, the PDI simply uses the difference between the total rainfall for the month and the median of that month.
2   In the next stage of the calculation, the index accumulates successive rainfall anomalies month by month but stopping and restarting the accumulation calculation every time a new negative anomaly appears in the context of a surplus run. As a result, preceding accumulated surpluses (that tend to be high because precipitation does not have an upper limit) do not mask new dry months.
3   Standardization is performed by assigning probabilities of exceedance of the monthly accumulated anomalies values calculated in the previous phase. Accordingly, the PDI index units are literally the empiric probabilities of exceedance of the deficit reached each month, calculated using the plotting position (*Weibull, 1939; Makkonen, 2006*). Values are therefore universally applicable and comparable between two variables or two different geographic contexts, if needed. The lower the PDI, the more difficult it would be to achieve a drought like the one experienced at that particular time (i.e., a drought of a higher return period) and, consequently, the drier the month is (*Limones, 2013 and Pita-López et al., in press*).

For this paper the PDI is applied to monthly rainfall and water table values in the study area in order to identify and evaluate the coincidence of the dry periods of each time series.

3 RESULTS

3.1 Precipitation and water tables. Regimes and time evolution

The evolution of rainfall data from 1980 to 2012 was analysed without finding any significant trend, although it is interesting to note that the interval from 2001 to 2006 registered lower values compared to the remainder of the period. (See *Figure 2*).

However, the analysis of the water levels historical series shows a very evident decreasing trend. It is a continuous and progressive drop until there is a specific

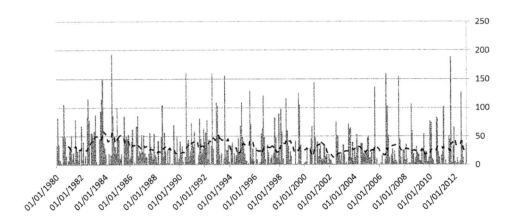

Figure 2 Monthly rainfall in Tucson in millimetres in blue and its 12 months moving average in black (data from Climate Research Unit 3.21 dataset).

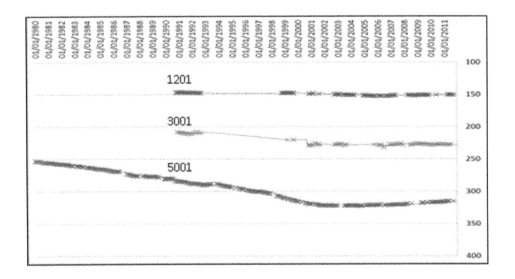

Figure 3 Evolution of the depth to water table in feet in three measurement points of the Upper Santa Cruz aquifer.

and very quick inflection point around year 2000 observed at the three wells (See *Figure 3*).

The 1990's were very humid years in general but, according to the TAMA water budget, more than 250,000 acre feet (308.25 cubic hectometres) were taken from the aquifer each year in that decade on average, which is the historical maximum. The quick depletion could have been triggered by such overexploitation. Another interesting fact is that from the year 2000 the values stabilise despite precipitation decreasing after 2001. Groundwater depletion does not continue because in Tucson, CAP water started replacing the groundwater abstracted, as mentioned in the introduction.

For this reason, henceforth some of the analysis will be divided into two time periods: before and after 2000 because there are two different water supply conditions and two different water table conditions. In fact, in view of these preliminary results, a new hypothesis was tested: after 2000 the system could have started restoring its natural equilibrium and rainfall and hydrogeology could be more connected than before that time.

With regard to the average regimes, we noticed clear differences in precipitation and water table monthly regimes. In fact, there is no coincidence at all between the peaks or the valleys in the two variables (see *Figure 4*). Precipitation is very low throughout the whole hydrological year except for monsoon rains, concentrated in July, August and, to a lesser extent, in September. However, this peak is not clearly perceived in water table records.

Another very significant finding was that the minimum and maximum values are not located equally in the three measurement wells throughout the year. There may be many causes for this result and this preliminary study cannot diagnose them in absence of more information, but it is likely that local fluctuations in the water tables may be caused by localised abstraction or recharge.

However, differences appear between the precipitation and the water tables between 2000 and 2012 (see *Figure 5*). First, the values of the standard deviation of the water table are in all the cases considerably lower than the prior period, which suggests a higher permanency of the intra-annual patterns identified from 2000, although this same pattern has not been recognized in precipitation. This could mean that there are other factors creating stability in the system. Second, the patterns of

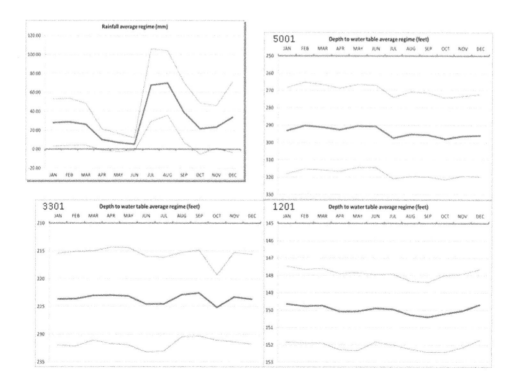

Figure 4 Rainfall and water table regimes 1980–2012.

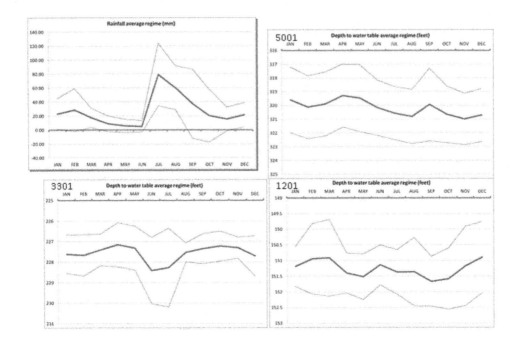

Figure 5 Rainfall and water tables regimes 2000–2012.

change among the months show concurrence with the precipitation ones but with some lag and attenuation with respect to it. Last, the patterns of the three measurement points are more parallel. These outcomes reflect a more natural behaviour of the aquifer after a decrease of pumping due to the arrival of CAP water and use of this alternative supply source in 2000.

3.2 Correlation of the monthly series of precipitation and water tables

According to these initial results and the opening hypotheses, a concrete response pattern of groundwater levels to rainfall should exist. To assess the correlation between rainfall and water tables, an initial correlation analysis between the rainfall time series and the three water table measurement points was established on a monthly basis. There is not significant evidence of dependence between the two variables, neither considering the entire reference period nor limiting the analysis to the interval 2000–2012. It is important to note that other nonlinear relationships were explored without improving significantly the results (*Table 1*).

The low correlation could be linked to the fact that the hydrological behavior is delayed in time. To test the hypothetical lags between variables there are many possibilities documented in literature, but the simplest consists in correlating the monthly values of groundwater levels with values from previous months of precipitation and analyzing the evolution of the correlation values as the correlation lag gets bigger. The coefficients resulting from this analysis are not significantly higher. Consequently, considering that there were no delay patterns in the hydrological response, the possible existence of accumulation patterns was assessed. For that purpose, the correlation between the months of water table levels and the sum of several consecutive months of rainfall (the particular month and the previous ones) was evaluated, again analyzing the evolution of the correlation values as the accumulation interval is broader.

This analysis produced significant results, showing that regardless of the measurement point or the interval considered, the accumulation of several months of rainfall yields substantially higher correlations with monthly values of water tables than without accumulation. Patterns differed between measurement points (*Figure 6*). An interesting fact observed is that if only the time period after 2000 is considered, the increase in correlation values changes significantly in comparison to the values for the whole period. In general, lower correlations between accumulated precipitation and water tables are obtained from 2000 to 2012, differing from what could be expected from the hypothetical more natural character of the groundwater behavior for that time period. In addition, the accumulation intervals that show the highest correlation between precipitation and water table levels are smaller between 30 and 82 months of accumulation. This implies that from 2000 to 2012 the water tables relate better with

Table 1 Dependence of rainfall and water tables in the three series.

Coefficient and water table measurement point	5001	3301	1201
Pearson coefficient using the whole period	0.048	0.121	0.104
Pearson coefficient for the interval 2000–2012	0.08	−0.284	−0.033

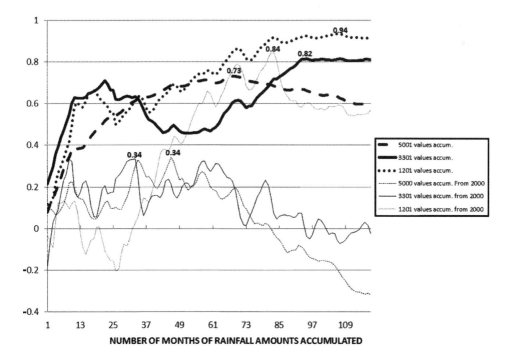

Figure 6 Diagram of the evolution of Pearson correlation versus accumulation interval length.

more immediate rainfall than encompassing the whole series, where the best correlation peaks fluctuate between 65 and 108 months of accumulation.

3.3 Results of the application of the PDI index in both variables

The results correlating precipitation and groundwater levels are inconclusive on an immediate time scale, but very significant when considering the conditions of the basin in the past. This result means that the dryness and humidity of the aquifer have "memory" and perhaps the analysis in the preceding section is reflected in a delay between the dry sequences identified in both variables.

The application of any drought index requires a frequent measure of the variable so that the persistence of the deficit can be assessed. Therefore, for this section only the point 5001 could be used because the other two measurement sites did not have enough historical continuity. However, despite the local differences found in previous sections between all the measuring points, the patterns identified were very similar and what will be found using the 5001 point may be sufficiently informative and a good starting point for future ideas.

It was necessary to de-trend the monthly median values for the two time periods separately as stated above. The series was split into two parts: the interval before 2000, a turning point occurred in that year, and the interval after that date. In the absence of such a division, the use of a single set of the twelve monthly medians, would have caused all the months before 2000 to appear as wet and all the final months as dry because they would have been systematically above or below the medians, respectively.

The series of the indices in precipitation and water tables for the period from 1980 to 2000show a systematic drop after the important rainfall drought that took place throughout 1987–88 and 1989–90 (*Figure 7*). This depletion does not relate to rainfall patterns but clearly shows the effect of increased groundwater use experienced during the decade of the nineties.

The results of the application of the index in precipitation and groundwater levels for the second interval show a counter-intuitive trend (*Figure 8*). Unfortunately, the length of the times series does not reflect the recent southwestern drought and this will have to be analyzed in coming studies; however, the drought from 2001–2005 was even more intense and prolonged for the particular case of the City of Tucson.

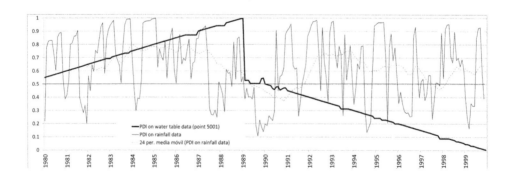

Figure 7 Evolution of the PDI in rainfall and in water tables from 1980 to 2000.

Groundwater dynamics: How is Tucson affected by meteorological drought? 333

In fact, that drought was one of the triggers for the promotion of a law and the Drought Preparedness and Response Plan for the city, in order to prevent future supply problems (*City of Tucson, 2012*). This rainfall drought is clearly identified in the PDI graph of *Figure 8*. On the other hand, the water table began to drop in 2004, reached the minimum in 2007, and did not begin to recover until 2010.

The correlation between the two index series is as low as obtained in the series of the raw values of both variables measured in the previous section, in this case yielding only a Pearson coefficient value of *0.14* from 1980 to 2000 and *0.08* from 2000 to 2012. However, a 24-month moving average was applied to the graph of the PDI index on precipitation and its shape is quite similar to the graph of the PDI index applied on groundwater levels (*Figure 8*), but with a separation of approximately three years. Again, this suggests that both phenomena might be mutually delayed. This delay is not clearly perceived in the previous interval (dashed line in *Figure 7*). Therefore, the same analysis as in the previous section was applied, generating new results. In the case of the period from 1980 to 2000 lagging the series of one PDI index with respect to the other produced a very slight increase in correlations that could not be considered statistically significant (solid line in *Figure 9*, left side). However, in the

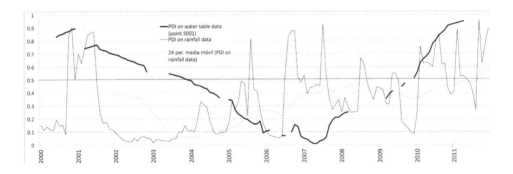

Figure 8 Evolution of the PDI in rainfall and in water tables from 2000 to 2012.

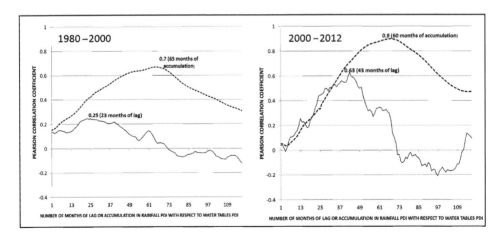

Figure 9 Evolution of Pearson correlation. The solid line shows the correlation as the two series of PDI are lagged. Dashed line is the correlation as a function of accumulation interval length.

analysis performed for the interval from 2000, a rapid and sustained increase in the correlations was found, reaching a peak in the 43 months lag (a little over the three years perceptible by the naked eye) when both series present a Pearson coefficient value of 0.63 (solid line in *Figure 9*, right side).

If instead of lagging the PDI series of both variables, the PDI on precipitation is accumulated in the same way as it was done in the previous section and compared to the PDI of water tables, we find that the correlations values reach a Pearson of 0.7 for a 65 months accumulation (dashed line in *Figure 9*, left), and reaching 0.9 for 60 months of accumulation for the time period after 2000 (dashed line in *Figure 9*, right). This indicates that the conditions of deficit or surplus that take place throughout the five preceding years determine the aquifer conditions at every particular moment and that this effect is more evident in the recent time period. These results indicate the possible long-term predictive value of precipitation anomalies on the status of the aquifers.

4 SYNTHESIS, CONCLUSION AND NEXT STEPS

On a monthly scale, there is a direct and linear correlation between rainfall and aquifer levels, although the values are low. Thus, both variables are different enough to be individually studied. Delays are seen in the response of water tables to rainfall. Specifically, this variable correlates much better with the sum of the volumes of several consecutive months of rainfall. This result is less manifest in the last part of the series beginning from year 2000. Moreover, the correspondence between the time series of the PDI drought index in precipitation and PDI in water levels at the point of measurement is direct and linear but also quite low, indicating that behaviors during dry sequences in both variables do not necessarily correspond better than the series of untreated measured values.

However, it has been found that in the most recent phase the behaviors of droughts in both variables are delayed between each other around three and a half years and, consequently, the beginning of a rainfall drought could allow us to anticipate the start of a hydrogeological drought in the Upper Santa Cruz. During the period when groundwater was the main water supply source in Tucson (1980–2000), this pattern is masked by human pumping and the response to dry and humid phases is less obvious but more immediate. In fact, in this period, the lag between the two series of PDI was approximately two years. Additionally, the accumulated values of meteorological drought over several years are also a good way to approximate the value of the drought in groundwater, although it is a more complex calculation and less intuitive than the simple delay of an index series with respect to the other.

Overall, the results confirm that the aquifer responds more proportionally to the anomalies in rainfall when groundwater is no longer extracted in large quantities, but it is important to delve deeper into this phenomenon and continue researching both phenomena with more data. Also, more exploration will be needed to develop a pattern of hydrological response to meteorological drought to develop new and different managerial approaches in the catchment.

Bearing in mind the significance of the preliminary findings demonstrated in this paper, the methodology should be used to test the results in the future on more measurement points to determine what occurs in other parts of the aquifer and, if possible, over longer time series that can also encompass the recent drought in the southwest (2012–2015). Some aquifer storage models of could also be used in the future to apply the same methods shown here, some of them are available via a web page called "Modeling Hydrology Unit" by Arizona Department of Water Resources (2015a). Another idea for future improvements could be applying other drought indices that could shed light on the status of groundwater without the need to perform the lag analysis, now that it has been shown that the connection is more evident in the long term which has been approximated to be between 43–60 months of delay. A good approach would be to compare the water table levels with long term SPI drought indexes such as SPI48.

The confirmation of a delayed linkage of drought in precipitation and groundwater drought could be valuable for management. Monitoring and predictive capability helps reduce vulnerability to droughts. Although socioeconomic activities are currently largely covered by the CAP water supply, and therefore there is little perception of risk among the population, groundwater has a high ecological value in the basin, as it still supports some natural riparian areas, and its dynamics in times of scarcity must be better understood.

Last, although we believe that the development and calculation of a hydrological drought index that is able to express the reality affecting the aquifer has sufficient magnitude to reflect on it in order to raise a valid solution from which to work further, we recognize the usefulness of widening this study to streamflow drought and to drought impacts. This will enhance the understanding of the connections between all the types of drought in the basin.

REFERENCES

Arizona Department of Emergency and Military Affairs (2013). *State of Arizona Hazard Mitigation Plan Risk Assessment.* Retrieved from: http://www.dem.azdema.gov/preparedness/docs/coop/mitplan/26_Flooding.pdf.

Arizona Department of Water Resources (2009). *Summary of water budget.* Retrieved from: http://www.azwater.gov/azdwr/WaterManagement/Assessments/documents/Tucson_AMA_Summary_Budget_Jan_6_2009_001.xls.

Arizona Department of Water Resources (2015a). *Hydrology. Modelling Unit.* Retrieved from: http://www.azwater.gov/AzDWR/Hydrology/Modeling/.

Arizona Department of Water Resources (2015b). *Tucson AMA Groundwater Conditions.* Retrieved from: http://www.azwater.gov/AzDWR/StatewidePlanning/WaterAtlas/ActiveManagementAreas/Groundwater/TucsonAMA.htm.

Anning, D.W. and Leenhouts, J.M. (2011). Conceptual Understanding and Groundwater Quality of the Basin-Fill Aquifer in the Upper Santa Cruz Basin, Arizona. In *Conceptual Understanding and Groundwater Quality of Selected Basin- Fill Aquifers in the Southwestern United States.* U.S. Department of the Interior – U.S. Geological Survey.

Barker, J. (2009). *Aquifers of the Tucson Active Management Area.* Retrieved from: http://academic.emporia.edu/schulmem/hydro/TERM%20PROJECTS/2009/Barker/John.htm.

City of Tucson (2012). *City of Tucson Water Department Drought Preparedness and Response Plan.* Retrieved from: http://www.tucsonaz.gov/files/water/docs/drought_plan_update_spring_2012_p_1-3.pdf.

Bazza, M. (2014). *Gaps in Drought Characterization and Monitoring in Countries.* [Power Point Presentation] Presented at the Building Drought Resilience World Bank Workshop (Washington, DC). November, 2014.

Bota, L. and Mason, D. A. (2006). Regional groundwater flow model of the Tucson Active Management Area Tucson, Arizona: Simulation and application. *Phoenix Arizona Modeling Report*,13. Arizona Department of Water Resources. Hydrology Division.

García Prats, A. (2006). *Sequía. Teoría y prácticas.* Editorial UPV. 145 p.

UNESCO/OMS (1992). *International Glossary on Hydrology.* 2° Ed. Retrieved from: http://webworld.unesco.org/water/ihp/db/glossary/glu/HINDES.HTM.

Garrido, N. et al. (2006). *Las sequías climáticas en la cuenca del Duero.* Instituto Nacional de Meteorología. CMT de Castilla y León. Retrieved from: http://www.unizar.es/fnca/duero/docu/p209.pdf.

Glennon, R.J. (1995). *The threat to river flows from groundwater pumping.* S.E.L. Associates.

Glennon, R.J. and Maddock, T. (1994). In search of subflow: Arizona's futile effort to separate groundwater from surface water. *Arizona Law Review, 36.* Arizona Board of Regents.

Glennon, R.J. and Maddock, T. (1997). *The concept of capture: the hydrology and law of stream/aquifer interactions.* Rocky Mountain Mineral Law Foundation.

Limones, N. (2013). *El estudio de la sequía hidrológica en el mediterráneo español. Propuesta de aplicación del Índice Estandarizado de Sequía Pluviométrica a las aportaciones hídricas.* University of Seville Publications.

Lloyd-Hughes, B. and Saunders, M.A. (2002). A drought climatology for Europe. *International Journal of Climatology* 22(13): 1571–1592.

Makkonen, L. (2006). Plotting positions in extreme value analysis. *Journal of Applied Meteorology and Climatology* 45(2): 334–340.

Marcos Valiente, O. (2001). Sequía: definiciones, tipologías y métodos de cuantificación. *Investigaciones Geográficas 26:* 59–80.

McKee, T.B., Doesken, J. and Kleist, J. (1995). Drought monitoring with multiple time scales. *Proceedings of the Ninth Conference on Applied Climatology, American Meteorological Society* (Dallas, TX): 233–236.

Morehouse, B.J., Carter, R.H. and Tschakert, P. (2002). Sensitivity of urban water resources in Phoenix, Tucson, and Sierra Vista, Arizona, to severe drought. *Climate Research* 21(3): 283–297.

National Drought Mitigation Center (2015). *What is Drought?* Retrieved from: http://drought.unl.edu/DroughtBasics/WhatisDrought.aspx.

Pita López, M.F. (1995). *Las Sequías: Análisis y Tratamiento.* Sevilla. Publicaciones de la Consejería de Medio Ambiente, Junta de Andalucía.

Pita López, M.F. (2000). Un nouvel indice de sécheresse pour les domaines méditerranéens. Application au bassin du Guadalquivir (sudo-uest de l'Espagne), *Publications de l'Association Internationale de Climatologie, 13:* 225–234.

Pita López et al. (In press). A new index for rainfall drought assessment: the Precipitation Drought Index (PDI).

University of Arizona (2014a). *Public Meeting in Tucson, AZ to Discuss Regional Groundwater Issues and Drought Planning.* Retrieved from: http://www.portal.environment.arizona.edu/events/public-meeting-tucson-az-discuss-regional-groundwater-issues-and-drought-planning.

University of Arizona (2014b). *Groundwater, Climate And Stakeholder Engagement (GCASE).* Retrieved from: https://wrrc.arizona.edu/GCASE.

Weibull, W. (1939). A statistical theory of the strenght of materials. *Ing. Vetensk. Akad.* 151: 1–45.

WWF (2012). Medidas para abordar la escasez de agua y la sequía en España. *WWF Informe 2012.* WWF España.

Chapter 17

Alternative water sources towards increased resilience in the Tucson region: Could we do more?

Kristin Kuhn
UNESCO-IHE, Institute for Water Education, Delft, The Netherlands

Aleix Serrat-Capdevila
UMI iGLOBES CNRS/Department of Hydrology and Atmospheric Sciences, University of Arizona, USA

Edward F. Curley
Pima County Regional Wastewater Reclamation Department (Retired), Arizona, USA

László G. Hayde
UNESCO-IHE, Institute for Water Education, Delft, The Netherlands

INTRODUCTION

The southwestern United States experiences currently what is said to be the worst drought in 500 years (Ingram, 2015). In July 2015 nearly 80% of the region has been classified as under moderate to exceptional drought conditions (*National Drought Mitigation Center, 2015*), threatening water supply in the state of California. While there are uncertainties about the exact amounts and the time frame of occurrence, supply and demand studies agree that new water supply options need to be developed in the future to meet increasing demand and mitigate the impact of climate change in the Colorado River basin (*ADWR, 2014, USBR, 2012, WRDC, 2011*).

Figure 1 shows the unbalanced state of water supply and demand in the service area of Tucson Water that is projected to occur by 2042 if no additional sources are utilized and based on the assumption that the current per capita demand stabilizes (482 LPCD/127.4 GPCD) (*Tucson Water, 2012*). The City of Tucson, located in the Sonoran Desert, is the second biggest metropolitan area in Arizona and highly dependent on Colorado River water in addition to its own aquifer. While a shortage in the Colorado might be declared as early as 2018 (*USBR, 2015a, USBR, 2015b*), it may not have an immediate impact on the municipal water supply in Tucson. Water managers do not expect cuts to municipal users before 2030 (*Tucson Water Department, 2015*). Nevertheless, with the current system, insufficient supplies may occur earlier than shown in the graph that indicates a reduction in CAP supply by 2040. The Santa Cruz River and its aquifer have historically enabled the development of what has become Tucson. A combination of irrigated agriculture, population growth

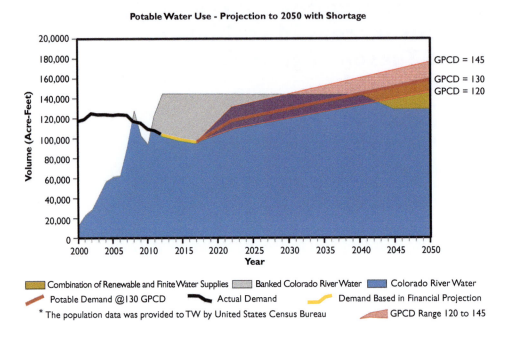

Figure 1 Projected supply and demand in Tucson until 2050 (*Tucson Water, 2012*).

and high per-capita water use resulted in the overexploitation of the groundwater. Riparian corridors along perennial streambeds died and disappeared once pumping in the last century lowered the water tables and effectively disconnected the riparian corridors from the groundwater that kept them alive during the long dry seasons.

"Tucson Water's need to develop renewable water supplies in order to reduce reliance on groundwater and meet projected future demand has long been recognized and is a critical goal" (*Tucson Water, 2004*). The water utility mainly relies on the Colorado River for future supply (blue and gray areas in *Figure 1*). Proposals to close the supply and demand gap include the acquisition and development of additional water supplies, the enforcement of stronger demand management and the increased effluent utilization (yellow area in *Figure 1*). Effluent is the "only locally generated renewable supply that grows with the community" and, accordingly, Tucson Water considers leasing or purchasing effluent entitlements from other water right holders. Furthermore, groundwater use is planned under a "hydrologically sustainable pumping rate" for a number of reasons, such as to cover peak demand and mitigate impacts of climate change on the availability of Colorado River water (*Tucson Water, 2012*).

This research aims at capturing an integrated view of the current water management by analyzing all available water sources for municipal and environmental use and the sustainability of their utilization. It focuses on the benefits that can be provided to the community from the use of 'new water sources' believed to be underutilized and that are successfully used here and in other semi-arid regions. It seeks to uncover the circumstances under which Tucson could enhance the resilience of its water management to water supply and environmental conservation.

1 RESEARCH METHODOLOGY

1.1 Sustainability assessment

In this work, the sustainability of water sources has been assessed based on criteria that are, in part, proposed by Gleick (*1998*) and Loucks and Gladwell (*1999*). Indicators of resilience, reliability and vulnerability fitted for the social-ecological conditions of Tucson were added (Loucks, 2000). Some criteria that identify unsustainability were included and phrased accordingly to enable a consistent method of categorization. Three scoring levels from unsustainable (−) to more sustainable (+) and sustainable (++) have been used in order to allow comparison and provide an appropriate level of detail. This set of standards is consistent with current water management approaches and is based on the assumption that water supply in the future will be provided by a mixture of different sources to create an overall more resilient system.

Criteria used to assess sustainability:

- The water source is renewable and its use does not exceed the capacity of the system to regenerate within human time frame;
- The water source is generated within the local watershed;
- The availability of this water source is reliable, considering long and short term supply;
- The water quality is adequate to meet potable demand, taking into account how much effort treatment would take in regards to feasibility, effort and cost;
- Use of the water source has no negative impact on water quality of the groundwater aquifer;
- Utilization of the source is independent of energy supply;
- No institutional settings or constraints are in place that could limit the use of the resource;
- It is feasible to utilize the source in terms of cost and effort.

Measuring sustainability has been identified as a somewhat fuzzy endeavor (*Loucks, 2000*), due to the complexity of the term and the different terminological approaches that are held by environmentalist and economists. Representing physical water scarcity, special focus was placed on the first three criteria that focus on the reliability, renewability and location of sources. The other criteria can be regarded as indicators of institutional water scarcity that depend to greater extent on the financial means, infrastructure and the institutional framework necessary to use water source.

The Ecosystem Services (ESS) concept is applied in order to review the benefits to the community that are generated from utilizing the different water sources and increasing urban greenery. This study focuses on the provisioning, regulating and cultural services and how they contribute to human well-being and sustainability (*see Chapter 12 Boyanova et al.: 197*).

1.2 Data collection

Interviews were carried out with 12 experts representing different stakeholder groups working in the fields of water management, water policy, hydrology and environmental management. The broad range of managers and stakeholders provides insights to the different perceptions held by individuals in the water sector. A semi-structured interviewing approach was followed, in which questions were phrased to provide a topic outline, while allowing flexibility without constraining discussions. The interviews complemented an extensive literature review including management and planning documents, peer-reviewed and secondary literature. Quantitative data were collected from the literature and through personal inquiries with local water agencies.

2 CHARACTERIZING AVAILABLE RESOURCES

Municipal demand in the Tucson Active Management Area (TAMA) is projected to reach 310 mm³ (251,018 AF) by 2025 based on the lowest reasonable water demand (*ADWR, 2012*). *Figure 2* indicates approximate estimates for the potential contribution (blue boxes) of all water sources to cover municipal demand in the TAMA.

Groundwater has historically been the predominant water source and, until the implementation of the Groundwater Management Act (GMA) in 1980, its use was widely unregulated (*Jacobs and Holway, 2004*). Model-derived estimates of annual natural recharge (from mountain front and streambed recharge) in the TAMA are in the order of 94.48 mm³ (76,600 AF) corresponding to 30.5% of demand in 2025. Groundwater management practices through the GMA were carried out to address overuse of the resources and to enable access to annually 3.45 bn m³ (2.8 MAF) of Colorado River water that are allocated towards Arizona (*Connall Jr, 1982, Jacobs and Holway, 2004*). Tucson holds the largest municipal CAP entitlement and the open canal transports 178 mm³ (144,172 AF) of water from Lake Havasu at the western border of the state 540 km (336 mi) through Arizona (*Tucson Water, 2012*). Under the assumption that shortage of Colorado River water will not influence municipal demand, 57%

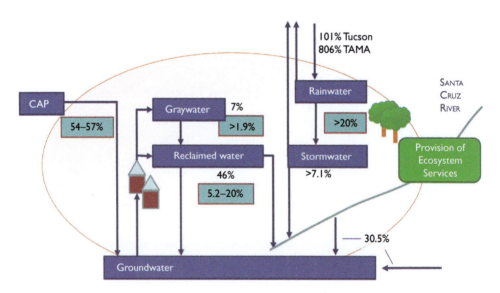

Figure 2 Approximate estimates for the potential contribution of all water sources to cover municipal demand in the TAMA 2025.

of demand can be covered with CAP water. Alternatively, if the shortages influence municipal users, this would result in 54% of demand coverage by CAP water.

Figure 3 indicates current demand in the Tucson Water service area and the quantities of annually available water sources. After recharge and recovery, CAP water is indirectly supplied as a blend with native groundwater. Because CAP supply currently exceeds demand, by one-third of Tucson's CAP allocation, the remaining water is recharged to the aquifer that now stores about the equivalent of two years' worth of municipal water demand.

Reclaimed water, also referred to as recycled water or effluent, fulfills the standards for landscape irrigation and certain industrial needs and is delivered to over 800 residential and commercial customers. Furthermore, additional reclaimed water is now discharged to the Santa Cruz River but future competition from human demands is expected. Tucson Water's Recycled Water Master Plan identifies the treatment of reclaimed water to potable standards as an option for the future (*Tucson Water, 2013, Tucson Water Department, 2015*). Based on projected amounts of reclaimed water in 2025 the total share of effluent from the wastewater treatment facilities of Pima County represents approximately 46% of total municipal demand in the TAMA. Up to 12.33 mm^3 (10,000 AF) of effluent supplied by Pima County and local water utilities have been assigned for environmental restoration through the Conservation Effluent Pool (CEP) (*Tucson & Pima County, 2010*).

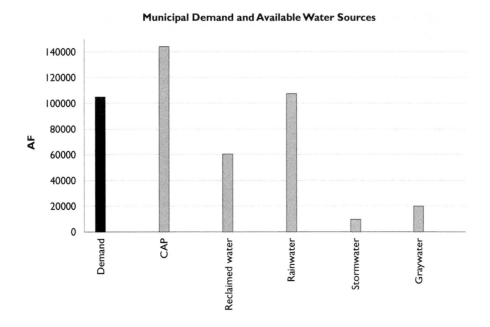

Figure 3 Municipal demand and available water sources in the Tucson Water service area (*NOAA, 2015, Pima County RWRD, 2013, RFCD, 2015, Sonoran Institute & Pima County, 2013, Tucson Water, 2012*).

The use of rainwater, stormwater and graywater has the potential to reduce the demand for potable water for landscape irrigation, therefore decrease financial dependence and support additional greenery along roadways and parking lots when combined with native vegetation (*Lancaster, 2013, Lancaster, 2015*).

Tucson Water estimates that 45% of the water provided to single-family homes, accounting for more than half of all water connections in Tucson, is used outdoors. The overall city average is 39% outdoor use where the lowest is in multifamily houses (26%) (*Tucson & Pima County, 2009*). Primary outdoor uses are landscape irrigation and swimming pools and water is, therefore, lost to the atmosphere through evapo-transpiration, and evaporation.

The use of graywater, defined as wastewater that does not originate from a kitchen sink, dishwasher or toilet, is widespread and often informally practiced in the United States (Crook and Rimer, 2009). It is estimated that 13% of Pima County residents, corresponding to 20,000 to 30,000 households, are using graywater to some extent (*Water CASA, 2004*). Accounting for 31% of all wastewater, a single-family house produces enough graywater to cover about half of all outdoor water needs. However, increased graywater use might have adverse effects on wastewater collection and treatment infrastructure, increasing operation and maintenance costs (*Graf, 2014*). Therefore, graywater use might be most feasible in rural areas without sewer connection. It is estimated that septic tanks serve about 20% of Arizona's population (*Graf, 2014*). In the TAMA this would represent 1.9% of the total municipal water demand in 2025, that could be covered without stressing sewer flows negatively, while increasing the life span of septic tanks.

The City of Tucson receives annually more rainfall (132.5 mm^3/ 107,455 AF) than the overall water demand. While light rain events evaporate without generating significant runoff, a higher quantity of stormwater is available than the Flood Control district measures. Using this water, as close to its origin as possible is therefore the most efficient way to reduce losses. Based on the approach to cover a significant part of landscape irrigation by rainwater and stormwater (Commercial Rainwater Harvesting Ordinance, Green Street Policy) and the great portion of potable water that is used outdoors, it is estimated that 20% of municipal water demand could be covered through these sources. A portion of the stormwater runoff discharges to the Santa Cruz River, supporting the aquatic and riparian ecosystem and recharging the aquifer. The amount of stormwater running to the river varies highly, depending on precipitation amounts and intensity (*Sonoran Institute & Pima County, 2013, Sonoran Institute & Pima County, 2014*).

Available water sources are sufficient for the present state; nevertheless, scarcity is projected to occur within the current planning horizon. If no measures are taken, water supply for human and environmental demand will eventually reach the carrying capacity and the system might become unable to mitigate impacts of climate change and population growth.

3 SUSTAINABILITY ANALYSIS

Sustainable water supply in Tucson is threatened on one hand by natural conditions such as droughts and climatic variability and, on the other hand, by institutional settings and socio-economic arrangements. Whereas future droughts in the region

Table 1 Sustainability characteristics of all available water sources.

	Renewable	Local	Reliability	Potable	No impact on water quality of the aquifer	Absence of instit. conflicts	In-dependent of energy intense system	Feasibility (cost)	Feasibility (effort)	Ease of use
Groundwater	−	+	−	++	+	−	+	+	+	++
CAP water	+	−	−	++	−	−	−	++	−	++
Reclaimed water	++	++	++	+	−	+	+	−	−	+
Graywater	++	++	++	−	++	+	++	−	++	−
Rainwater	++	++	+	++	++	++	++	−	++	+
Stormwater	++	++	+	−	++	+	++	−	++	−

are likely inevitable, the condition of scarcity might be a question of adaptive management. Water managers will face an increasing number of challenging decisions between urban and environmental use in the future if no sustainable measures are taken.

A reliable, flexible, interconnected water resources system with buffers to cope with variability is essential for the resilience of Tucson as a socio-ecological system in the semi-arid region of the Santa Cruz River. Sustainability of Tucson depends on the availability of water resources of sufficient quality to support the human society now and in the future, while preserving the integrity of the ecological system. Results of the sustainability assessment are presented in *Table 1* and explanations are provided in the following section.

3.1 Traditional water sources

3.1.1 Groundwater

Pumping rates in the last decades have significantly exceeded natural recharge, and have thus significantly depleted the aquifer. Since the arrival of the CAP water and its recharge into the aquifer, pumping is now less than the sum of natural and artificial recharge, and thus within safe yield. However, safe yield is not necessarily sustainable in the long term, as it does not account for the replenishment of groundwater flow outside the basin and evapotranspiration needs from natural riparian ecosystems. The spatially imbalanced distribution of pumping and recharge within the TAMA may likely continue to degrade existing riparian ecosystems if unchanged (*see Chapter 15 Cabello et al.: 291*).

Groundwater utilization requires only very little treatment to meet the standards for direct potable supply (*Tucson Water, 2004*). However, groundwater quality decreased in the past due to seasonal and diurnal recharge of reclaimed water and long-term storage of CAP water, both characterized by high salt content (*Tucson Water, 2004, WRRC, 2011*).

Salt treatment due to further mineral input will eventually be inevitable and management of Total Dissolved Solids (TDS) has been only postponed due to the economic crisis (*USBR & Tucson Water, 2004*). The demand for energy to pump groundwater from current water levels ranges between the amount of energy needed to transport CAP water and treat reclaimed water, but is much higher than the energy needed for the utilization of rainwater, stormwater and graywater (*Lancaster, et al., 2011*). Pumping costs and technical effort increase with dropping groundwater levels and the possible utilization of brackish groundwater from deeper layers would require more energy and therefore be cost intensive (*Lancaster, et al., 2011*).

3.1.2 CAP

CAP pumps water nearly 900 m (3,000 ft) uphill to Tucson (*see map 4, p. 22*) and is therefore the largest single energy consumer in the state. Approximately 90% of the power used to transport the water is provided by a coal-fired power plant (*Eden, et al., 2011*) that is one of the most polluting power plants in the country (*Lustgarten, 2015*).

The importation of Colorado River water relieved pressure on the groundwater system, enabled meaningful groundwater regulations and increased, at first sight, sustainability of the Tucson system by diversifying its water supplies. The substitution buffered the negative impact of the last two decades and gives momentum to develop a utilization scheme for sustainable sources. However, Tucson's dependency on CAP water changed the spatial focus from local to regional scale. Depending on management decisions and climatic conditions on a bigger spatial scale brings new uncertainties. The Colorado River was over allocated in the Colorado River Compact (*1922*) and water demand that exceeded supply in the last decades made the water level of the main reservoirs drop to the lowest state since construction (*Grafton, et al., 2013, Hutchinson, 2015*). Water users are interconnected on local, state and federal level. The development in the fast growing neighboring cities and the agricultural sector, as well as interstate negotiations on Colorado River water, for which Arizona does not hold a senior water right, will influence the future of water in Tucson.

3.1.3 Reclaimed water

Reclaimed water can be considered a sustainable source that is local, renewable and reliable. The amount of reclaimed water can be influenced by the extent of graywater reused. Reclaimed water is a relatively drought resistant and secure source that depends on the quantity and type of water used. While potable water used outdoors is lost for the reclaimed water system, water used indoors results in a nearly equal amount of sewage produced and, the amount of reclaimed water therefore grows with the community.

The supply of reclaimed water depends on rather energy-intensive and therefore comparatively expensive treatment (*Lancaster, et al., 2011*). The feasibility (effort) factor for reclaimed water depends on proximity to Tucson Water's reclaimed water distribution system, which is currently constructed to serve primarily large volume customers such as golf courses, parks and schools. To expand this system to allow residential reuse throughout the community would require an enormous investment in infrastructure, so that cost for community-wide implementation would be a deterrent. The future plan is to treat reclaimed water to potable standards, which are associated with increasing energy demand and costs (*Tucson Water, 2012*).

The Lower Santa Cruz River is fully effluent dependent and acts as a managed Underground Storage Facility that accrues water credits at a rate of 50% of total volume recharged. Most parties have plans to put the water in uses that are more beneficial in an economic sense. Treatment plant upgrades increased water quality and enhanced infiltration rates. Consequently, with higher recharge rates, effluent water reached a shorter stretch of river, causing a negative impact on the length of the riparian area, for habitat and wildlife. (*Sonoran Institute & Pima County, 2014*). Currently, it is unknown how much of the CEP water will be allocated towards environmental conservation. Grassroots organizations such as the Community Water Coalition are advocating for the discharge of effluent in reaches upstream from shallow groundwater areas supporting existing natural riparian areas (such as in the Tanque Verde sub-watershed).

An intensified utilization of effluent for human demand may or may not have a negative impact on the environmental system as it could offset in the long run the water pumped from the aquifer. The effects on the social-ecological system as a whole will depend on the nature and spatial distribution of water use allocations in the future. The prevailing community sentiment should be to consider preservation of the natural environment over restoration efforts.

3.2 Alternative water resources

Rainwater, stormwater and graywater are, like reclaimed water, renewable and locally generated resources.

3.2.1 *Graywater*

As reclaimed water, graywater is a consistent water source throughout the year without seasonal fluctuation in its amount. Its supply is therefore much more reliable than rainwater and stormwater, and independent from climatic conditions. Transport and treatment of graywater is centralized through the reclaimed water system. Decentralized reuse of graywater from showers and washing machines can add an additional cycle of use for example, by using it for toilet flushing or irrigation. For the public, the primary concerns for the use of graywater are threats to public health. While the quality of graywater is variable and dependent on its exact origin, no risks for human health have been detected that exceed existing regulations and no negative effects have been reported in the State of Arizona (*EPA, 2012, Huang, 2015*).

Regulations in Arizona are considered the best statutory framework for residential graywater use in the United States (*Ludwig, 2012*). The Arizona Department of Environmental Quality sets a number of guidelines and commonsense best management practices that correspond with obtaining a permit on a residential level. A Residential Graywater Harvesting Ordinance (*City of Tucson Ordinance No. 10579*) requires installation of separate drain lines and plumbing for all new housing developments that enable future homeowners to implement graywater systems. Developments with higher risk than residential users, such as systems producing over 1,514 LPD (400 GPD), multi-family, commercial, and institutional systems require a standard permit (*Arizona Revised Statutes*).

Apart from domestic use, there is a growing interest in graywater utilization for commercial properties where the high increase in graywater use is expected to cause resistance from the reclaimed water utility, as wide spread graywater use has the potential to influence the performance of the reclaimed water infrastructure and treatment scheme significantly (*Graf, 2012, Huang, 2015*).

3.2.2 Rainwater

Collection and utilization of rainwater for beneficial use has been a common practice in many parts of the world and in parts of the southwestern United States. Common Rainwater Harvesting (RWH) practice include the use of rainwater tanks and cisterns to store rainwater for periods of low rainfall (*Figure 4a*) and rain gardens, where earthworks slow down runoff and encourage infiltration by directing the water towards vegetation basins and mulched soil (*Figure 4b*).

The most obvious limitation for the utilization of rainwater, as well as stormwater is the very high variability of rainfall: occasional storms providing abundant water during the winter rains and summer monsoons, and long dry periods in between posing a challenge for storage. RWH is dependent on the availability of rainwater in appropriate intervals and quantity. A lack of water during dry periods obviously prevents RWH and too much water during rainy seasons exceeds cisterns' capacity and excess rainfall goes unused.

RWH systems can be implemented and maintained with relatively low technical knowledge and efforts. Its decentralized supply makes it independent from distribution infrastructure. However, because RWH must be implemented and managed at each individual site, it is difficult to facilitate community-wide use. While Arizona has one of the most RWH- friendly legislatures and its residential use does not face legal restrictions, rainwater is a water source that lacks the institutional support of a managed utility to ensure its success (*Gaston, 2010, Lancaster, 2013*).

Rainwater quality in Tucson is considered to be good and it meets drinking water standards (*Lancaster, 2013*). However, due to the open collection system, some type of filtration and disinfection is necessary for potable use, while a number of non-potable uses can be covered (i.e. irrigation, washing, cleaning, flushing). The potential development of mosquito habitats needs to be avoided as they act as vectors for water-borne diseases (*Zinser, et al., 2004*).

The energy need to utilize rainwater, stormwater and graywater is zero if it is untreated and distributed by gravity and comparatively low when using pumps for local distribution. Using the water where it is generated and tailoring its use towards the level of treatment has the lowest demand for power supply.

3.2.3 Stormwater

The utilization of stormwater is highly restricted by water quality issues. Stormwater is considered the most contaminated type of water reviewed, containing non-point source pollutants such as heavy metals and oils from traffic surfaces. Consequently, stormwater use is rather limited to be utilized for landscape irrigation.

Figure 4: a) Residential rain water tank (*Lancaster, 2013*); b) rain garden (*Lancaster, 2013*).

Large-scale stormwater projects have the potential to cause conflicts between holders of existing water rights. The legal boundary for stormwater use is not clearly defined (*Gastélum, 2012*). The implementation of Green Infrastructure (GI, i.e. in this case: landscape water collection, passive storage and infiltration, for flood mitigation and watering of urban vegetation) on neighborhood scale is not expected to interfere with water rights, as an estimated 95% of precipitation is lost to evaporation and therefore only a small percentage actually reaches streams (*Cleveland, et al., 2015*).

Figure 5 a) Illustration of green infrastructure with curb cut (after Watershed Management Group, 2015a) b) photo of implementation *(Lancaster, 2013)*.

Stormwater utilization on a neighborhood scale with curb cuts and GI basins as shown in *Figure 5(a) and (b)* are currently led by grassroots movements and non-profit organizations. Efforts of local activists and environmental groups have been significant drivers to date. A city regulation, the Green Street Policy requires that stormwater-harvesting features capturing at least the first 12.7 mm (0.5 in) of a rainstorm be integrated in all new street developments and redevelopments that are publicly funded *(TDOT, 2013)*.

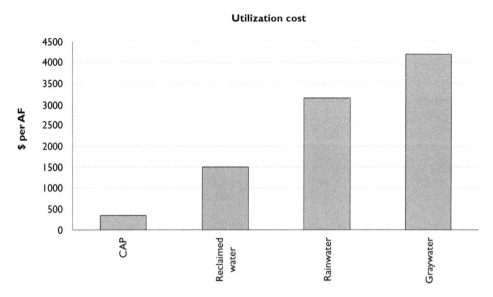

Figure 6 Cost estimation per unit of CAP water, reclaimed water (*Tucson Water, 2012*), rainwater and graywater (*USBR, 2012*).

Responsibility for stormwater management lies with the Pima County Regional Flood Control District and the City of Tucson Department of Transportation. Although the Flood Control District is funded by property taxes, neither of these agencies has an established funding mechanism for stormwater management that could better enable the development of such projects.

3.3 Cost assessment

Cost estimations per unit of water are presented in *Figure 6*. Significantly higher costs for the utilization of rainwater and graywater are published by Tucson Water based on the Rain Water Harvesting and the Single Family Residential Gray Water Rebate Program, but do not take long-term benefits into account (*Tucson Water, 2012*). CAP, marketed on an equal postage stamp rate for all users, is the cheapest available water source due to subsidies. In case of shortage, the price of CAP is expected to rise per unit due to the division of fixed costs over a lower amount of water purchased.

Economic feasibility is currently the main obstacle for augmented utilization of rainwater, stormwater and graywater. However, these approaches present less environmental risks and do not depend on the implementation of large technical infrastructure. While capital intensive, large scale projects such as Hoover Dam or the CAP have been the key solution to previous water management problems. In the Southwest, administrative leaders consider the time for great investments to be over (*Hirt, et al., 2008*).

Utilization of graywater requires separate collection from sewage water, which was not part of standard plumbing practice before the Residential Graywater Harvesting Ordinance. Costs for treatment and reuse of graywater through the reclaimed water system are much lower per unit of water (*DaRonco, 2012, Graf, 2015*). While graywater reuse might be cost effective when installed into new homes, ranging between several hundred and a thousand dollars, costs for retrofitting existing plumbing are at least three to four times higher (*DaRonco, 2012*).

Implementation costs for RWH systems are highly dependent on the type of system, whereas earthworks are cheap in comparison, but provide a lower ease of use than cisterns. (*Watershed Management Group, 2015b*). The payback time depends on the size, type and complexity of the system. Recent estimations for residential rain gardens predict a benefit-cost ratio of 4.4, with a breakeven period of approximately eight years under the current water pricing scheme (*Watershed Management Group, 2015b*).

When estimating the economic feasibility of stormwater use all benefits beyond water provision need to be taken into account. Neighborhood GI projects show evidence that they are cost effective considering their long-term benefits, taking into account the ESS provided (*Ogata, 2015, Wise, 2015*). Applicable cost estimations for stormwater use exist only by area of green street feature, whereas rain gardens with curb cuts have modelled benefit-cost ratios of 2.7 (*Watershed Management Group, 2015b*).

4 BENEFITS OF ALTERNATIVE WATER RESOURCE USE FOR URBAN GREENERY

Green infrastructure and urban vegetation can provide significant ecosystem services, modulating the partitioning of rainfall into evaporation, infiltration and runoff, as well as temperatures and energy use in the hot summer months.

4.1 Provisioning services

The use of rainwater, stormwater and graywater through appropriately designed systems can reduce the amount of potable water used for landscape irrigation. Furthermore, water is indirectly provided through savings in outdoor water use of homeowners who live in close proximity to greenery, such as a golf course, natural preserves and riparian areas (Halper, 2011). Green neighborhoods are expected to directly influence the amount of outdoor water use of single-family households because they can substitute potable water use for residential landscaping (Halper, et al., 2015). GI features, tend to reduce evaporation by slowing down flow and increasing infiltration rates which will replenish the groundwater aquifer and contribute to the safe yield goal (*Potter, 2000, Shuster, et al., 2007*).

The Sonoran desert has over 450 drought-tolerant, food-bearing plants such as mesquite, palo verde, ironwood and hackberry and it is estimated that stormwater drained from one mile of street in a residential neighborhood can support the growth of up to 400 native trees (Davis, 2015). A study has shown that desert trees do not take up and store heavy metals in the eatable parts and can therefore be also watered by stormwater (*Lancaster, 2015*).

4.2 Regulating services

Green infrastructure also provides urban heat mitigation and flood regulation, much needed services during Tucson's hot summers and intense storms.

Tucson's population and urban growth in the last century increased minimum temperatures in the city. Urban growth caused a temperature increase of about 3 °C (5.4 °F), of which 2.1 °C (3.8 °F) occurred during the previous three decades (*Comrie, 2000*). The reduction of vegetation cover and its substitution by impervious surfaces and heat absorbing materials (concrete, dark pavement) enhanced the urban heat island effect. This effect increases energy demand and costs for air conditioning, and has negative impact on human health, causing heat-related illness and mortality. As the installation of air-conditioning systems or shading vegetation are greatly dependent on economic potential, heat mitigation is a topic of social justice. Low-income areas of Tucson's metropolitan area have a much less dense tree cover.

Vegetation can regulate thermal conditions by providing shade and evaporative cooling, while also reducing evaporative losses to the atmosphere and keeping moisture in the soil longer for plant use. Positive effects are reduced expenses for energy, prolonged life of the asphalt and lowered maintenance costs through shading and benefits for human health (*McPherson and Muchnick, 2005*). GI also helps reduce flood volumes and decrease peak flows during extreme storm events (*Watershed Management Group, 2015b*). The water is stored in basins, delayed, and contributes to infiltration instead of surface runoff. Flood regulation reduces flood damage costs and has positive impact on property values. GI also intercepts and retains pollutants and heavy metals in the GI basins. As a result, potential pollutants such as E. coli and copper, are not transported with the runoff to washes and watercourses (*Watershed Management Group, 2015b*).

Vegetation in urban settings reduces air pollution and sequesters carbon. Urban trees and shrubs are a valuable strategy to improve air quality because they trap significant amounts of various pollutants such as aerosols, nitrogen dioxide, sulfur dioxide and carbon monoxide (*Nowak, et al., 2006*). As ozone levels increase with rising temperatures, urban vegetation can mitigate its impact through cooling (*Jackson and Kochtitzky, 2002*).

Green infrastructure and urban greenery also decrease noise level through softer and multifaceted surfaces (*Bolund and Hunhammar, 1999*). This benefits human health through decreased levels of stress and increased opportunities for recreation. In addition, urban greenery provides habitat for local species such as birds, bees and other insects that provide regulating services through pollination and seed dispersal (*Kandziora, et al., 2013, Maes, et al., 2013*).

4.3 Cultural services

The cultural services provided by GI and urban greenery include recreation, tourism, educational systems and improved health and quality of life. Sustaining and increasing native vegetation in the urban environment along roadways and open spaces provides societal benefits by reconnecting the community with its natural and cultural heritage. Appreciation and understanding of the local desert environment can enhance the sense of belonging to a community and the level of environmental and cultural awareness.

Numerous studies show the interconnection between human health and the environment and have identified a direct positive impact of urban greenery, especially in public places, on human well-being (*Jackson, 2015, Kardan, et al., 2015*). Kaplan

(*1984*) shows that natural environment in an urban setting can provide a restorative experience and decrease the stress from life in urban agglomerations, thus increasing quality of life. Apart from decreasing stress levels, urban trees have been shown to improve self-perceived health and result in an overall measurable better physical health, thereby reducing expenses for public health (*Kardan, et al., 2015*).

In the United States, heat is the major reason for death associated to weather conditions and, the state of Arizona has the highest rate of heat mortality (*Centers for Disease Control and Prevention, 2005, Kalkstein and Sheridan, 2007*). Occurring especially in low-income neighborhoods, cases related to heat mortality and morbidity are believed to be significantly underreported. The mitigation of heat islands and the provision of shade on pathways and public places can reduce heat stress significantly and increase the time people spend outdoors, which in turn can improve their physical health and fitness through exercise (*Kuo, et al., 1998*).

Furthermore, greenery can have a positive impact on mental health by providing recreational opportunities, locally grown food and an aesthetic environment. Studies have shown that vegetated neighborhoods increase the likelihood that people interact with their neighbors and care about the local community (*Coley, et al., 1997, Jackson, 2015, Kuo, et al., 1998*). Furthermore greenery increases the feeling of safety in urban environments (*Kuo, et al., 1998*), reduces incivilities and crime rates (*Kuo and Sullivan, 2001*) and leads to more reasoning and less physical violence when dealing with conflicts in private homes (*Sullivan and Kuo, 1996*), and thus "enhance residents' psychological resources for coping with poverty" (*Kuo, 2001*).

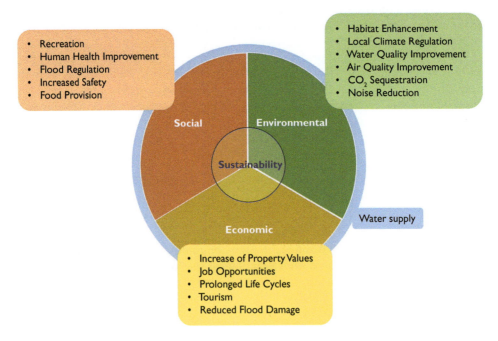

Figure 7 Benefits of urban greenery to the social, economic and environmental system.

It has been shown that homeowners value vegetation and especially prioritize riparian areas, resulting in increased property values (*Bark, et al., 2009, Colby and Wishart, 2002*). Beyond general greenery, this refers also to the biodiverse vegetation of the riparian forest provides great visual contrast to the desert vegetation. The implementation of urban greenery not only increases property values, but has also been shown to increase commercial benefits and success (Wolf, 2004). The local economy profits from tourism that is related to wildlife and bird watching and environmental restoration projects and urban greenery can provide nesting areas and habitat (*Tucson Audubon Society, 2013*).

Figure 7 illustrates the potential benefits of ESS supplied by urban greenery, based on the integrated concept of sustainability, combining the triple-bottom line of economic, social and environmental aspects.

5 INSIGHT FROM THE WATER COMMUNITY

The insights presented here illustrate the diverse views on water management in the Tucson region. They were obtained during interviews with expert agency officials and stakeholder representatives. The public opinion on Tucson's situation concerning water resources is divided. On the one hand, environmental groups and other non-profit organizations perceive Tucson as defined by a rather traditional water management view, focusing mostly on gray infrastructure, designed with relatively little innovations, and which could do significantly more to adjust its water supply and demand scheme to become more sustainable. These interviewees understand water scarcity as a reason to worry and point out that, given the circumstance of a desert community and the depleted aquifers, more efforts to utilize sustainable sources should be carried out combined with intensified conservation and monitored growth.

On the other hand, government representatives and water agency officials view Tucson as having an ensured water supply for a minimum of 25 years, without any changes in practice or policy, and are therefore ahead of numerous other communities. Comparisons are often drawn between regulations and practices in Arizona and those in drought-affected California. Comparisons are also made between Tucson and other regions in Arizona that either have no access to CAP water or less underground storage and recovery capacity such as Phoenix. In such comparisons, the current situation and the future of Tucson's water supply is described as sufficient and sustainable.

5.1 Social dimension

The United States Bureau of Reclamation (USBR) projects that the Santa Cruz region faces a substantial risk for insufficient water supply by 2025 and stated that a "water supply-related crises will affect economies and resources of national and international importance" (*USBR, 2005*). However, the main message from official institutions is that water supply presents no reason to worry. Groundwater and CAP supply were perceived to be abundant and the official position is that these conditions can surely be expected to occur in the future. In July 2015, the mayor of Tucson published an article in a local newspaper with the headline *"Tucson's water strategy ensures a secure, sustainable future"* (Rothschild, 2015). He points out the city's attractiveness for business establishment due to its long-term water supply. The common message is, *"it's better than in a lot of places."* A government representative expressed the following statement, *"We don't need the water right now, that's the thing. We don't have to catch every drop of rain so that we have something to eke out to live on this planet. We are not there. We are so very well planned."* An interviewee from the environmental community reports the opposite, saying, *"We are in a desert and we can't afford to lose this water. We should capture every drop."* The topic of future scarcity has been mentioned by a stakeholder from an environmental organization as *"the elephant in the room"* and the inaction of the past years is understood as hubris as *"not only is Arizona feeling good about Arizona, Tucson is feeling good about Tucson."*

Government officials understand water conservation and the residential utilization of rainwater, stormwater and graywater as a life-style choice. It has been emphasized that the significant use of these sources is very unlikely, too expensive and that *"people don't really want to do that. They want to make it work the way it works right now."* Environmental stakeholders stress the idea that the use of these sources does not mean draconian measures and a loss of life quality. Conservation efforts are not seen as a sign of shortage and scarcity, but as *"just the right thing to do"* concerning the natural conditions in a desert environment.

While policies such as the Commercial Rainwater Harvesting Ordinance are a viable measure for commercial developments, it is not a common practice in Arizona to limit private landscaping and irrigation through administrative or regulatory practice (*Jacobs and Holway, 2004*). The interviewed stakeholders agree widely that given the importance of individual freedom, well-placed incentives are more appropriate and under the equality approach, the community as a whole should be supported rather than just individuals. Incentives should be complemented by projects focusing on education and awareness building. Nevertheless, according to the environmental community it should become standard practice to cover at least a part of the landscape irrigation by rainwater and *"every house owner should be responsible to use the water that falls on the property and enhance greenery; and be responsible to green the desert."* It was highlighted that all projects on private, neighborhood or regional level should be examples for good practices and illustrate appropriate behavior and infrastructure that reduce water demand or increase water efficiency should be installed to support a cultural shift.

Interviewees from the environmental community have stressed the fact that cultural aspects have been widely ignored in the past and that awareness of good practices and the cultural value of the desert environment can introduce change into the system that results in changes in legislation. In Tucson the bottom-up approach that strongly engages the community and shapes changes in legislation through practice has a long history and grassroots movements and community led non-profit initiatives have been significant drivers for environmental projects and local conservation initiatives. Water management policy is a question of political will and activism of city council members, whereas water pricing and allocation are politically dangerous topics. The development and implementation of new policies are a way to promote innovation in the community. Individual initiatives creating peer pressure and influencing regulations and the top-down change in policies are mutually influencing.

5.2 Economic dimension

The economic crisis of the past decade and the lack of funding have resulted in relatively little action within the water sector. Water providers faced poor revenues and citizen involvement and action opposing water rate hikes has increased. Discussions on water rate increases are classified as an unpopular topic, while most interviewees agreed, *"it all goes back to the price of water."* Water supply is considered underpriced by numerous individuals considering it scarce. As water bills do not account for the intrinsic value of water, but only the cost of treatment and delivery, other costs such as scarcity/opportunity, environmental, and generational costs, are not reflected in the water price. Representatives of NGOs expressed the hope that *"we will come to understand the value of water better"*.

Benjamin Franklin said in 1774, *"when the well's dry, we know the worth of water."* In the case of Tucson, the implementation of the CAP avoided this realization to this point. Considering the high energy demand and the infrastructural efforts of supply, a representative of a local non-profit organization calls CAP, *"ridiculously cheap"* in comparison with the other available sources. Its outstanding performance is based on federal intervention and subsidies. The CAP had been planned as a means to provide irrigation water for the highly influential agricultural sector, represented by a local manager saying, *"Water flows uphill towards the money."* It is expected that the first shortage declaration will cause an increase in cost and *"water is about to get a lot more expensive in some areas."* Government officials know that reclaimed water treatment will be used as potable supply or the treatment of CAP to reduce its salt content is inevitable in the future (*Tucson Water, 2004*). The unperceived urgency (*"human supply is not setting off any alarm bells"*), high investment costs (*"until there is financial recovery we won't see a lot new great infrastructure"*) and the presumed politically unpalatable topic of 'toilet to tap' are likely to delay their implementation by decades.

A topic of controversial discussion is the Rain Water Harvesting Rebate program offered by Tucson Water. The limited level of participation in the utility's promotion of these resources indicates how difficult it will be to implement the decentralized approach throughout the community. To date the rebate program has not achieved the intended goal of reducing water demand and has been criticized for its inefficiency and inequality. It was pointed out that similar investments could lead to much more water savings and support low-income customers by setting incentives for the implementation of low-flow toilets and showers instead of RWH systems (*Steller, 2015*). Stakeholders from the environmental community interpret the outcome as valuable data on which to build and ask for a broader perspective as new landscape goals are accomplished benefiting the community. Several interviewees reported personal experience of drastically reduced water consumption because of the rebate program. However, it has been mentioned by numerous interviewees that rebates for earth works and rain gardens might hold more potential than active systems in terms of cost-effectiveness and equality. Furthermore, the opinion was expressed that increased educational support on the use of RWH systems will be necessary in order to reduce outdoor demand for potable water.

A government representative admitted, "*once the water price goes up, there will be less green*" referring to reduced irrigation in the case of cuts in funding. It all goes back to the value of urban greenery and the financial support responsible agencies are willing to give. Additionally an interviewee associated with water management on regional level said, "*The question is, will we still have the natural amenities to draw people, will people come here for the environment? Is the environment an economic driver, or just a nice thing to have? And that really depends on where we are in comparison to other places.*" Some stakeholders support and emphasize the potential of green infrastructure for the development of Tucson as a lush oasis and numerous interviewees identified urban greenery as a pulling factor for people and therefore economic development.

5.3 Environmental dimension

Overexploitation of groundwater resources in the past led to the disappearance of many valuable and biodiverse riparian corridors. It is questionable which environmental state should be taken as a baseline to discuss water management approaches, pristine or current conditions? On one hand, interviews with official water resource managers gave the insight that it is not considered possible to reverse this ecological damage and manage water resources to return to pristine conditions. On the other hand, members of the environmental community mention the recovery of riparian areas and the reestablishment of the Santa Cruz as a natural flowing river as an ideal development. In addition, environmental stakeholders are concerned that a crisis is needed to do things differently. Representatives from environmental groups and non-profit organizations identify a climate of denial on the implications of water scarcity in Tucson.

Existing riparian vegetation is either dependent on shallow groundwater areas or the allocation of effluent towards the Santa Cruz River and are therefore directly threatened by human actions. Water managers indicated that water is currently only recharged to the stream as a result of the momentary lack of financial capacity to put the water in different use. In addition, managed recharge facilities are assigned only 50% water credits while constructed facilities result in credits equal to the amount of water recharged (Arizona Revised Statutes.) (Chapter 3.1, Article 4, 852.01.C.1). While these regulations support a greater cut to the aquifer and therefore support replenishment of the aquifer, there are no significant incentives in place to discharge effluent to the channel to support the aquatic and riparian ecosystem. Therefore, alternative use is planned by all parties for the future and solutions need to be developed to ensure flow in the channel to keep the riparian corridor. Interviewees expressed the view that a reduction in effluent discharge to the river that caused further environmental degradation would likely be prohibited due to a required environmental impact assessment after the National Environmental Policy Act (NEPA).

The Conservation Effluent Pool- (CEP) Agreement, assigning effluent for the environment has been called, "a *good idea gone wrong*" by a government official and its realization is subject to criticism. The official reason for the current proposal to allocate a great part of the annually 12.33 mm^3 (10,000 AF) CEP water to the Santa Cruz River is the possible depletion of the ecosystems. However, the fear was sensed that if not put into use, the CEP could be under debate again and utilized in a different manner. In addition, it was described by an interviewee how the economic sector reacted with concern to the allocation of such a significant share of effluent towards environmental restoration instead of industrial development. The use of water and allocations for different users has great potential for dispute and conflict on a local and regional scale that increases with scarcity.

Another means to preserve riparian ecosystems and support urban greenery is the 'Conserve to Enhance' program that has been implemented to provide funding for environmental restoration through participant donations, based on water savings. It is widely acknowledged as very successful by the environmental community. This can be traced back to the fact that it provides funding that otherwise would have been not available. However, criticism has been expressed about residents being made responsible for tasks, which the government or local agencies should undertake, and provide sufficient financial means for such restoration programs.

Environmental group representatives emphasize that there should be RWH carried out to support urban greenery whenever there is reasonable redevelopment, similar to the green street policy. Criticism has been expressed by environmental stakeholders that not all new streets and retrofits comply with this regulation and that the implementation of GI along medians and on roundabouts is not the default. Lancaster (2015) advances the view that the ultimate purpose of GI is to design, implement and maintain oases that are thriving on natural, local resources without overuse. Interviewees emphasized that the water management should mimic a natural hydrological cycle, where water is cycled instead of quickly drained. Environmental representatives emphasized that *"we need to recycle what we have multiple times and do it in the lowest-energy, highest-productive way"* meaning that it should become standard practice to implement landscapes that harvest water instead of draining water.

Local non-profit organizations and professionals from the environmental sector see much greater potential in GI projects stating, *"Stormwater should not be understood as a threat, but a wasted resource."* Local water managers, in contrast, state that *"What should we do with the most contaminated water around?"*. The message from the stakeholders involved in the implementation of RWH systems and GI is that *"everything is under capacity."* Representatives from the environmental group and some government officials support the approach to develop a payment mechanism for stormwater utilization and management.

6 CONCLUSION

The goal of the work presented here was to evaluate the significance of rainwater harvesting, stormwater capture and greywater reuse practices in Tucson, in the broader context of traditional resource use such as groundwater and CAP transfer water, and assess their current and potential future contribution to sustainability. Significant efforts exist to date: effluent recharge and reuse (Agua Nueva and Tres Rios Water Reclamation Facility), multi-purpose stormwater capture and effluent facility at KERP, rebate programs, city ordinances and community initiatives for rainwater and street stormwater harvesting, grey water reuse, etc.

However, to date, rainwater, stormwater and graywater use are not measured and represent a relatively small percentage of the water budget. Much more could be done in the future. This research concludes:

- Increased groundwater use in the future will be necessary if water demand develops as projected, calling for the intensification of conservation efforts. While CAP supply is threatened on the long term by climate change and institutional arrangements, Colorado River water will be the predominant source in the next decades (>54%). Intensified use of reclaimed water and treatment to potable standards will likely be necessary in the future in order to cover demand without further groundwater over-exploitation.
- The utilization of rainwater and stormwater has the potential to reduce pressure on the aquifer and cover more than 20% of municipal water demand in the TAMA. Graywater re-use could contribute up to 2% without affecting sewer flows. These sources can have significant impact on both water provision and re-greening of urban areas. Their utilization can be understood as an adaptive approach to reduce vulnerability and improve resilience of the social-ecological system.
- High implementation cost is currently the main constrain for the utilization of these alternative sources. Efforts of water managers, stakeholders from the environmental community and city planners should be coordinated to intensify research on best practices and cost-effectiveness.
- The integration of ecosystem services and the contribution of both riparian ecosystems and urban greenery to human well-being should be better integrated in cost-benefit analyses and water resources planning. Having reached safe yield in the TAMA, it remains to be seen if Tucson as a community will be capable of maintaining and conserving the last existing natural riparian systems that depend on shallow-groundwater areas in the fringes of the basin.
- The implementation of a specific stormwater utility that provides funding and service for municipal GI projects may be needed to utilize rainwater and stormwater in a significant and meaningful way.

Whether Tucson's water management is sustainable in the long term will depend on the measures that are put in place to meet water needs of the community and the environment under the constraint of further population growth, potential CAP shortages and local impacts of climate change. The development of alternative water resources, increasing conservation, embracing green infrastructure, and tailoring water uses to the water quality level, will contribute towards sustainability and resilience.

REFERENCES

ADWR (2012). Tucson Active Management Area: Water Demand and Supply Assessment: 1985–2025. Arizona Department of Water Resources.
ADWR (2014). Arizona's Next Century: A Strategic Vision for Water Supply Sustainability: Executive Summary. Arizona Department of Water Resources.

Arizona Revised Statutes. Title 45: Waters. http://www.azleg.gov/ArizonaRevisedStatutes.asp?Title=45. Cited September 15, 2015.

Arizona Revised Statutes. Title 49: Environment. http://www.azleg.gov/ArizonaRevisedStatutes.asp?Title=49. Cited September 15, 2015.

Bark, R.H., Osgood, D.E., Colby, B.G., Katz, G. and Stromberg, J. (2009) Habitat preservation and restoration: Do homebuyers have preferences for quality habitat? *Ecological Economics* 68: 1465–1475. doi: 10.1016/j.ecolecon.2008.10.005.

Bolund, P. and Hunhammar, S. (1999). Ecosystem services in urban areas. *Ecological Economics* 29: 293–301. doi: 10.1016/S0921-8009(99)00013-0.

Centers for Disease Control and Prevention (2005). Heat-Related Mortality: *Arizona*, 1993–2002, and United States, 1979–2002. MMWR: Morbidity and mortality weekly report 54: 628–630.

Cleveland, J., Eden, S., Moxley, J. and Audrey, A. (2015). Charting a path toward integrated water resources: Using water harvesting as a community-scale problem solver. *Stormwater: the Journal for Surface Water Quality Professionals*, 32–43.

Colby, B.G. and Wishart, S. (2002). Quantifying the influence of desert riparian areas on residential property values. *Appraisal Journal* 70: 304–308.

Coley, R.L., Sullivan, W.C. and Kuo, F.E. (1997). Where does community grow? The Social Context Created by Nature in Urban Public Housing. *Environment and Behavior* 29: 468–494. doi: 10.1177/001391659702900402.

Comrie, A.C. (2000). Mapping a Wind–Modified Urban Heat Island in Tucson, Arizona (with Comments on Integrating Research and Undergraduate Learning). *Bulletin of the American Meteorological Society* 81: 2417–2431. doi: 10.1175/1520–0477(2000)081<2417:MAWMUH>2.3.CO;2.

Connall, Jr. D.D. (1982). History of the Arizona Groundwater Management Act. *Arizona State Law Journal* 14: 313–322.

Crook, J. and Rimer, A. (2009). Technical Memorandum on Graywater. Environmental Engineering Consultant, Boston, Massachusetts.

DaRonco, D. (2012). City Council orders review of gray-water requirement. *Arizona Daily Star*, December 12, 2012.

Davis, T. (2015). Tucson's rain-catching revolution: In the Sonoran Desert, rainwater harvesting is finally going mainstream. *High Country News*, April 27, 2015.

Eden, S., Scott, C.A., Lamberton, M.L. and Megdal, S.B. (2011). Water-energy interdependencies and the Central Arizona Project. In: Kenney, D.S., Woilkinson, R. (eds) The water-energy nexus in the American West: 109–122.

EPA (2012). Guidelines for Water Reuse: U.S. EPA/600/R-12/618. U.S. Environmental protection agency, Washington D.C.

Gastélum, J.R. (2012). Analysis of Arizona's Water Resources System. *International Journal of Water Resources Development* 28: 615–628. doi: 10.1080/07900627.2011.625524.

Gaston, T.L. (2010). Rainwater harvesting in the southwestern United States: A policy review of the four corners states. University of arizona Water Resources Research Center, Tucson.

Gleick, P.H. (1998). Water in crisis: paths to sustainable water use. *Ecological Applications* 8: 571–579.doi: 10.1890/1051-0761(1998)008[0571:WICPTS]2.0.CO;2.

Graf, C. (2012). Large-scale non-household gray water resuse in Arizona. *AZ Water Association* spring 2012: 18–19.

Graf, C. (2014). Regulating Gray Water in Arizona: Lessons learned. presented at the national research council committee on the On-site reuse of graywater and stormwater, marina del rey, CA, January 21, 2014.

Graf, C. (2015). Personal Communication: Cost-effectiveness of graywater use.

Grafton, R.Q., Pittock, J., Davis, R., Williams, J., Fu, G., Warburton, M., Udall, B., McKenzie, R., Yu, X., Che, N., Connell, D., Jiang, Q., Kompas, T., Lynch, A., Norris, R., Possingham, H. and Quiggin, J. (2013). Global insights into water resources, climate change and governance. *Nature Climate Change* 3: 315–321. doi: 10.1038/nclimate1746.

Halper, E.B. (2011). Do open space resources shape residential outdoor water consumption? http://citeseerx.ist.psu.edu/viewdoc/summary?doi=10.1.1.551.5458. Cited 23. 07. 2015.

Halper, E.B., Dall'erba, S., Bark, R.H., Scott, C.A. and Yool, S.R. (2015). Effects of irrigated parks on outdoor residential water use in a semi-arid city. *Landscape and Urban Planning* 134: 210–220. doi: 10.1016/j.landurbplan.2014.09.005.

Hirt, P., Gustafson, A. and Larson, K. (2008). The Mirage in the Valley of the Sun. *Environmental History* 13: 482–514. doi: 10.1093/envhis/13.3.482

Huang, L.Y. (2015). Water Quality Characteristics of Greywater and Lessons from the Arizona Experience. University of arizona Water Resources Research Center, Tucson.

Hutchinson, N. (2015). 21st century dams. *Geodate* 28: 9–11.

Ingram, B.L. (2015). To cope with California's drought, policymakers must go beyond water conservation and rationing. *American Politics and Policy Blog*, http://eprints.lse.ac.uk/62696/. Cited June 08, 2015.

Jackson, R. (2015). Green Infrastructure benefits for personal and community health. Presented at the Second Green Border Infrastructure Forum, Tucson, AZ, May 21, 2015.

Jackson, R.J. and Kochtitzky, C. (2002). Creating A Healthy Environment: The Impact of the Built Environment on Public Health. Centers for Disease Control and Prevention, Washington D.C.

Jacobs, K. and Holway, J. (2004). Managing for sustainability in an arid climate: lessons learned from 20 years of groundwater management in Arizona, USA. *Hydrogeology Journal* 12: 52–65. doi: 10.1007/s10040-003-0308-y.

Kalkstein, A.J. and Sheridan, S.C. (2007). The social impacts of the heat–health watch/warning system in Phoenix, Arizona: assessing the perceived risk and response of the public. *International Journal of Biometeorology* 52: 43–55. doi: 10.1007/s00484-006-0073-4.

Kandziora, M., Burkhard, B. and Müller, F. (2013). Interactions of ecosystem properties, ecosystem integrity and ecosystem service indicators—A theoretical matrix exercise. *Ecological Indicators* 28: 54–78. doi: 10.1016/j.ecolind.2012.09.006.

Kaplan, R. (1984). Impact of urban nature: A theoretical analysis. *Urban Ecology* 8: 189–197. doi: 10.1016/0304-4009(84)90034-2.

Kardan, O., Gozdyra, P., Misic, B., Moola, F., Palmer, L.J., Paus, T. and Berman, M.G. (2015). Neighborhood greenspace and health in a large urban center. *Scientific Reports* 5: 1–14. doi: 10.1038/srep11610.

Kuo, F.E. (2001). Coping with Poverty: Impacts of Environment and Attention in the Inner City. *Environment and Behavior* 33: 5–34. doi: 10.1177/00139160121972846.

Kuo, F.E., Sullivan, W., Coley, R.L. and Brunson, L. (1998). Fertile Ground for Community: Inner-City Neighborhood Common Spaces. *American Journal of Community Psychology* 26: 823–851. doi: 10.1023/A:1022294028903.

Kuo, F.E. and Sullivan, W.C. (2001). Environment and Crime in the Inner City: Does Vegetation Reduce Crime? *Environment and Behavior* 33: 343–367. doi: 10.1177/0013916501333002.

Lancaster, B. (2013). Rainwater Harvesting for Drylands and Beyond: Volume 1: Guiding Principles to Welcome Rain into Your Life and Landscape. Rainsource Press, Tucson.

Lancaster, B. (2015). Thoughts on Green Infrastructure: An invitation to growing oases. Presented at the Second Green Border Infrastructure Forum, Tucson, AZ, May 20.

Lancaster, B., Lane, L., Hartman, M., Lellou, B. and Strassberg, V. (2011). The Energy Costs of Water (ECW). http://www.harvestingrainwater.com/water-energy-carbon-nexus.

Loucks, D.P. (2000). Sustainable water resources management. *Water International* 25: 3–10. doi:10.1080/02508060008686793.

Loucks, D.P. and Gladwell, J.S. (1999). Sustainability criteria for water resource systems. Cambridge University Press, Cambridge.

Ludwig, A. (2012). Create an oasis with greywater: choosing, building and using greywater systems, includes branched drains. Oasis Design, Santa Bárbara, CA.

Lustgarten, A. (2015). End of the miracle machines: Inside the power plant fueling America's drought. ProPublica, June 16, 2015.
Maes, J., Teller, A., Erhard, M., Liquete, C., Braat, L., Berry, P., Egoh, B., Puydarrieux, P., Fiorina, C. and Santos-Martín, F. (2013). Mapping and Assessment of Ecosystems and their Services-An analytical framework for ecosystem assessmentsunder action 5 of the EU biodiversity strategy to 2020. European Union, Luxembourg.
McPherson, E.G. and Muchnick, J. (2005). Effect of street tree shade on asphalt concrete pavement performance. *Journal of Arboriculture* 31: 303–310.
National Drought Mitigation Center (2015). U.S. Drought Monitor. http://droughtmonitor.unl.edu/. Cited June 28, 2015.
NOAA (2015) Monthly and Daily Normals (1981–2010). plus Daily Extremes (1895–2015) for Tucson, Arizona. http://www.wrh.noaa.gov/twc/climate/tus.php. Cited 30. 05. 2015.
Nowak, D.J., Crane, D.E. and Stevens, J.C. (2006) Air pollution removal by urban trees and shrubs in the United States. *Urban Forestry & Urban Greening* 4: 115–123. doi: 10.1016/j.ufug.2006.01.007.
Ogata, I. (2015). Economic studies of green infrastructure in Tucson presented at the second green border infrastructure forum Tucson, AZ, May 21, 2015.
Pima County RWRD (2013). 2013 Effluent generation and utilization report. Regional Wastewater Reclamation Department, Tucson.
Potter, K.W. (2000). Final report: Field evaluation of rain gardens as a method for enhancing groundwater recharge. University of Wisconsin-Madison, Madison, Wisconsin.
RFCD (2015). Dataset on flow measurements in KERP. Personal Communication.
Rothschild, J. (2015). Tucson's water strategy ensures a secure, sustainable future. Arizona Daily Star, July 26, 2015.
Shuster, W., Gehring, R. and Gerken, J. (2007). Prospects for enhanced groundwater recharge via infltration of urban storm water runoff: A case study. *Journal of Soil and Water Conservation* 62: 129–137.
Sonoran Institute & Pima County (2013). A living river: Charting wetland conditions of the lower sanata cruz river. Sonoran Institute & Pima County, Tucson.
Sonoran Institute & Pima County (2014). A living river: Charting wetland conditions of the lower santa cruz river. Sonoran Institute & Pima County, Tucson.
Steller, T. (2015). Trend towards rainwater harvesting meets adversary. Arizona Daily Star, March 15, 2015.
Sullivan, W.C. and Kuo, F.E. (1996). Do trees strengthen urban communities, reduce domestic violence? Urban Forestry Center for the Midwestern states, Illinois.
TDOT (2013). Active practices guidelines: green streets. City of Tucson, Department of Transportation, Tucson.
Tucson & Pima County (2009). Water and wastewater infrastructure, supply and planning study: Phase 1: Chapter 2. City of Tucson & Pima County, Tucson.
Tucson & Pima County (2010). Intergovernmental agreement between pima county and the city of tucson for implementation of the conservation effluent pool. City of Tucson & Pima County.
Tucson Audubon Society (2013). Economic contributions of wildlife viewing to the Arizona economy: A county-level analysis. Tucson Audubon Society, Tucson.
Tucson Water (2004). Water Plan 2000–2050. City of Tucson Water Department, Tucson.
Tucson Water (2012). 2012 Update: Water plan: 2000–2050. City of Tucson Water Department, Tucson.
Tucson Water (2013). Recycled water master plan: Volume 1. City of Tucson Water Department, Tucson.
Tucson Water Department (2015). Personal communication: Water supply data.
USBR (2005). Water 2025: Preventing crises and conflict in the West. U.S. Bureau of Reclamation.
USBR (2012). Colorado river basin water and demand study: Executive summary. U.S. Bureau of Reclamation.

USBR (2015a). Colorado river basin update and status: September 2015. U.S. Bureau of Reclamation.

USBR (2015b). The Colorado River System: Projected future conditions 2016–2020: August 2015. U.S. Bureau of Reclamation.

USBR & Tucson Water (2004). Reverse osmosis treatment of central Arizona: Project water for the city of Tucson: desalination research and development program report No. 36. U.S. Department of the Interior, Bureau of Reclamation; City of Tucson Water Department.

Water CASA (2004). Residential graywater reuse study. Water Conservation Alliance of Southern Arizona, Tucson, Arizona.

Watershed Management Group (2015a). Green streets primer. https://watershedmg.org/learn/resources/green-streets-primer. Cited 14. 06. 2015.

Watershed Management Group (2015b). Solving flooding challenges with green stormwater: infrastructure in the airport wash area. Watershed Management Group, Tucson.

Wise J.S. (2015). Evaluation of GI/LID Benefits in the desert Environment-AutoCASE. Presented at the sustainability director's network, Green Infrastructure Group, April 9 2015.

Wolf K.L. (2004). Trees and business district preferences: A case study of Athens, Georgia, US. *Journal of Arboriculture* 30: 336–346.

WRDC (2011). Water resources development commission final report: Volume I. Water Resources Development Commission.

WRRC (2011). Salinity research is wide and varied. https://wrrc.arizona.edu/awr/s11/salinity. Cited August 12, 2015.

Zinser, M., Ramberg, F. and Willott, E. (2004). Scientific note culex quinquefasciatus (Diptera: Culicidae) as a potential West Nile virus vector in Tucson, Arizona: Blood meal analysis indicates feeding on both humans and birds. *Journal of Insect Science* 4: 1–3. doi:10.1093/jis/4.1.20.

Chapter 18

Differentiated approaches of groundwater management: A comparison between the TAMA and the San Pedro basin

Susan Harris
Department of Hydrology and Atmospheric Sciences, University of Arizona, USA

INTRODUCTION

Due to the environmental and economic consequences of long term, substantial groundwater use, Arizona and its stakeholders have had to adopt and implement new policies to address groundwater pumping. Survival of riparian ecosystems depends upon the existence of consistently high groundwater levels that are within reach of vegetation in riparian corridors. The effects of pumping caused by a quickly growing population's increasing demand for water threaten both existing riparian ecosystems and the long term water security of that same population.

In two groundwater basins in Southern Arizona, two very different approaches have been taken to manage water to achieve this elusive balance between economic activity and the environment. The Groundwater Management Act enacted by the Arizona Legislature and the influx of the water from the Central Arizona Project (CAP) financed by the federal government drive many of the policies and practices implemented by institutional water managers in the Tucson Active Management Area (TAMA) (see *Figure 1*). In contrast, the Groundwater Management Act does not influence much of the water management in the Upper San Pedro Basin because very little of that legislation applies to that basin. Similarly, CAP water has no impact on Upper San Pedro Basin policies because no CAP water arrives in the basin (*Figure 1*). In addition, unlike the TAMA, the ecosystem of the Upper San Pedro Basin includes a national riparian habitat that requires a steady flow of water from the aquifer and thus groundwater levels close to the riverbed. Also unlike the TAMA, the Upper San Pedro Basin has received millions of dollars in federal funds that have been spent on scientific research to provide knowledge on which to base sound water management practices. The funding has also supported the operations of a stakeholder group of representatives from the federal, state, and local governments and environmental groups to address the impact of pumping by the basin's main cities, the military base and other users. Accordingly, an examination of the TAMA and the Upper San Pedro Basin provides insights into the relative efficacy of the management of water resources using a "top down" regulatory approach compared to a "bottom-up," collaborative, community-based approach between stakeholder organizations and agencies and the scientific community.

Figure 1 Map of the Central Arizona Project from the Colorado River to Pima County, Arizona. Source: Central Arizona Project.

1 DEVELOPMENT OF TWO MANAGEMENT APPROACHES

Arizona, historically a rural, sparsely populated state reliant on agriculture and mining, has long depended upon groundwater to satisfy a significant portion of its water needs (*Saliba & Jacob, 2008*). During its formative years, the Arizona courts applied and further developed the common law that governed the allocation and use of groundwater. Under the common law doctrine of "reasonable use", landowners could pump unlimited amounts of groundwater "for a reasonable, beneficial use of the land from which the same is taken" (*Bristor v. Cheatham 1953; Pima Farms v. Proctor, 1926*). The *Bristor* court stated:

> A great majority of the states which in recent years have been presented with this problem adhere to the principle that the owner of lands overlying ground waters may freely, without liability to an adjoining user, use the same without

limitation and without liability to another owner, providing his use thereof is for the purpose of reasonably putting the land from which the water is taken to a beneficial use. *(75 Ariz. at 235, 255 P.2d at 178.)*

Unlike with surface water, controversies rarely arose over title to groundwater or its apportionment among users *(Schaffer, 2010)*. Abundant groundwater supplies historically made it difficult to deplete the common supply or to interfere with groundwater use by others. In time, however, the growing demand for groundwater, improvements in pumping technology, and the economic benefits to be garnered from accelerated groundwater use throughout portions of Arizona led to serious overdrafts in the aquifers underlying the State's major urban areas and the agricultural areas *(Maguire, 2007)*.

By the 1960s, Tucson began to acquire land outside the city limits solely for the purpose of pumping groundwater to satisfy its growing demand. It purchased more than 21,000 acres of farmland in the Avra Valley, approximately 16 to 18 miles outside the city limits, and proceeded to pump and transport the groundwater to meet municipal demand. Similarly, during this period, mining companies transported water away from rural areas, threatening the water supplies of agricultural producers *(Schaffer, 2010)*. Legal challenges to these actions resulted in a series of ruling issued by the Arizona Supreme Court in *Jarvis v. State Land Department I, II, III (1969, 1970, and 1976)* and *Farmers Investment Company v. Bettwy (1976)*. These decisions imposed constraints on the transport of unlimited amounts of groundwater and threatened to disrupt both economically important mining operations in Arizona and municipal deliveries of water to many thousands of residential and commercial water users *(Staudenmaier, 2007; Pearce 2007)*. By 1976, groundwater withdrawals had exceeded natural recharge from rain and snow by 2.2 million acre feet *(Glennon and Maddock, 1994)*. The Arizona legislature succinctly captured the escalating problem in its 1980 assessment of the water situation:

> The people of Arizona are dependent in whole or in part upon groundwater basins for their water supply and that in many basins and sub-basins withdrawal of groundwater is greatly in excess of the safe annual yield and that this is threatening to destroy the economy of certain areas of this State and is threatening to do substantial injury to the general economy and welfare of this State and its citizens. *(Arizona Revised Statutes (A.R.S.) §45-401(A))*.

In 1980, two landmark events occurred to address the significantly depleted aquifers in the state: the Arizona legislature passed the Groundwater Management Act and Congress approved funding for the CAP. The Groundwater Management Act and the CAP primarily impacted Arizona's largest urban areas and agricultural areas affected by the groundwater overdrafts. The Groundwater Management Act designated these geographic areas as Active Management Areas (AMAs) *(A.R.S. §45-402(2))*. By establishing the AMAs, the legislature sought to craft tools to help the newly created state agency, the Arizona Department of Water Resources (ADWR), to more proactively govern the use of groundwater in high use or populated areas *(A.R.S. §45-411(A))*.

At that same time, Congress provided the funding to construct the CAP to bring Colorado River water to three of the AMAs to meet the demands of the rapidly

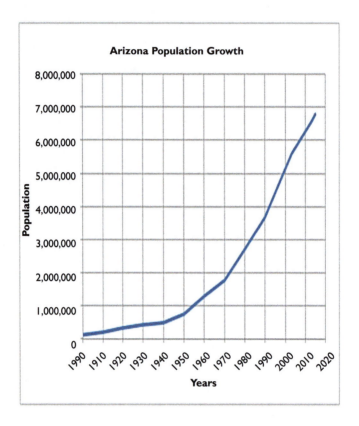

Figure 2 Source of Data: U.S. Census Bureau.

growing population. State law provided the governance structure for the CAP when it established the Central Arizona Water Conservation District (CAWCD), a state entity known as a "municipal corporation" with taxing authority and an elected board. By design, neither the Groundwater Management Act nor the CAP have a direct impact on the rural areas. Thus, Arizona has effectively managed the use of its groundwater and the demands of its growing population (*Figure 2*) using two basically different approaches and philosophies: Areas designated as AMAs are controlled by the Groundwater Management Act and supplied with CAP water (the Prescott AMA in north central Arizona does not receive CAP water). Areas outside of the AMAs remain subject to the common law of reasonable use applicable to groundwater and receive only the water provided by the ecosystem.

2 THE TUCSON ACTIVE MANAGEMENT AREA

In the case of the TAMA, Arizona has implemented a more traditional top down regulatory model for groundwater management. The Tucson Active Management Area consists of two groundwater sub-basins, which are hydrologically distinct bodies of groundwater located in Avra Valley and in the Santa Cruz Valley under the City of Tucson. To support its growing population and sustain its historic agricultural operations, the TAMA relies on two sources of water: groundwater and CAP water.

The Arizona Legislature recognized both that agricultural operations would continue in AMAs, and that the use of water for agricultural purposes contributed more to the depletion of existing groundwater resources than any other use. Thus, the Groundwater Management Act generally prohibited the irrigation of any land that had not already been irrigated during the preceding five years. This provision had the effect of banning the use of water to irrigate land in TAMA unless the land had been irrigated at some time during 1975 through 1979 (*Megdal, 2008*). Further, the Groundwater Management Act created a system of groundwater rights and permits for most groundwater pumpers, and required owners of large wells to meter or measure groundwater pumping and to report groundwater withdrawals. The new law also defined a management goal for TAMA, which must be met by January 1, 2025 (*A.R.S. § 45-562(A)*). This legislative mandate states that, on a long-term basis, TAMA's annual withdrawal of groundwater must not exceed the amount of water naturally and artificially recharged each year (*A.R.S. § 45-561(12)*). This long-term balance of withdrawals against recharge is known as "safe yield". The Groundwater Management Act further details an incremental process to achieve this management goal, consisting of a series of five management plans that establish conservation regulations for the municipal, industrial, and agricultural sectors.

The Groundwater Management Act also had the potential to restrict new development of commercial and residential subdivisions in TAMA. Before development of this type can proceed, the developer must demonstrate the availability of a 100-year water supply from renewable sources, also known as an "assured water supply" (*A.R.S. §45-576*). This requirement for a developer to substantiate the existence of an assured water supply entails a showing by the developer to the Arizona Department of Water Resources that: (1) sufficient groundwater, surface water, or effluent of adequate quality will be continuously available to satisfy the water needs of the development for at least 100 years; (2) the projected use is consistent with the applicable management plan; and (3) the developer has the financial capability to construct the water facilities necessary for the water supply (*A.R.S. §45-476(J)*).

As shown by this brief summary of the Groundwater Management Act, the state regulatory structure prohibits new agricultural land use in TAMA, restricts groundwater pumping, mandates conservation measures, and may limit new development. Enforcement and implementation of this law in TAMA resides in the state under the auspices of the Department of Water Resources and a Groundwater Users Advisory Council, a five member board appointed by the governor, to represent the TAMA groundwater users. TAMA's efforts to attain its management goals are also aided by the federally-financed CAP, managed by the Central Arizona Water Conservation District, another state entity. Beginning in 1992, the CAP has provided a significant new source of water to TAMA, and more specifically to its aquifer. As shown in *Figure 3*, the CAP has provided a steadily increasing source of recharge to the aquifer which is then used to provide groundwater for municipal, agricultural and industrial purposes.

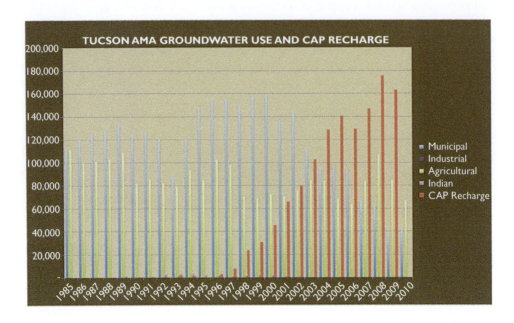

Figure 3 Source of data: Arizona Department of Water Resources, Tucson AMA Historical Assessment 1985–2010.

3 UPPER SAN PEDRO BASIN

In contrast to the regulation of groundwater in the AMAs in general, and TAMA in particular, state regulation does not govern groundwater management in rural Arizona. State law does not require farmers, municipalities and industries outside of an AMA to adopt management plans or conservation measures or to achieve management goals. It also does not impose a similar obligation on developers of commercial and residential subdivisions to demonstrate the existence of an assured 100 years water supply as a prerequisite to building. Also unlike in TAMA, no CAP water supplies the needs of the Upper San Pedro Basin and constraints have been imposed on the use of existing surface water in the San Pedro River.

The Upper San Pedro Basin holds approximately 20 to 26 million acre-feet in storage (ADWR, 2005) so it does not appear, at least initially, that existing groundwater presents any cause for concern. As of 2005, when approximately 126,459 people lived in the basin, ADWR found that the current primary water demand sectors were agricultural and municipal with relatively little demand from the industrial sector. In the intervening years, the population in the area has continued to expand. As of 2014, approximately 130,000 people live and work in seven incorporated towns and several unincorporated communities in the two counties within the Upper San Pedro River Basin with the population expected to grow by another 45% by 2050 (*Figure 4*).

As they continue to pump, the City of Sierra Vista, Fort Huachuca, industry, owners of exempt wells, private water companies and farmers deplete the groundwater in the Upper San Pedro Basin. During the period 1990 – 2001, water levels declined

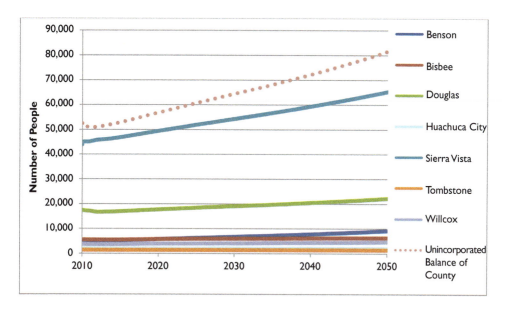

Figure 4 Source of Data: Arizona Department of Administration, Office of Population and Statistics).

in some areas around Sierra Vista by almost 15 feet (*ADWR, 2005*). Groundwater depletion continues to occur, as evidenced by the groundwater budgets prepared for the basin between 2003 and 2011 that report annual storage deficits ranging from 8,000 to 20,000 acre-feet per year (*Richter, 2014*). For the period 2006 through 2013, more than 81,700 acre feet of groundwater was pumped from wells subject to ADWR reporting requirements (*see Figure 5*). This number does not include the number of acre feet pumped from the more than the 6,000 exempt wells in the area for which there are no reporting requirements. Exempt wells have a pumping capacity that does not exceed more than 35 gallons per minute. At maximum capacity, a single well pumping at a rate of 35 gallons per minute translates into 56 acre-feet per year or 18,247,680 gallons each year.

While a depleted aquifer in TAMA poses a risk to available water supplies for its population, a declining aquifer in the Upper San Pedro Basin poses a twofold risk to the existing population and to the San Pedro River and its riparian ecosystems. Declining groundwater levels strongly affect riparian ecosystems where the aquifer system has a large volume of water in storage, but a relatively small rate of natural annual recharge and discharge (*Richter, 2014*). The aquifer in the Upper San Pedro Basin, is recharged at the mountain front and perennially drains down to the San Pedro River, thereby "*sustaining a lush riparian ecosystem year-round. Because of this linkage between the aquifer and surface water, riparian vegetation can easily tap ground water along the river*" (*Serrat-Capdevila, 2011*). Unrestricted pumping of groundwater in the Upper San Pedro Basin intercepts this discharge and, because the basin is connected to the San Pedro River, the pumping depletes the streamflow and associated riparian habitats (*Glennon & Maddock, 1994*). As such the current projected

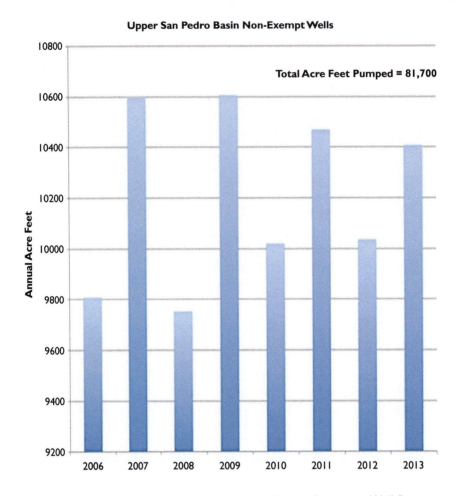

Figure 5 Source of Data: Arizona Department of Water Resources Well Registry.

decline in groundwater levels threatens the remaining water levels and baseflow in the San Pedro River (*Serrat-Capdevila, 2011*). In fact, the average streamflow in the river has declined over the past 80 years (*Randle & Meixner, 2013*) as pumping and population in the basin have increased.

This decline in streamflow has national importance because it threatens the continued viability of the San Pedro Riparian National Conservation Area (SPRNCA), the first riparian national conservation area of its kind in the country (*Arizona-Idaho Conservation Act of 1988, Pub. L No 100-696, 102 Stat. 4571*). Congress established SPRNCA to protect the riparian area and the aquatic, wildlife, archeological, paleontological, scientific, cultural, educational and recreational resources of the public lands surrounding the San Pedro River. SPRNCA runs along 40 miles of the San Pedro River and encompasses more than 56,000 acres in Cochise County, Arizona. Extending from the Mexican border to a few miles south of St. David, SPRNCA represents the most extensive, healthy riparian ecosystem remaining in the desert southwest, supporting approximately 390 species of birds, 83 different types of mammals and at least three endangered species, the south western will flycatcher, the Huachuca water-umbel and the jaguar (*Glennon, 2002*).

The decline in streamflow also has the potential to cause serious economic disruption in the area. The U.S. Army installation at Fort Huachuca, also located in the Upper San Pedro Basin, represents a major driver for southern Arizona's economy as the largest employer in the region, and it contributes approximately $2 billion annually to Arizona's economy (*Richter, 2014*). Fort Huachuca also pumps groundwater from the aquifer which may affect the viability of the SPRNCA. Due to the National Environmental Policy Act, the future viability of SPRNCA may affect the continued existence of Fort Huachuca (*Serrat-Capdevila, 2009*) as federal law could require the closure of the military base if its use of groundwater imperils SPRNCA.

The twin threats to SPRNCA and Fort Huachuca resulted in the 1988 Memorandum of Understanding (MOU) which formed the Upper San Pedro Partnership (the Partnership) funded by the federal government with millions of dollars. The Partnership, a key stakeholder organization, is a consortium of federal, state and local governmental entities, business and non-profit organizations.[1] The Partnership's purpose is to meet the long-term water needs of both the SPRNCA and the area's residents (*Richter, 2014*). According to the Partnership's mission statement, this purpose is to be accomplished through the identification, prioritization, and implementation of policies and projects related to groundwater conservation and (or) enhancement. Effectively, therefore, groundwater management in the Upper San Pedro Basin has taken the form of a "bottom-up," collaborative, community-based approach between stakeholder organizations and agencies and the scientific community to address the issues generated by its declining aquifer and increasing population (*Browning-Aiken et al., 2003*). This approach is markedly different than the top down framework imposed by the Groundwater Management Act.

As has been done in the TAMA, the Partnership defined a management goal for the basin. Unlike TAMA, the goal is "sustainable yield" rather than "safe yield". The Partnership's definition of "sustainable yield," incorporates the concept of environmental flow needs with human needs in the water balance in the aquifer – in contrast to the "safe yield" concept that assumes it is acceptable for consumptive human uses of water to equal groundwater recharge inflows (*Richter, 2014*) without accounting for any consumption or demand by the ecosystem (here the riparian habitat of the San Pedro River). In other words, safe yield exists when consumptive human uses of groundwater equal groundwater recharge. Safe yield is only safe for humans; it can be lethal to riparian habitats.

---

1 The members are: Arizona Department of Environmental Quality, Arizona Department of Water Resources, Arizona State Land Department, Audubon Arizona, City of Bisbee, City of Tombstone, City of Sierra Vista, Cochise County, The Cochise Water Project, Fort Huachuca, Friends of the San Pedro River, National Parks Service, Natural Resource Conservation District, The Nature Conservancy, Southwest Arizona Association of Realtors, Town of Huachuca City, US Bureau of Land Management, US Bureau of Reclamation, US Fish and Wildlife Service, US Geological Survey and USDA Agricultural Research Service.

Table 1 Source of data: U.S. Army Environmental Command, 2007.

Huachuca	Funds obligated	Acres protected	Military funds expended	Partner funds expended	Total expended
In FY12	$1,250,000	1,811	$3,068,692	$9,886	$3,078,578
Through FY12	$17,488,200	5,894	$11,699,008	$2,826,154	$14,525,162

Over the long term, the amount of recharge to a groundwater basin equals the amount of discharge. It only stands to reason that the amount of water entering a basin from rain, snow, sleet, underground flow from another aquifer, and discharge from streamflow must equal the amount of water leaving the basin via evaporation, transpiration, flow to another aquifer, and discharge into a stream. Absent this equilibrium, the basin storage will change: it will either eventually flood or dry out. The introduction of pumping into this setting upsets the established equilibrium. Clearly, while pumping does not affect evaporation that occurs at or near the surface of the basin, it does affect the remaining discharge from the basin. More pumping means that water that would have otherwise been available to SPRNCA is instead used to satisfy human needs (i.e., the concept of "capture"). When viewed from this perspective, it becomes clear that groundwater basins are a zero sum game, which leads to the conclusion that the groundwater pumping must be carefully managed so that sufficient water remains to support riparian habitat.

Limited regulatory action has occurred to address groundwater use in the Upper San Pedro Basin as compared to the broad sweep of the Groundwater Management Act. No state-mandated restrictions have been imposed on the irrigation of land. Yet, local policy makers and private water utilities have implemented conservation measures in the city of Sierra Vista to reduce water demand. Effluent has been treated and used to recharge the aquifer. Cochise County has adopted assured water supply rules similar to those that exist at the state level that govern new commercial and residential development (*Cochise County, Az. Subdivision Regulations* § *408.03; see A.R.S.* §*11-823A*). In addition (see *Table 1*), members of the Partnership, namely the Bureau of Reclamation and The Nature Conservancy have acquired and retired agricultural land in the Upper San Pedro Basin (*Richter, 2014*).

5 COMPARISON OF RESULTS

Neither water management approach adopted by the basins stopped population growth; both basins experienced robust population increases that created more demand for water. Cochise County has grown from 86,500 people in 1980 to more than 130,000 in 2014. The population of the cities in the Upper San Pedro Basin (which does not include the unincorporated areas of Cochise County) grew from 17,841 to 56,051 in 2014. The population of the five incorporated cities in TAMA has increased from 301,255 people in 1980 to 645,095 in 2014. According to the most recently available information about TAMA as a whole, the 1985 population was less than 600,000 and had increased to almost 1,000,000 by 2010.

The water management approach adopted in TAMA, as shown in *Figure 3* and discussed more completely in *Chapter 15*, had little impact on the agricultural sector. Land use and irrigation devoted to the agricultural sector rose between 2000 and 2010. Water use remained relatively constant. In contrast, the efforts of the members of the Upper San Pedro Partnership between 1985 and 2002 contributed to a 45% decline in water demand by agriculture. Through the cooperative efforts of the Partnership, agricultural land was purchased, retired and subjected to conservation easements (*ADWR, 2005*).

Although TAMA and the Upper San Pedro Partnership used different approaches to the management of municipal demand, both methods obtained positive results. In TAMA, the number of gallons used per capita per day fell during the period 2000 to 2009 (*Figure 6*). Similar results were obtained in key locations in the Upper San Pedro Basin. Fort Huachuca reduced its water use by almost 45% between 1993 and 2002 due to a combination of irrigation efficiency, installation of low water use plumbing fixtures, replacement of high water use landscaping and education. Although, not as dramatic, the City of Sierra Vista implemented water conservation programs and ordinances and reduced use measured in terms of per gallon per day per capita rate (gpdc). Specifically, water use declined by 14% since 2005 from 155 gpdc to 133 gpdc (according to the *City of Sierra Vista, Department of Public Works, Environmental Services Division Area Water Use*). This rate of reduction compares favorably with the TAMA results and results reported by the Phoenix and Pinal AMAs during the same time period (*Figure 6*).

The governance of new development in TAMA differs from the rules governing the construction of new subdivisions in the Upper San Pedro Basin. As discussed above, a developer seeking to construct new housing in the Phoenix and Tucson AMAs that

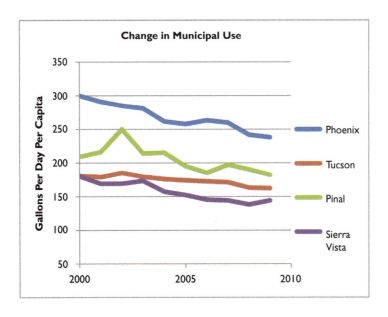

Figure 6 Source of Data: City of Sierra Vista, Department of Public Works, Environmental Services Division Area Water Use; ADWR.

relies on groundwater must be able to recharge with the aquifer with approximately the same amount of surface water that was pumped to meet the standard of "safe yield". According to the development community, this requirement essentially prohibited development in certain areas of the AMAs because replacement surface water was not economically available. To address the concerns of the real estate community, the Arizona Legislature authorized the formation of the Central Arizona Groundwater Replenishment District (CAGRD) to acquire water for its member developers to recharge the aquifers pumped to support the new development (*Maguire, 2007*). Membership in the CAGRD essentially deems that a member's development will comply with the management goals within the AMA. A total of 26,529 homes have been built in the Tucson AMA between 1995 and 2013, with more than 7,000 of those houses built since 2007 (*Central Arizona Groundwater Replenishment District 2015 Plan of Operation*). Thus, while the state imposed significant restrictions on the use of groundwater by the development community, it also acted in response to their concerns to add provisions to the law that created a new state obligation to accommodate the need for recharge.

Membership in the CAGRD is not available to developers in the Upper San Pedro Basin, but no requirement exists that the developers must comply with a "safe yield" management plan in connection with their planned use of groundwater to supply new subdivisions. Nevertheless, Cochise County has implemented regulations applicable to the City of Sierra Vista that prohibit approval of a new subdivision plat unless a developer proves the existence of an adequate water supply (*Cochise County, Az. Subdivision Regulations § 408.03; see A.R.S. §11-823(A)*). For Cochise County purposes, an "adequate water supply" consists of either a determination by the director of ADWR that there is an adequate water supply or a written commitment of water service from a city, town, or private water company designated by ADWR as having an adequate water supply. The necessary determination articulated in the state laws, which Cochise County has adopted, requires the director to find: (1) Sufficient groundwater, surface water or effluent of adequate quality will be continuously, legally and physically available to satisfy the water needs for the proposed use for at least one hundred years; and (2) financial capability to construct the water facilities necessary to make the supply of water available for the proposed use, including a delivery system and any storage facilities or treatment works (*A.R.S. §45-108(I)*).

The impact of these regulations adopted by Cochise County, a member of the Upper San Pedro Partnership, became apparent in a recent lawsuit involving a developer seeking to build a subdivision with 6,959 homes in Sierra Vista. Although the director of the ADWR determined that the developer had demonstrated a physical and legal access to a 100 years supply of water, the Arizona Superior Court determined that ADWR erred in making a determination that Pueblo del Sol had an adequate water supply that was legally available. The case is currently on appeal to the Arizona Court of Appeals.

As can be seen from the above discussion, the two governing approaches have had different results in the areas of agriculture and development and a similar result in terms of municipal use. Another measure of the impact of the two types of governance may be the economic health of the relative populations in TAMA and the Upper San Pedro Basin. As one of its prime goals, the state legislature enacted the Groundwater Management Act to protect the economic health of Arizona. The Upper

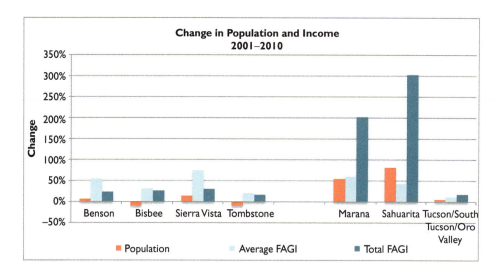

Figure 7 FAGI = Federal Adjusted Gross Income. Source of Data: Arizona Department of Revenue.

San Pedro Partnership and its members also took action in the Upper San Pedro Basin to protect the local economies as well as to shield SPRNCA. The economic results, as measured by the average reported income of the people living in the TAMA and the Upper San Pedro Basin, show mixed results for the period 2001 to 2010 within each of the two areas (*Figure 7*). The average incomes in Tombstone and Tucson, Upper San Pedro Basin and TAMA, respectively, increased but not in excess of the accumulated inflation rate of 23.2% for the time period based on the consumer price index published by the Unites States Department of Labor. The growth of the average federally adjusted gross income of residents of Benson and Sierra Vista in the Upper San Pedro Basin and in Marana and Sahuarita, in the TAMA, exceeded the inflation rate. Given the absence of a consistent pattern within each basin, no conclusion can be reached from this data about the relative merits of the governing tools vis-à-vis economic growth.

6 CONCLUSION

As demonstrated by the different approaches taken in TAMA and the Upper San Pedro Basin, groundwater management can take very different forms. In both cases, municipal demand measured in gallons per capita per day declined and, therefore, the management measures can be deemed successful by that assessment. In contrast, the changes in agricultural use in response to the Partnership's actions and the Groundwater Management Act vary dramatically. Little change has been observed in agricultural demand in TAMA, but it has been reduced significantly in the Upper San Pedro Basin. It is questionable, however, whether the approaches used in the areas surrounding the San Pedro River can be duplicated in TAMA.

Similarly, the results in the two basins dealing with new development show different results. The CAGRD effectively allows developers to proceed with subdivision development in TAMA. The effects of funding a deemed assured water supply for projects through CAGRD need to be more fully considered, particularly if drought conditions continue. The enactment of new regulations governing development in the Upper San Pedro Basin may create new challenges for future development.

No first effort by policymakers to address the needs of stakeholders should ever be viewed as complete. The Groundwater Management Act needs to be reevaluated to determine whether its provisions can achieve the policy goals of reduced agricultural use and a guaranteed long-term supply of water to new development. In addition, this law should be further evaluated to determine whether a "safe yield" standard achieves the best results for the state. The determination of local policy makers in the San Pedro to employ a different metric – sustainable yield – is an example of the incremental improvement that might be incorporated into an amended law.

Policymakers and stakeholders should be ready to revisit the solutions that they have crafted and revise them so that they better address existing issues. Groundwater use is no exception. In the case of TAMA, it appears that reliance on the groundwater hundreds of feet below its land has been replaced by reliance on water from a river hundreds of miles away that must be transported using pumps that are the single largest consumer of power in the state of Arizona. In the Upper San Pedro Basin, the streamflow in the San Pedro River continues to decline.

REFERENCES

Arizona Department of Water Resources (2005). Upper San Pedro Basin Active Management Area Review Report. Retrieved from http://www.azwater.gov/AzDWR/StatewidePlanning/WaterAtlas/SEArizona/Hydrology/UpperSanPedro.htm.

Browning-Aiken, A., Varady, R.G., Goodrich, D., Richter, H., Sprouse, T. and Shuttleworth, W. (2003). *Hydrology and Water Law – Bridging the Gap: A Case Study of HELP Basins.* London: IWA Publishing.

Glennon, R. (2002). *Water Follies. Groundwater Pumping and the Fate of America's Fresh Waters.* New York: Island Press.

Glennon, R. and Maddock, T. (1994). In Search of subflow: Arizona's Futile effort to separate groundwater from surface water. *Arizona Law Review,* 36: 567–610.

Maguire, R.P. (2007). Patching the holes in the bucket: safe yield and the future of water management in Arizona. *Arizona Law Review,* 49: 361–383.

Megdal, S., Smith Z.A. and Lien, A.M. (2008). Evolution and Evaluation of the Active Management Area Management Plan. Report of the Water Resource Research Center. Retrieved from http://www.azwater.gov/azdwr/WaterManagement/AMAs/documents/AWI_evaluation__Jan08.pdf.

Pearce, M.J. (2007). Balancing Competing Interests: The History of State and Federal Water Laws. In B.G. Colby and, K.L. Jacobs (Eds.) *Arizona Water Policy: Management Innovation in an Urbanizing, Arid Region.* New York: Routledge, 26–44.

Richter, H.E., Gungle, B., Lacher, L.J., Turner, D.S. and Bushman, B.M., (2014). Development of a shared vision for groundwater management to protect and sustain baseflows of the upper San Pedro river (Arizona, USA). *Water,* 6: 2519–2538.

Randle, N. and Meixner, T. (2013). Modeling the Impact of Increased Pumping from the Pueblo Del Sol Water Company in the Upper San Pedro Basin, http://www.biologicaldiversity.org/

programs/public_lands/rivers/san_pedro_river/pdfs/well_modeling_20131216_FINAL_REPORT.pdf.
Saliba, G. and Jacobs, K.L. (2008). Science, collaboration and water sustainability in Arizona. *Environment*, *50*(2): 30–43.
Schaffer, R.G. (2010). Davis v. Agua Sierra resources: Bringing some clarity to groundwater rights in Arizona. *Arizona Journal of Environmental Law & Policy*, 1: 25–46.
Serrat-Capdevila, A., Scott, R., Shuttleworth, J. and Valdes, J.B. (2011). Estimating evapotranspiration under warmer climates: Insights from a semi-arid riparian system. *Journal of Hydrology*, 399: 1–11.
Staudenmaier, W.L. (2007). Between a rock and a dry place: The rural water supply challenge for Arizona. *Arizona Law Review*, 49: 321–338.
U.S. Army Environmental Command Report for Fort Huachuca, Arizona (2007). ACUB approval date: 23 February 2007. Retrieved from http://www.aec.army.mil/Portals/3/acub/AZ-Huachuca.pdf.

DOCUMENTS, CASES AND LEGAL REFERENCES

Bristor v. Cheatham, 75 Ariz. 227, 255 P.2d 173 (1953).
Chino Valley v. City of Prescott, 638 P. 2d 1324 (Ariz. 1981).
Farmers Investment Company v. Bettwy, 558 P.2d 14 (Ariz. 1976).
Jarvis v State Land Department, City of Tucson I, II, III 456 P.2d 385 (Ariz. 1969); mod. 479 P.2d 169 (Ariz. 1970); injunction mod. 550 P.2d 227 (Ariz. 1976).
Pima Farms v. Proctor, 30 Ariz 96, 245 P. 389 (1926).

Stakeholders' perspectives

Chapter 19

Presentation

Aleix Serrat-Capdevila
UMI iGLOBES CNRS/Department of Hydrology and Atmospheric Sciences,
University of Arizona, USA

Edward F. Curley
Pima County Regional Wastewater Reclamation Department, Arizona, USA

Alba Ballester Ciuró
Institute for Government and Public Policy, Autonomous University
of Barcelona and University of Seville, Spain

In the spirit of the SWAN Project's transdisciplinary approach to science and in line with the attempt to involve practitioners from non academic sectors in new forms of collaborative research, we also consider it essential to include the perspectives of professionals from different sectors and stakeholder representatives. This chapter is devoted to their collective views and voices on water issues. The request to share their perspective was sent to a broad range of institutions. All of the seven individuals that agreed to share their perspectives – featured here – have been of invaluable assistance to the SWAN Project and to the students and researchers personally. We invited them to contribute a page on their impressions and views on the following issues:

- Your vision of water management in the Tucson Basin, Arizona, and the Southwest in the coming years.
- Will water be a limiting factor for growth?
- How do we reconcile growth and sustainability?
- How do we ensure safe yield in the future? The importance of sustainability (safe yield?) and how to ensure it in the future.
- How do we achieve public awareness and involvement of the public for water planning for the future?
- What is the role of ecosystem services in this water future?
- How do we make the environment a stakeholder in water planning?
- What new demands, what future supply.
- Fears, power, (un)certainties, change, risks.

We requested them to pick a few issues of their interest and develop some thoughts around them. We are very pleased to be able to include this chapter with their contributions and views.

Chapter 20a

Claire L. Zucker
Program Director, Water Sustainability Program, University of Arizona, USA
Previously: Director of the Sustainable Environment Program,
Pima Association of Governments

Water is in the news more than ever before and yet, despite journalistic exploration, it remains a topic that is difficult to fully grasp. This is, in part, because water management has moved beyond the purview of the water institutions and regulations of the last century. Today, thorough water management must also include emerging themes such as the water/food/energy nexus, water marketing opportunities, and riparian habitat needs. We need new tools to address the broader format including flexible water management, new organizations to help us integrate stormwater and reclaimed water with other supplies, and more inclusive planning to ensure that natural systems do not bear the brunt of our water management choices.

As in previous decades, comprehensive water management understanding must not only include knowledge of groundwater, surface water, reclaimed water, and human consumptive and agricultural demand, but must also include statewide management efforts such as the Central Arizona Project (CAP), water banking, recharge and recovery, and the various regulatory mechanism of the Groundwater Management Act and subsequent legislation. In my view, statewide water management has been in a holding pattern over the last ten years because of a severely underfunded Arizona Department of Water Resources. Although the regulatory engine may have stalled, climate change continues to impact our water resources and people's involvement and conceptualization of water management continues to evolve. For example, stormwater is now commonly considered a viable resource, reclaimed water is approaching acceptability, many rural areas face actual water crises, and each year managers around the state hold their breath to find out if CAP shortage will be declared.

To calm the public's fear about diminishing water availability, as our population grows and climate change fears build, Arizona's water managers have consistently provided the message "don't worry, we have it under control". If we remain complacent, no matter how well existing practices are managed, we risk losing opportunities. We need to accelerate efforts to increase thoughtful context-sensitive flexibility in water management. There are early examples of this as ADWR holds discussions about creating modified Irrigation Non-expansion Areas (INAs) to include some agricultural expansion for high value crops and possibly moving away from a "use it or lose it" philosophy. In Active Management Areas (AMAs), new models of sub-basin pumping/recharge impacts might replace basin-wide safe yield calculations. Also, water markets and agricultural water leasing are gaining traction around the state. I don't see how we can make timely progress on these and many other state-wide issues if we continue to underfund ADWR.

It is my hope that we can shift our administrative mind-set to incorporate new ideas and to develop a more holistic and sustainable water management future. One of the most significant shifts would be to fully utilize rainwater/stormwater as a water resource. If the state and municipalities would support creating stormwater utilities to fund region wide green infrastructure projects, breaking down barriers with departments of transportation and parks management, this may be achievable. Unfortunately, without system-wide municipally-driven implementation, it is difficult to envision how stormwater can be fully harnessed or how it can achieve the many societal ancillary benefits such as greening-up inhospitable urban areas and addressing urban heat island and environmental justice issues. There are planners throughout municipal governments working hard on these issues, implementing numerous pilot projects and ready to go into full action should financing become available.

As Arizona continues to grow, competition for water resources builds. Rural areas dependent on groundwater and without access to imported CAP water will likely be first to incorporate water resources into land use planning decision making. These areas may also be the first to turn to potable reuse as a solution for growing water instabilities. In contrast, Arizona's large urban centers have diverse water and financial resources that they can use to offset short term shortages and maximize long term options. They also count growth as the primary economic indicator and therefore, I think it is unlikely that water availability will significantly challenge growth in our larger metropolitan areas.

A sustainable societal future must necessarily include allocating water to support our natural environment. If we continue on the current course, we will very likely lose the low elevation streams and critical habitat that currently rely on shallow groundwater aquifers. In these areas, there are no impediments to private wells and rising temperatures will only add to the problem. We need to prioritize monitoring and preservation of these systems instead of waiting until they are lost and, only then, undertaking costly and often in-effective restoration efforts. I believe that the plight of our mountain-edge streams has the potential to capture people's imagination and motivate people to action, serving as an entrée into water management issues, much as rainwater harvesting has done for the Tucson region.

For the environment to truly be a stakeholder in water management discussions, it cannot be the purview of environmentalists and researchers alone. Ecosystem sustainability will need to be recognized for the significant societal and economic value it provides. Mainstream economic discussions currently disregard support of natural systems even as their public relations documents rely on environmental imagery and tout recreation as a primary regional attractant. It is my hope that collaborative research, demonstrating the societal and economic benefits of water for the environment, will bolster implementation of projects throughout the state. However, it is up to our political and business leaders, as they seek a more resilient water future, to recognize that using water for growth of trees and societal balance is integral to growing jobs and economic vitality.

Chapter 20b

Wally R. Wilson
Chief Hydrologist, Water Resources Management, Tucson Water

As water resource uncertainties become more focused and visible to the general public, water managers and provider organizations must continue to develop and/or implement sound strategic plans to meet current and future water demands. A sound strategic plan, at a minimum, must provide policy guidance in three major areas: Resource Planning, Capital Plan Implementation and Asset Management. Demand management often termed conservation programs are an important part of strategic planning, but this really falls under Resource Planning in this context. If you don't have a plan for the major strategic initiatives, the small measures afforded by conservation will not matter in the long view.

Resource planning has been critical to the State of Arizona and many larger communities for many decades. The result is all too clear when one compares the water conditions in the State of California to those in Arizona. A lack of planning and no regulation of groundwater have placed California in an epic water crisis. In contrast, Arizona has had 30 years of sound groundwater management and regulation that focused on the use of renewable supplies rather than groundwater dependence. Municipal water providers in Central Arizona have had to plan for diversity and renewable water supply portfolios. Coupled with well executed conservation measures and effective changes in plumbing codes, potable water demands have steadily dropped making renewable supplies more sustainable and preserving groundwater supplies. It must also be noted that a good resource plan is a snapshot in time built on information and forecasts available at the time. Updating and evaluation of goals is required for a long view plan to be implemented.

Good planning should be coupled with sound investment in water infrastructure. Implementation of a capital program that satisfies the goals and recommendations of a resource plan is the path to a reliable water supply for water providers in all sectors (i.e. Municipal, Industrial, Agriculture and Mining). Without improvements to water infrastructure the goals of any plan cannot be achieved. This results in increasingly unreliable treatment and delivery systems which can lead to both financial and supply crises for a water provider.

Asset management involves the inspection, repair and replacement of a provider's water delivery systems. This can include wells, pipelines, pumping stations, treatment facilities, recharge basins, metering equipment and whole system communication/operation hardware. Without a robust management of all of these systems a water provider begins to fail the goals of a strategic plan in multiple ways. Conservation targets for resource management become difficult if there is growing levels of lost or unaccounted water due to leaks and faulty metering equipment. Water quality targets are difficult to achieve if treatment equipment is not maintained appropriately. Many other examples can be used to illustrate the need for asset management that must be in place to accomplish any long term vision set by a resource plan.

Chapter 20c

David Modeer
General Manager—Central Arizona Project (2009–2015)
President—Colorado River Water Users Association (2014–2015)

I began my career in the water industry in 1972 and at that time we in the industry took pride in being the "silent service". We strove to provide quality water, quality service, when it was needed and as much as was needed. What a difference 40 years makes.

Today the matter of drinking water is one of the highest profile issues throughout the world. High quality and sufficient supply of drinkable water is no longer assumed. This fact manifests itself here in the southwest portion of the United States as well or better than most places in the world. From the drought of record in California to the nearly 16 years of below normal precipitation in the Colorado River Basin to water quality threats from previous industrial activities, the challenges to providing this life giving are unparalleled.

The effects of the issues related to water supply and water quality manifest themselves in almost every area. From our quality of life to our economic prosperity water is a key aspect. The majority of our population resides and the majority of our economic activity occurs not where water presents itself naturally. The needed water must be moved from where it occurs naturally to these areas. Given the Southwest's naturally arid climate, this has made for challenges even in years of abundant precipitation. The challenge is now much greater given the years of drought and the ensuing aspects of climate change. The challenges in the future for maintaining sufficient water supply for our quality of life and economy are daunting.

While the challenge is difficult it is not one that can't be met successfully. It is my belief that in order to meet this challenge we must as individuals recognize that we are no longer in an era of abundant water supplies and we must change the way we value water. Even our success to date in modifying our use of water will not be enough with a significant alteration in the value we place on water. While scarcity alters the value and hence cost of other natural resources while driving down demand it is unlikely that the cost of such a life giving resource could be altered to drive down demand without injuring segments of our population. Rather we must change the manner in which we place value on water. If we do this we will be able to make good decisions every day about how we use water and the proper role it plays in our quality of life. It will take each one of us on an individual basis to make this change and contribute to building a sustainable supply for the communities where we live.

At the same time we in the industry, in government and public service must do more to educate the public and lead them to understand the role we can play in addressing the future challenges to a sustainable supply of water.

The Southwest has one of the most dynamic economies in the world and water plays no small part in this fact. Recent studies by our Arizona Universities have shown that for Arizona over 50 per cent of all economic activity happens due to the presence of sustainable supply of water. Much of the water used through the Southwest is moved to where it is used from elsewhere. A sustainable supply is not just an individual state's issue, it is a regional issue and as such the solutions to today's challenges and those in the future must be solved regionally. The conditions that lead to water scarcity and other impacts of climate change recognize no boundary. We must get beyond our local, state and other biases to reach acceptable solutions. It will be of vital importance for all participants, water managers, regulators and government entities to understand the different needs and perspectives we all have. In doing so we must act with courage, integrity and honesty. There are differences in priorities, regulations and intra-state relations for all of us but there is also a common need. A sustainable supply of quality water is what binds us all together. By recognizing our common goal, working with each other in an atmosphere of trust we can be successful in meeting the challenges of tomorrow. Silence in services will not be sufficient for the future.

Chapter 20d

David Godlewski
President, Southern Arizona Home Builders Association

The place we call home: Managing water in Tucson and Arizona

The Colorado River Basin is now in its sixteenth year of drought. While there have been some encouraging signs given a particularly wet May and June, there is still a chance of an official declaration of shortage in the Colorado River water supply for 2017. The counties of Pima, Pinal and Maricopa are dependent on the Central Arizona Project (CAP) and thus the Colorado River for most of their water.

Under shortage conditions CAP will be subject to the first reductions in Colorado River water because CAP holds a "junior" priority water entitlement to Colorado River water among the Lower Basin states (Arizona, California, and Nevada).

Under the current guidelines, shortage will directly reduce or eliminate deliveries to the CAP excess supply for underground storage, and severely reduce the volume of CAP water available for agriculture. The initial reductions are not projected to impact deliveries of municipal supplies but should be taken seriously nonetheless.

The Lower Basin uses about 1.2 million acre-feet more annually than it receives from Lake Powell. It is the responsibility of all Lower Basin states and water users and the United States to take action to close the structural deficit. Despite the "junior" status, it should not be the burden of CAP users alone. Each lower basin state must adopt measures to reduce system losses and keep Lake Mead at the critical level to prevent shortage declaration.

Reducing the imbalance in the short-term and aggressively seeking long-term augmentation solutions are efforts we all must work on. This effort starts at home. New homes built today have significantly reduced the amount of water used by a family compared to homes built only a few years ago. Further, many SAHBA members have become leaders in the installation of water-efficient household devices and water reuse technologies including high-efficiency faucets, grey water and rain water harvesting. Effective conservation depends not only on individual commitment but increased municipal and state incentives and commercial and residential rebates.

Augmentation of the river is an important long-term solution and includes diverse ideas, many of which will require multi-state legislative action, funding and construction. Nonetheless, projects such as cloud seeding, brackish water and sea water desalting plants and structures to capture run-off are vital to the health of the river. Reactivation of the Yuma Desalting Plant alone could result in as much as 100,000 acre-feet of water retained behind the dam at Lake Mead, water that now runs un-accounted for into Mexico.

We are not coming to this situation unprepared. Working together over the last decade on ideas like water banking, ground water replenishment and conservation measures have put us in a good position to develop more solutions. Mahatma Gandhi, said "Your values become your destiny." During times of tough decision-making it is tempting to blame others and discount their value. Municipal, environmental, agricultural and economic needs *must all be* considered.

It is our responsibility to engage water planners and policymakers to ensure that decisions are made with the best information. It is also our job to make sure that worthy projects or efforts are fully evaluated. SAHBA has been a leader in the Tucson Regional Water Coalition (a group of business organizations that focuses on water policy) and actively engaged on local and statewide water issues. We will look to expand this effort so that the policies and actions required in the future balance the needs of all.

Chapter 20e

Julia Fonseca
Pima County Office of Sustainability and Conservation

During the Sonoran Desert Conservation Plan in early 2000, the saying was "we will run out of water and money before we will run out of land [for growth]." This phrase framed a debate around where urban expansion might be most appropriate, and enabled people to see that protecting landscapes around Tucson would not be at odds with economic development.

Today it seems we may have run out of money before running out of water. Most people realize that neither the federal nor the state governments will be financing major new water importation projects, at least for southern Arizona. The cost of energy needed to move and treat water has risen while domestic per-capita consumption has dropped. A stable or declining revenue base must now finance the higher costs of aging infrastructure, higher water-quality standards, and more costly energy.

Regardless, state water policy provides for growth at all costs. For instance, Arizona allows water companies to pump an aquifer until it is 1000 feet below the land surface in the Active Management Areas. The state refused measures to restrict pumping to protect farming in southeastern Arizona, leaving rural communities with no tools to ensure a sustainable economy. Pumping for new mines and development supported by exempt wells remain unrestricted.

In contrast, the flows of streams and springs have diminished, and state water rights do not prevent further reductions. Agua Caliente spring has ceased to flow. Miles of streams that used to flow year-round now dry out during months of hot weather. Wetlands along the San Pedro have disappeared, and the canopy of trees along Tanque Verde Creek and Cienega Creek has visibly suffered. Without concerted local actions, the reliability of flows at streams and springs will diminish further.

We need spatial water management to achieve specific ecological outcomes. That is, managing water to achieve localized outcomes such as groundwater levels that would support mesquite or cottonwood trees, or flowing streams and springs in locations with favorable geology. There are many areas were groundwater declines will never harm an ecosystem: Avra Valley is a good example. But the Tucson basin has places where relatively small shifts in groundwater use or groundwater recharge could make a difference in ecosystem health. The paper by Violeta Cabello and others in this book is one of the first to explicitly recognize these places and join them to a more localized water-management framework.

Spatial water management is already being used successfully along the San Pedro River to reduce or prevent groundwater pumping near Sierra Vista from harming the river. Despite drought and continued pumping in the city, there are places where the river flows farther now during the driest time of the year, in part because of specific local actions to manage water use and recharge the aquifer.

Local citizens are asking for new tools to manage water for ecological benefits. In Tucson, the City Council responded to those who live along Tanque Verde Creek by locally reducing groundwater pumping and bringing reclaimed water to the area. At San Xavier, the Tohono O'odham are replenishing the aquifer and have secured a buffer area that restricts new pumping.

The land conservation successes of the Sonoran Desert Conservation Plan came through collaboration to meet needs for a wide variety of people. In a similar fashion, managing water with objectives that differ by location will require close collaboration among municipal leaders, water companies, land owners, scientists and citizen groups. In eastern Pima County, we are perhaps closer to this kind of collaboration than any other part of the state, in part because we need a more fine-grained spatial resolution to our water management anyway—it's needed to address other problems such as subsidence, costs of pumping, reliability of wells, and water quality needs. By recognizing that ecologically sustainable water management can also meet other community needs, it can become a standard part of good governance now and for the future.

Chapter 20f

Randy Serraglio
Southwest Conservation Advocate, Center for Biological Diversity

Whenever the discussion turns to water—whether it's sustainability, security, supply, demand or whatever—the assumed default perspective is "water for human consumption." Even when water for the environment is the topic, it's generally considered only in relation to humans—how it conflicts with human uses or needs, how ecosystem services could be quantified and valued with regard to their benefits to human life, how we must find a way to carve out some water for nature from the assumed dominion of humankind.

Leaving aside the absurdity of this notion, and the truth that we're really doing just the opposite—carving out water from its natural place for our own (often foolish) uses—this dynamic is not just narrow-minded, short-sighted, and selfish, it's also counterproductive. Being unwilling or unable to escape the mentality of exceptionalism that allows us to blithely subdue, or even sacrifice nature for our needs—or moreover, *wants*—ensures that we will repeat this behavior indefinitely, to the point of crisis.

Let's say we sacrifice the Chiricahua leopard frog on the pyre of extinction because we can't bring ourselves to share "our" water with the living beings that are closest to its source and most dependent on it. Let's say we decide that the few hundred jobs provided by an open-pit copper mine are more important than the continued survival of the Gila topminnow. Let's say we prioritize human consumption over the needs of Huachuca water umbel, Gila chub, and yellow-billed cuckoo, or we fail to take adequate steps to ensure that we can coexist with these and hundreds of other aquatic and riparian dependent species. Will this ensure, or even aid, our survival?

No, it almost certainly will not. In fact, by evading responsibility for living sustainably—which is to say, living in a way that sustains ecosystems and maintains the integrity of the web of life—we merely ensure that someday we, too, will suffer the same fate as the frogs and fish. We humans will run out of water, or make war on each other over the inadequate quantities that remain. In any event, our proud civilization, with all of its technology, creativity and progress, will suffer and possibly collapse.

It would be far better to draw the line in the mud where the turtle lives. If we do what it takes to ensure the turtle's survival—e.g., rein in our consumption, curb our ridiculously unsustainable population growth, use our ingenuity and scientific prowess to determine exactly where and when the line should be drawn to guarantee the *turtle's* well being—then we almost certainly will ensure our own survival in the process. And we'll very likely do it for far less cost and long before it becomes a crisis of apocalyptic proportions.

Developing the foresight and will to protect species that depend directly on aquatic and riparian habitat is step one toward sustainability and water security, placing well ahead of (and possibly eliminating the need for) pipelines, dams, canals, desalination plants, and other boondoggles that are mere emblems of human irresponsibility.

Chapter 20g

Eve Halper
Natural Resources Specialist, U.S. Bureau of Reclamation (Phoenix Area Office)

COLLABORATION CAN ACHIEVE MULTIPLE GOALS WITH LIMITED WATER RESOURCES

As water grows scarcer in the West, it is also becoming more expensive. Higher prices can be an incentive for prudent management, but they also have a down side. As water becomes a commodity, it tends to be diverted from uses without easily valued benefits. While natural amenities such as urban vegetation, streams and riparian areas are appreciated by the public, supplying the water they need can be difficult to justify while there are more fundamental needs (drinking and sanitation), as well as more profitable ones (industry and agriculture). There is a risk that water-supported natural amenities, such as green space that provides shade and cooling, will become accessible only to those who can afford to purchase water.

Multi-benefit projects can address the need to protect natural ecosystems and provide public green space while fulfilling other objectives. Such projects require engineers, geologists, hydrologists and biologists to collaborate and innovate. A good example is a groundwater storage project that supports vegetation and wildlife habitat. If sufficient, uncontaminated land is available, ephemeral streambeds can be used as groundwater recharge facilities. As water flows down the streambed, most of it infiltrates through the coarse sediments, but a small portion supports vegetation, which generates shade, reduces evaporation, and provides high-quality riparian habitat. Another type of project uses wetlands to polish treated wastewater and while providing wildlife habitat and recreational wildlife viewing facilities.

Such areas can become attractions for tourists and economic drivers in themselves. In addition, these projects have the potential to attract public support and generate the financing necessary for continued operation. Both these types of projects have been supported by the Bureau of Reclamation and operate successfully on a long-term basis.

Local governments play a key role in water provision in the West. While reliable and affordable water supplies are essential for modern communities, providing hospitable places to live and work can support the local economy and attract new businesses. Communities can realize multiple benefits from limited resources by addressing water supply and green space needs in the same projects. For example, new construction could be designed to take advantage of the stormwater it generates, supporting on-site vegetation while decreasing storm water runoff. These measures would increase the up-front costs of construction, but they lower the long-term costs of flood damage, pollution control and landscape irrigation. This will require collaboration between departments that do not typically consult each other, such as wastewater treatment, flood control and urban planning.

While this is a very challenging time for Tucson, the state of Arizona and the West in general, crisis can bring innovation, as well as an appreciation of our precious water resources. As water supplies dwindle, inter-disciplinary collaboration is essential to maintain critical water uses, as well as the quality of life provided by natural ecosystems and urban green space. At least in the Tucson area, professionals in water resources, wastewater treatment, flood control, urban planning, parks and recreation and environmental conservation are beginning to work together and accomplish multiple objectives with limited resources, creating a better future for all.

Chapter 20h

Tom Buschatzke
Director of the Arizona Department of Water Resources

Arizona has had a long history of sound water management. Because we live in an arid environment we cannot take water supplies for granted. Starting with irrigation projects and culminating with the completion of the Central Arizona Project delivering Colorado River water to central Arizona, we have robust infrastructure and supplies for our agricultural and population centers. From a water management perspective, the passing of the Groundwater Code in 1980 provided water management tools including mandatory conservation requirements for all water use sectors in the most populous areas of the state, and later mandated water conservation statewide through water conservation plumbing requirements. The combination of programs and requirements has been successful; Arizona now uses less water than we did in 1957 even with robust economic and population growth.

Successes

In recent years, the Tucson area has attained Safe Yield, and for that Tucson should be congratulated. Reduction in mined groundwater use has been significant. In 1985 Tucson was 100 percent reliant upon mined groundwater; by 2013 the proportion of groundwater had been reduced to only 24 percent of municipal sector use. By 2009 less than half the volume of groundwater was used than in 1985. In total, there have been recharged over 2.8 million acre-feet of renewable supplies in the Tucson Basin, which represents a significant investment in sustainability. Tucson also continues to be a leader in conservation. With the successes of the region, the question is how do we maintain the Safe Yield status moving forward?

Drought has always been a part of life in the arid Southwest. In the drought we are currently in, it was recognized early that state wide efforts were needed. In 2000 and 2003 gubernatorial water commissions were established to examine statewide water issues and drought impacts; many resulting recommendations were adopted. On a local level, particularly in rural Arizona, many of the communities are small and reliant upon groundwater. Assistance with infrastructure, and drought planning and mitigation, are offered by state agencies including the Arizona Department of Water Resources (ADWR). Water providers statewide are required to have drought response plans, and the state itself has a drought response plan. As drought has progressed on the Colorado River system, multiple in-state and out of state entities have stepped up to conserve supplies and bolster water levels to avoid shortage declarations. All of these efforts need to continue.

Challenges

The basic strategy of the Groundwater Code was simple although the implementation was not. The strategy was to increase water efficiency through conservation, shift to the sustainable use of renewable water supplies (surface water, Colorado River water, and reclaimed water), store excess renewable supplies when available, and reduce our reliance on mined groundwater saving that supply as emergency drought backup. The implementation of this strategy is on a basin-wide basis, which may allow situations to develop where local groundwater declines may be seen even though the basin overall is in Safe Yield. Local area management may be needed to further encourage the direct use of renewable supplies, or help facilitate the storage or replenishment of renewable supplies closer to the point of withdrawal. In addition, a certain amount of residual, unreplenished groundwater pumping is still allowed, which may create local issues.

Drought will continue to challenge rural Arizona particularly for smaller communities that have fewer resources and limited flexibility. Ongoing drought and structural deficits will need to be addressed on the Colorado system as well. The Arizona Department of Water Resources 2014 Strategic Vision for Water Supply Sustainability study identified areas of concern as well as common solutions, including: continuing conservation, forest and range management, macro-level rainwater harvesting, brackish groundwater desalination, weather modification, and facilitating resolutions of water right settlements. However, Arizona will ultimately need to augment its water supplies. In October of 2015, Governor Doug Ducey announced his water initiative for Arizona and directed ADWR to reach out to local communities, rural and metropolitan areas alike, to address local challenges and needs. Ultimately, there may be the need for additional supply importation into the state, most likely involving sea water desalination either through direct use or exchange.

Conclusion

Chapter 21

Bringing all the stories together: Beyond the Tucson case study

*Aleix Serrat-Capdevila, Violeta Cabello,
Kremena Boyanova, Franck Poupeau,
Nuria Hernández-Mora, Hoshin Gupta,
Zhao Yang, Natalia Limones, Brian O'Neill,
Sergio Segura Calero, Eliza Benites-Gambirazio,
Rositsa Yaneva, Murielle Coeurdray, Joan Cortinas,
Kristin Kuhn, Susan Harris, Maria Sans-Fuentes,
Edward F. Curley, Leandro del Moral, Juan Valdes,
Graciela Schneier-Madanes, Owen King,
Dulce B.B. Rodrigues & Gloria Salmoral*

This chapter contains three sections describing the integration of (1) people and the development of a collaborative research focusing on the Tucson Region; (2) disciplinary fields and methodological approaches; and (3) research results and their relevance for planning and management.

1 THE BIRTH OF A COLLABORATIVE RESEARCH PROCESS

Coming together: As may be expected at the beginning of any collaborative research effort, we spent considerable amounts of time during the first few months of the SWAN project (initiated in April 2012) learning about each partner's work. This period was characterized by long discussions about how to work across disciplinary divides, similarities and differences in methods and perspectives, and the seemingly unavoidable challenges of dealing with the diversity of academic jargon. During the spring of 2013, a group of SWAN-affiliated scientists based at the University of Arizona, began to meet once a week with the rotating group of *"extended-stay"* international students to develop a cooperative approach oriented towards trans-disciplinary research. The initial discussions focused on establishing strong connections among the different disciplines represented, the development of a holistic vision of water-related research, and a search for common frames of analysis. The differences between disciplines and frameworks, especially between the basic notions of hydrological modeling and ecosystem services approaches, were carefully analyzed during several interactive working sessions.

The seeds of integration: These discussions resulted in an initial proposal for project integration, that was realized in the form of a poster at the April 2013 Progress Meeting of the SWAN project held in Tucson, from which the foundation for our ongoing collaborative engagement became much more clear. With the value of weekly student/researcher meetings now extremely apparent to the group, the practice was unanimously institutionalized as a concrete means to advance scientific cooperation.

Equally important, it was also decided that the student group would collaborate on the development of a working paper to connect the variety of scientific disciplines, analytical frameworks, and models of understanding, thereby providing a practical mechanism for trans-disciplinary collaboration[1]. Within this framework, it became apparent that a *'place-based'* case study would provide the vehicle for moving from the abstract to the specific, by focusing everyone's research efforts towards a specific location.

The power of place: After discussing several possibilities, a focus on the Tucson Basin emerged as the natural choice to anchor the variety of methods and approaches to a contextual transdisciplinary reality. Besides being the location of the University of Arizona where the visiting students would physically spend much of their time, the setting of the Tucson Basin presented us with a unique combination of characteristics: a specific ecological system, a history and a strong *know-how* in terms of water management technologies and institutions, a very organized water community with public debates and information meetings, and a long-standing tradition of collaboration between institutions, research centers and professional associations.

Guides to development: As described in the Introduction to this book, the SWAN team drew upon the critique of positivist science and the development of the Post-Normal Science approach (*Funtowitz & Ravetz, 1994*) as a guide for the development of collaborative relationships with the local stakeholders. This approach emphasizes the uncertainty of model predictions and the incommensurability of complex issues due to differences in stakeholder values and perceptions, and deems the involvement of non-academics to be essential in the definition of the research problems and in the collaborative research process.

Stakeholder involvement: During the course of the project (and especially since the spring of 2013), the SWAN team in Tucson progressively developed congenial relationships with a number of local water managers and stakeholders, thereby establishing an essential basis for this research. Many of these persons were invited to participate in one or more of the weekly meetings, to present seminars, and to engage in discussions of how the research being conducted by scientists could be made more relevant to the water challenges faced by the Tucson Basin. This dialogue led to the formulation of several research questions, and a workshop with local stakeholder experts and members of the broader academic water community in Tucson was then held (in October 2013). The workshop accomplished several goals including:

a   Identification of key management challenges in the Tucson basin region
b   Evaluation and prioritization of pre-defined research questions
c   Identification of knowledge gaps and proposal of new research questions
d   An initial map of relevant regional stakeholders, and
e   An open roadmap for future collaboration.

A key feature of this workshop was that it enabled an extensive conversation with a set of very knowledgeable and experienced stakeholders, facilitated an in-depth exchange of ideas, and resulted in a rich and productive working session. Learning from this, the team decided that future interactions with stakeholders would be

---

[1] For more details, see: https://swanproject.arizona.edu/sites/default/files/Deliverable_5_Supplement1_web.pdf.

organized to occur in small groups to maximize information exchange. The water challenges, priorities and knowledge gaps elicited during the stakeholder meeting (through a number of structured participatory exercises) became the main motivations that significantly influenced the research activities presented in this book.

Genesis of the book: Eight months later, during the June 2014 SWAN progress meeting that took place in Seville, Spain, a total of 11 research topics (including abstracts and paper outlines) were proposed by the project participants, all of them informed by the above-mentioned efforts to understand and characterize water resources challenges in the Tucson basin. Since then, these efforts have advanced and evolved, with the explicit intent to coordinate, acknowledge and benefit from each other's research.

Learning from a case study: The following SWAN progress meeting, held in Tucson (USA) in November 2014, dedicated a full three days to discussing the Tucson Case study, and included presentations from each research team about advances of their research and links between papers, overlaps between spatial scales and challenges, and overlaps between stakeholders relevant to each research endeavour and efforts to coordinate interactions with them. From these presentations and discussions emerged a debate regarding the definition of the research *"object"*. A number of collaborative tools were employed during these sessions, including spatial diagrams with scale boundaries, pie charts showing types of connections between papers, network analysis of connections, sharing of perspectives and exploration of common storylines. These tools helped to bridge the gaps across the national and disciplinary backgrounds of the people involved. One important fruit of that collaborative effort was the emergence of the idea of developing this book, as a way of documenting and cementing together the different disciplinary approaches into a transdisciplinary perspective; this idea finally took on a mature form at the subsequent progress meeting in Sofia, Bulgaria (April 2015), where initial drafts of various chapters were presented and discussed by the participants.

2 SCIENTIFIC LESSONS: METHODS INTEGRATION AND A HOLISTIC APPROACH

This section discusses how the effects of growth and human demands can be successively tied to demands on ecosystem services, to water budget components, to hydrologic processes and functions, to climate, and finally to feedbacks between climate and land use cover, which again is strongly influenced by spatial planning and social uses of water (*Figure 1*). In other words, it describes a potential integration of the different approaches presented in this book, in which each methodology poses feedbacks from/to the others, not only between variables and indicators but also between concepts. This integration can help the reader to understand the synergies and overlaps among disciplinary fields. The goal of a transdisciplinary and integrative methodology is to combine physical- and water-centric modeling with understanding arising from the social sciences and the knowledge of the controversies among various social agents', thereby providing the quantitative and qualitative research framework required for the analysis of water management in socio-ecological systems.

Atmospheric and Hydrologic Modeling: Within this integrative effort, the role of atmospheric and hydrologic modeling is to provide critical information regarding the

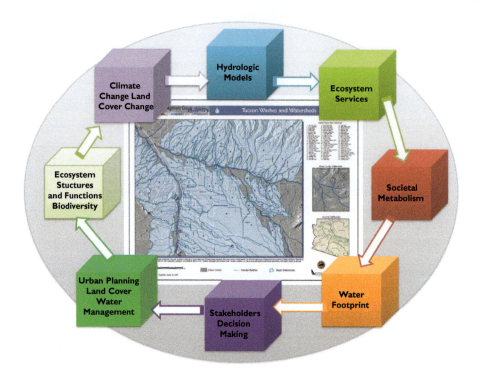

Figure 1 System components relating to water in the SWAN Project's integrative research efforts (linkages are simplified for clarity).

functioning of (and constraints to human actions provided by) the physical environment. Atmospheric variables such as precipitation, temperature, specific humidity, snow cover and others are the basic input data used to run hydrologic simulation models in order to generate estimates of the spatio-temporal evolution of various hydrologic variables such as evapotranspiration, streamflow, groundwater recharge, and soil water content. These data (including the model simulated estimates) are important to being able to relate changes in water availability to human well-being. By processing projections of future climate change through hydrologic models (*Rajagopal 2014; Serrat-Capdevila et al. 2007, 2013*), the trajectory of future states of the hydrologic system under climate change conditions can be generated. These can then be used for water management purposes such as drought planning, development of water supply planning scenarios, connections to agricultural and other use activities, and evaluation of management options that optimize water availability.

Ecosystem Services: Hydrological models generate quantitative data at specific spatial and temporal scales, and can usually also provide qualitative information regarding various hydrological attributes relevant to the ability of a hydrological system to supply ecosystem services. Such an ecosystem services assessment can then allow a manager to draw upon a comprehensive background of information during decision-making about land use and water related services (i.e., visualization of flood regulating areas and level of flood risk).

To properly quantify water-related ecosystem services, an appropriate set of indicators must be derived from the simulations provided by hydrological models. These indicators must reflect the ecosystem properties, functions and services of interest, while also reflecting the role of land management as a main force driving land use changes (*Burkhard et al., 2012*). For example climatic indicators (e.g. precipitation, temperature, albedo) can provide information regarding the ecosystem service *local climate regulation*. Similarly, potential indicators for *water flow regulation* include groundwater recharge rate, infiltration, runoff and peak flow. For *water purification* one can draw upon a variety of water quality indicators such as sediment load, total dissolved solids, nitrates, phosphorous, and others. The freshwater *provisioning ecosystem service* can be accounted for in terms of the withdrawal of freshwater (*Kandziora et al., 2013*). In comparison with methods for assessing supply, approaches for assessing the demand for ecosystem services are much less developed.

Social Metabolism: To better understand coupled human-environmental systems, the concepts of societal metabolism and water footprint assessment have proved to be useful, and the ecosystem services framework provides a suitable way to connect water-related ecosystem services to the societal metabolic demand for those services. A complete analysis of water metabolism (*Madrid and Giampietro, 2015; Cabello et al., 2015*) requires the integration of information regarding both the climate and the ecological status of water bodies. By following the different flows of water (taken from the ecosystems) through the social structure (using demographic, labor and economic information) one can assess how these are combined with labor and other resources to produce goods and well-being. Water planning scenarios can then be used to assess the trade-offs associated with different solutions and how they might contribute to a sustainable balance between human-use and ecosystem health. And in all cases, one must consider the institutional configuration of water rights and management plans to arrive at a proper definition of the constraints on each scenario.

'Social metabolism' and 'Ecosystem services' are complementary approaches to understanding how human activities interact and overlap with ecosystem functioning. The two approaches follow different conceptual metaphors, and highlight (or conceal) different perspectives and objects of analysis. While ecosystems services focus on the societal benefits obtained from ecosystems, societal metabolism is based on the concept of autopoiesis (i.e., societal requirements to maintain and reproduce itself where ecological thresholds can't be surpassed to guarantee this reproduction) in systems (*Maturana and Varela 1980, Giampietro et al. 2012*).

Water Footprint: There are three types of water footprints: 'green' (water from rainfall stored in the soil, evaporated or used by plants), 'blue' (water taken from surface or groundwater bodies for human activities), and 'grey' (freshwater required to dilute pollutants to meet specific water quality standards). Minimal levels of blue water flows (in terms of quality, quantity and temporality) are typically required to achieve environmental goals. Meanwhile, green water flows from vegetated areas constitute the total consumptive water used in biomass production [*Falkenmark and Rockström, 2006*]. The consumption of each of these kinds of water (blue and green) has been incorporated into the Water Footprint concept developed by Arjen Hoekstra [*Hoekstra & Hung, 2002; Hoekstra et al., 2011*].

The green and blue water footprints (WF) associated with agricultural and natural areas will necessarily vary based on the amounts of precipitation and evapotranspiration over those areas; estimates of these can be obtained using models as discussed above. Similar to the ecological and carbon footprints (*Rees, 1992; Wiedmann and Minx, 2007*), the concept of WF addresses the appropriation of water resources by humans and introduces the metaphor of virtual water (water embedded in a product), which facilitates analyses of the role of water in trade, the study of how consumer choices can impact ecosystem functioning and health, and the investigations of topics such as water equity and food security.

Water Management Systems: Alternative management systems can be compared and contrasted in terms of their relative success in facilitating the sustainability of resource exploitation. A successful management system must also take into account existing and potential conflicts, the ways in which institutions magnify or ameliorate inequalities in resource access, value systems, and possibilities for conservation. In the specific case of water, a management plan must reflect clear goals, account for future scenarios of water use, and include plausible strategies for managing and responding to new water demands. Given that land planning is the main driver of land use change, realistic scenarios (along with an assessment of their social viability and biophysical feasibility) necessarily require detailed analyses that consider the interdependence of both water and land in the context of human activities.

Participatory Processes: To maximize the relevance of research for the benefit of management, it is essential that participatory processes are employed between decision-makers and the research community. Participatory planning and research approaches have been developed to help provide guidance during the definition and structuring of problems to be analyzed, the identification of relevant stakeholders, and the establishment of a fruitful collaborative process. As it was done for many chapters of this book, ensuring that scientific questions are vetted by a stakeholder community from the very beginning, and providing a framework for continuous dialogue and feedback, will increase the relevance of research insights and their usefulness to inform planning and decision-making.

The integrated approach presented here illustrates how human water use, planning and management are related to specific components and footprints of the water budget, to ecosystem functions, to changes in climate, land use and social parameters, and to environmental impacts. The coupling of human-centric and physical-centric approaches enables the analysis of feedbacks and linkages between fields of research that have been largely disconnected. This methodological integration is the foundation from which many of the chapters in this book have developed, focusing on specific transdisciplinary research questions tied to the Tucson Basin and its stakeholders. Acknowledging the need for an evolving science with new schools of thought to analyze environmental issues, the work presented here is an effort to provide an integrative analysis to keep up with observed growing levels of complexity in social-ecological dynamics. More effective strategies are needed to deal with present and soon to come human-ecological problems.

3 RESULTS AND INSIGHTS FOR WATER MANAGEMENT IN TUCSON

Given the arid nature of the American West, water is a limited resource. Consequently, the history of water management in the West has been one that is both intricate and convoluted, characterized by power struggles, antagonistic coalitions and lobbies, strategy, drama and intrigue. This history gives rise to a complex and rather unique context by which the American West stands apart from the rest of the United States and most of Europe. The body of work presented in this book has investigated the great puzzle of water in the context of culture, power, money, people and the environment, and the pieces of this puzzle fit together in interesting ways. While a completely holistic view is beyond the possibilities of this work, the chapters of this book have helped to illuminate and understand several of the interlocking pieces.

The struggle for water: The history of water struggles in Southern Arizona during the past century is explored in depth in *Chapters 3-7* of this book (*Poupeau et al., Cortinas et al., Coeurdray et al., O'Neill et al.*), placing it in the context of the broader Southwest and the Colorado River Basin. By analyzing the power dynamics among coalitions, advocacy groups and government institutions, those chapters describe the evolution of water management from a more technocratic approach to a more inclusive one that now accounts for environmental and social concerns, and one that has made considerable progress towards the democratization of information and decision making. The key factors that have determined the outcomes of past water struggles have been more than technical, being especially affected by institutional and political characteristics, and until recently have mainly focused on the development and control of water supplies to satisfy the increasing demands imposed by economic growth. However, in the last two decades, demand management and conservation have emerged as important dimensions, with water professionals shifting and expanding their focus from infrastructure development towards policy and governance. The current water management landscape in the West, bound as it is by its legal framework and interpretation, is a result of both the historical precedence and the evolution of a society dealing with this arid context. It therefore needs to be understood both in relationship to the past, and to its multiplicity of spatial scales and interconnections—from the local (Tucson) to the regional, the State (and its politics), the Colorado Basin, and ultimately the Federal levels.

Engaging the beneficiaries: Given the contentious and conflicted history of water in the West, water managers are understandably concerned with maintaining a positive public perception. For this reason, their aim is often to antagonize the least number of people when designing and implementing a project, and to work towards building a consensus. While there are many ways to quantify the economic value of water, its environmental, social, cultural and generational values need to be reflected in the management and planning processes through both the participation of all relevant social groups and by the acknowledgement of their economic, cultural and other kinds of preferences. A better understanding of growth and its impacts on land use, ecosystem services, the water budget and human well-being, can help to better inform participatory processes, and to contribute toward a scientifically sound, socially equitable and environmentally sustainable planning process.

Growth and Urban Sprawl: Drawing upon a historical and qualitative investigation, *Benites-Gambirazio* (*Chapter 8*) has sought to uncover the roots of beliefs, representations and actions around water and growth, analyzing to what extent they shape the current situation of water and urban development – urban sprawl – in Arizona cities. Sociological research can enable a necessary understanding of the positions of actors and institutions involved, the ensuing dynamics, and the elements that can lead to potential resistance and conflict regarding any proposed project. Without such understanding it can be difficult, if not impossible, to be able to change certain norms and practices and to balance competing interests.

Pressure for growth: In this regard, the *"demand for suburbanization"* is not so much the result of cultural preferences for the privacy that characterizes suburban life as it is the consequence of institutional and private actions at both national and local levels. Highlighting the importance of the real estate industry in the overall economy, it is clear that a *"pro-growth"* coalition exists, composed of public and private actors. These actors have not been found to express much concern regarding the future of water supply, and instead tend to highlight conservation through increased efficiency in residential water use. They adopt the position that technology will solve future water problems, thereby legitimizing the *"no concern"* discourse.

However, even if that were to be the case, it must be acknowledged that in spite of high efficiencies, continued growth will inevitably lead to the full allocation of resources and the progressive elimination of buffers to cope with variability, ultimately resulting in what has been termed *brittle solutions*.

Factors enabling growth: Looking at the patterns of urban growth to try and understand how the community has developed, *Schneier-Madanes et al. (Chapter 9)* found that the key factors controlling development are the cost of land, the availability of water, and the existence of infrastructure for wastewater treatment. When all three factors exist for a property, dense development is both feasible and very profitable. In some cases, the developer chooses to completely fund the development of public water and wastewater infrastructure to enable proceeding with development. The three main development models can be termed *"urban expansion"* (characterized by steady growth); *"leap frog"* (dense development that leaps from the urban fringe into a new area); and *"wildcat"* development (characterized by one-acre lots, no public water or sewer connections, domestic wells, manufactured homes, septic systems, propane tanks and connections to the electrical grid). Each development model is aimed at a different kind of customer need and socio-economic levels.

Interestingly, in Tucson, only about two thirds of new developments overall are connected to the public sewer system, even in prosperous developments like the *Foothills*, with one third relying on septic tanks. As we move to the future, the use of new technologies such as time lapse remote sensing, can enable detailed studies of the patterns of urban growth, and can also provide a means for communities to shape and mold their patterns of growth through thoughtful planning and management practices (such as the inclusion of green infrastructure, the development of alternative water resources, and the integration of land and water planning).

Achieving Safe Yield: The goal of *"safe yield"* has been achieved in the Tucson Active Management Area by the arrival of CAP water and its use to recharge the groundwater aquifers. *Cabello et al.* analyzed the evolution and effectiveness of different elements of Arizona's institutional arrangements to ensure long-term water supply. They found that the municipal sector has been the most adaptive one, reducing overdraft by replacing more than half of its groundwater consumption with CAP-recovered water, and by reducing per capita consumption, thereby stabilizing overall demand in the face of continuing population growth. However, a thorough understanding of the effect of the stagnation of urban development as a result of the global financial crisis and the potential effects of the reactivation of the sector is necessary.

In fact, the growth projections for this sector remain unaltered; they have simply been set back ten years because of the economic crisis.

The agricultural sector drives inter-annual variability in overall water demand and overdraft in the TAMA, mirroring weather patterns and agricultural market vagaries. Overall, Arizona's CAP allocation, including the partial substitution of agricultural groundwater pumping by CAP in lieu, is vulnerable to droughts in the Colorado basin. The impacts of the current Colorado drought and potential shortage declarations remain to be seen on CAP allocations and on the achievement of safe yield in the Tucson AMA in the next decades.

The Central Arizona Groundwater Replenishment District (CAGRD) is a state agency that buys and delivers water to replace excess groundwater pumped by certain developers and water companies, thus providing a mechanism to circumvent restrictions from the Groundwater Management Act (GMA) on new groundwater uses for development.[2] Current water purchases by the CAGRD include excess CAP water, long term storage credits, water from Indian Tribes, and effluent, and it is investigating acquiring additional sources of water such as desalinization. The costs of CAGRD are initially borne by the developer (in the case of new urban development) and then by property owners through property tax assessments. While this growth-enabling mechanism has allowed developers to proceed with subdivision development even in regions were groundwater levels continue to drop, the CAGRD does not necessarily replenish the water where it was pumped. The impacts and spatial implications of granting a deemed assured water supply for new projects through CAGRD need to be more fully considered within the local water budget over the long-term, if sustainable management is a true goal.

Overall, the future of safe yield in the Tucson Active Management Area is based on the consideration that CAP Colorado River water is a renewable resource. However, in spite of the fact that the Colorado River is currently experiencing an extended period of severe drought, over-exploitation and extremely vulnerable ecological status, the reliance on CAP water continues to enable economic growth without questioning the growth model that is resulting in water scarcity.

The role of Water markets: The emerging Long Term Storage Credits (LTSC) market and the future reclaiming of the groundwater credits is an issue to be looked at in detail. While anyone recharging renewable water supplies in permitted storage facilities can create and sell LTSC, Native Tribes are increasingly important players due to the use of their CAP water allocations for recharge, thereby accruing LTSC's which are marketed to generate revenue. To date no integrative plan exists to govern the future pumping of credited water, and the distribution of its spatial effects on the water budget and the environment has not yet been analyzed. The main buyers of LTSC to date have been Investment Firms, followed at a significant distance by CAGRD, and then municipalities (*Figure 2*).

Limitations of the Groundwater Management Act: The current limitations of the Groundwater Management Act (GMA) need to be addressed for it to achieve its policy goals, as discussed by *Harris (Chapter 18)*. Any water sector practitioner knows that "safe yield" is not sustainable, as it does not account for environmental needs

---

2 CAGRD: http://www.cagrd.com/index.php/water-supply-program.

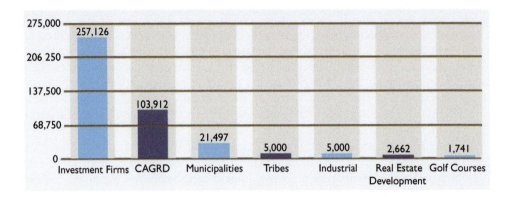

Figure 2 Total volume of Long Term Storage Credits (LTSC) acquired by buyer type during the period 2008–2014 in the State of Arizona. Adapted from: Water Market Insider (2014) West Water Research, Llc, Arizona Q3, 2014, at http://www.cagrd.com/documents/news/FNAL_2_14–0905_WWInsider-6 singles.pdf (last accessed on November 20th, 2015).

(i.e., evapotranspiration) or groundwater outflow. Thus, if pumping equals recharge, the aquifer volume continues to decrease. "*Safe yield*" is a laudable intermediate goal, but "*sustainable yield*"—as adopted in the nearby San Pedro river basin—will come closer to long term sustainability as a legal amendment to the GMA.

Environmental Considerations: The lumped safe yield water accounting in the TAMA conceals the uneven spatial distribution of recharge areas and pumping locations. Groundwater levels continue to decrease in many areas away from CAP recharge sites, with mines and new developments acting as primary drivers of this continued aquifer pumping. As discussed by *Cabello et al.*, ecologically valuable riparian and shallow groundwater areas (that act as local hotspots of biodiversity) overlap to some extent with areas of decreasing water table levels. Because such areas are in danger of disappearing in the near future due to their disconnection from the water table, they should be carefully monitored.

Conservation of riparian areas: If the last remaining riparian areas (with their phreatophyte ecosystem biodiversity) are to be conserved, it is critically important that the spatial distribution of safe yield be taken into account, so as to reflect environmental considerations through proper spatial distribution of recharge and pumping locations. Safe yield accounting should be spatially sensitive, by being split into sub-regions that more accurately reflect the locations and sources of groundwater being pumped; i.e., water pumped from groundwater aquifers near recharge sites (real CAP-recovered) should be differentiated in the water budget accounting from groundwater pumped far from recharge areas that is being offset elsewhere through the CAGRD. Groundwater recovered outside the area of influence of recharge sites should be accounted for as that area's overdraft. The current efforts to create "Water Accounting Areas" are a good step in this direction. Spatial mismatches are being addressed via agreements to transport CAP water through currently available infrastructure (such as through the Tucson Water network to the City of Vail), as well as through the construction of new infrastructure (such as to Sahuarita), thereby bringing CAP water to areas experiencing groundwater table decreases. Pumping and recharge disconnects are also being addressed through trades involving reclaimed water where possible.

While groundwater levels in regions affecting riparian areas continue to decrease, the investigation by *Limones* (*Chapter 16*) shows there can be a lag of between 2 and 5 years (depending on location and events) before recharge at the surface has an effect on groundwater levels. Therefore, time is of the essence in balancing these spatial mismatches, and the current progress of such efforts deserves to be praised.

Ecosystem services provided by nature: Natural areas provide a range of services that benefit society. The work by *Boyanova et al.* (*Chapter 12*) describes the implications of growth and land use change scenarios in the Upper Santa Cruz Basin, by combining hydrologic modeling with an ecosystem services framework to assess current and future water related ecosystem services (WRES): provisions of freshwater, water flow regulation, and water purification. Results have shown high and very high supply across all services only from evergreen forests (oak and conifers) and forested wetlands (including riparian areas), and medium to very low supply from all other land use types. Even though having a limited spatial extent, evergreen forests and forested wetlands still have very high relevance for the supply of WRES in the basin, in addition to being habitat and biodiversity hotspots. Deciduous forests (i.e. mesquite bosques) show lower but still relevant supply of freshwater provision and water flow regulation; and barren areas show relevant supply of water purification. Urban areas show low supply of all three WRES. An assessment of three different future scenarios of land use change show decreases in the supply of almost all services due to the projected urbanization and decrease in forested areas. The biggest decrease is projected for the scenario reflecting current urbanization trends.

This clear projection of a decrease in freshwater supply and other services is relevant and concerning for the future of safe yield in the Tucson AMA. Especially since they do not include the potential effects of climate change, and the fact that similar urbanization processes may be taking place in the broader Colorado Basin, whose current shortages and CAP supplies are already an issue of debate. Future planning in the basin will have to consider the tradeoffs between how growth is done and its impacts on biodiversity (as discussed by *Sans and Meixner* in *Chapter 14*) and water related ecosystem services, using relevant indicators.[3] As discussed by *Yaneva* (*Chapter 13*), maps that link land use cover types with their associated ecosystem services can be very helpful in promoting understanding and informing planning and decision making processes.

Combined impacts of Urban Growth and Changing Climate: The uncertainties related to climate and global change will impact both the CAP allocations from the Colorado, as well as the hydrology of the Tucson basin itself, modulated by land planning choices and green versus grey infrastructure choices at scales ranging from individual homes to city and county projects. The research on regional climate as influenced by land use and urban sprawl scenarios by *Yang et al.* (*Chapter 11*) found that urbanization results in significant increases in daily minimum (night time) temperature. It will likely not impact the magnitude of daily maximum temperature,

---

3 The Federal Resource Management and Ecosystem Services Guidebook can be a useful resource and can be found at https://nespguidebook.com/.

but will increase the mean temperature by not cooling off during the night, as concrete and pavement store heat from the day. The increase in magnitude and duration of night time temperature will likely increase the risk of heat-related health issues. Precipitation is also likely to decrease because of urbanization. Due to the urban island heat storage (during the day) and release (night) the maximum and minimum temperatures are delayed—about an hour for the maximum and 30 minutes for the minimum temperatures. Sensible heat flux (dry air heat) will increase during the night time and decrease during the day. Latent heat flux (heat taken up as water vapor) is likely to decrease dramatically throughout the day, resulting in less evaporation over urban regions and downwind mountainous areas. Ground heat flux (heat absorbed and released by the ground) mainly stores during the day and releases during the night. Water and energy demands for future urban expansion scenarios up to 2050 (using past business-as-usual trends) and temperature changes indicate that water and energy demands will significantly increase to meet future water supply and air conditioning needs. As of today, the Tucson AMA relies on Colorado River water to meet safe yield: water that is pumped and transported from 300 miles away, making the CAP the single largest electricity consumer in the state of Arizona. Comprehensive planning will face significant challenges to meet the demands for water and energy associated with the growth scenarios modeled here.

While the uncertainty of climate change projections is too large to accurately determine future storm and flood frequencies, climate variables will influence future human well-being, ecosystem services and hydrologic partitioning in the basin, governing aquifer recharge, water availability and floods. Thus, climate and the consequences of its interplay with how urban growth occurs need to be taken into account in current planning processes. The urban heat island effect associated with urban expansion will have major implications on temperature and affect the health and well-being of the Tucson's inhabitants. It is therefore imperative that urban planners consider the interplay of climate and urban growth.

Comprehensive planning: As discussed by *Segura* (*Chapter 10*), comprehensive planning in Pima County and the City of Tucson has positively influenced water management. Administrative coordination, cooperation and public participation have increased over the years, yielding specific efforts, actions and policies regarding water management issues. The first regional joint Water & Wastewater Infrastructure, Supply & Planning Study 2008–2010 is a good example of this cooperation between the County and the City. Comprehensive planning is usually limited to coordination of sector specific policies and plans rather than their integration, as is the case of water policy aspects relating to the safe yield objective in Pima County's comprehensive plans. As is common with integrated planning, the main challenge is who is responsible for the funding and effective implementation of specific plan components. There have been important efforts to incorporate implementation measures for proposed policies in recent Pima County plans. The County and the City comprehensive plans have mechanisms to monitor progress towards its goals, but not to assign specific responsibilities to guarantee their accomplishment. Public participation stands out as one of the biggest achievements of the latest comprehensive planning process for Pima County, especially in relation to water in this arid region. A continuation of these efforts to effectively engage all sectors of society in comprehensive planning and water management processes can only benefit the community as a whole.

Because comprehensive planning is the integrative process by which the many dimensions of economic development and urban growth must come together, it is where the rubber meets the road in terms of building a better community with human well-being, as well as social, environmental and economic sustainability and resilience. The solutions to the water and environmental challenges we face require integrated efforts and coordination among institutions for effective implementation. Aspects such as land use cover, growth and building codes, energy efficiency, new water use permits, local environmental impacts, no pumping zones, water re-use, alternative water sources and green infrastructure, are all topics that overlap and influence each other through feedbacks that contribute to a sustainable future.

The effective implementation of measures to spatially balance the accounting of safe yield and conserve or restore valuable natural areas will have to be tied to comprehensive planning in the basin. While enhancing natural or artificial recharge in strategic areas within the Water Accounting Areas is one option, urban planning, zoning, design, policies, incentives and buyouts can also be effective mechanisms to regulate existing and new pumping affecting sensitive areas. In addition to overall conservation efforts, a special emphasis reflecting the local sensitivities of pumping, land and water use needs to be integrated in urban planning.

Alternative Water Sources and Green Infrastructure: As shown by *Kuhn et al.* (*Chapter 17*), the systematic integration of green and grey infrastructure in future planning efforts offers promising potential to address future water, energy, and environmental challenges. The development of alternative water resources is not new to Tucson, with facilities that produce effluent for recharge and re-use (old Roger Road and Randolph, and new Agua Nueva and Tres Rios reclamation facilities), large scale storm-water capture (Kino Environmental Restoration Project), and ordinances, incentives and efforts towards rainwater and stormwater harvesting at the home and street level, and others. It is far, however, from reaching its potential: rainwater and stormwater can cover more than 20% of municipal demand in the TAMA, and greywater could contribute 2% without negatively affecting flows to reclamation facilities. Effluent, which meets high quality standards, is becoming a precious resource, with the possibility of a future shift from indirect to direct potable reuse.

Both active rainwater harvesting measures (cisterns & piping) as well as passive (landscaping basins, curb-cuts) can all contribute to offsetting aquifer pumping through water provision for drinking and the outdoors. While active rainwater harvesting has initial investment costs that can vary depending on the size and nature of the storage tank and system, it has been demonstrated that a home in Tucson with a garden and a pool can be 100% reliably supplied with rainwater.[4] Thus, rainwater harvesting is clearly under-utilized in Arizona's desert cities, which enjoy two rainy seasons. Passive rainwater and stormwater harvesting—where berms and basins in the urban landscape slow down water and promote infiltration and soil storage—has much lower costs and can self-support native urban vegetation.

As shown in this book, green infrastructure in the urban landscape provides important services such as heat regulation (decreasing the urban heat island effect,

---

4 See Watershed Management Group Spring 2015 newsletter, pages 10–11, at: https://watershedmg.org/sites/default/files/documents/2015-may-watershed-management-group-newsletter.pdf.

and energy demand), resilience to drought through water savings, air dust regulation, a healthy environment and shade, outdoor spaces for recreation, and well-being. At municipal scales, stormwater harvesting can provide water for supply or recharge, as well as being a flood control measure. The ecosystem services of green infrastructure need to be better integrated in benefit-cost analyses of planning processes, and mechanisms need to be found to develop the potential of alternative water sources.

Need for a new institution: It has been raised by researchers and practitioners alike that an effective harnessing of stormwater resources at the regional level can only be achieved through the creation of a new institution, able to work across department and agency boundaries (county, city, transportation, parks management, ADWR). This new "stormwater utility" could fund green infrastructure projects, well integrated in current and future urban planning, and building resilience by saving groundwater for times of drought and scarcity, balancing the heat island effect, providing flood control and enhanced recharge, and contributing to environmental justice through green areas in poor neighborhoods. Perhaps taking advantage of higher public awareness and the current drought crisis in the Colorado basin and California, the state and municipalities can take advantage of the opportunity to bring these ideas to fruition.

In parallel, a fuller accounting of hidden costs of development, in land use change (such as negative footprint, i.e., a decrease in supply of ecosystem services) or increased water use, can be enforced in environmental impact assessments and costs be passed on to users. While this can be carefully done in the public sector without placing additional burdens on vulnerable socio-economic populations, it is especially relevant for the business and industrial sector, encouraging good practices and linking conservation and development.

On Research and Governance: The work presented in this book has already been of interest and benefit to the stakeholder community by illustrating key issues regarding the relationship between urban growth, land use, drought, ecosystem services, biodiversity, green infrastructure, water source development and reuse, policy and planning. The findings of this research are especially relevant in the current context of potential shortages in the Colorado River, drought and the uncertainties arising from climate and global change.

However, none of the ideas and initiatives discussed above will be effective without strong institutions with more power to pass policy and enforce regulations. If continued in the future, the severe underfunding of the Arizona Department of Water Resources in the last years can represent a challenge to the proper monitoring and transparent reporting of evolving water budgets. This would undermine science- and information-based decision making, as well as the evolving needs of good water resources management practices when trying to keep up with the always moving target of sustainability.

As current official projections indicate growth, it will be interesting to see what are the new institutional and policy arrangements, land use planning, green and grey infrastructure, technology, use practices and behaviors that will take place to accompany such growth projections in the future. In this region that prides itself of having the most innovative and progressive water policies, some of the best water research institutions, and an environmentally concerned private sector, it remains to be seen whether it will manage to maintain its last natural riparian areas. While the aquifers—

and the CAP water we are storing in them—are a good temporal buffer to scarcity and drought, the future of the last riparian shallow groundwater areas will be a good indicator of how sustainable is Tucson's water management in the decades to come.

REFERENCES

Burkhard B., Groot R. de, Costanza R., Seppelt S.E. and Potschin M. (2012). Solutions for sustaining natural capital and ecosystem services. *Ecological Indicators* 21: 1–6.

Cabello, V., Willaarts, B., Aguilar, M. and del Moral, L. (2015). River basins as social-ecological systems: linking levels of societal and ecosystem water metabolism in a semiarid watershed. *Ecology and Society* 20(3): 20. http://dx.doi.org/10.5751/ES-07778-200320, http://www.ecologyandsociety.org/vol20/iss3/art20/.

Dumont, A., Salmoral, G. and Llamas, M.R. (2013). The water footprint of a river basin with a special focus on groundwater: The case of Guadalquivir basin (Spain). *Water Resources and Industry*, 1–2, 60–76. doi:10.1016/j.wri.2013.04.001.

Funtowicz, S. and Ravetz, J.R. (1994). The worth of a songbird: ecological economics as a post-normal science. Ecological Economics 10(3) 197–207.

Kandziora M., Burkhard B. and Müller F. (2013). Interaction of ecosystem properties, ecosystem integrity and ecosystem service indicators – A theoretical matrix exercise. *Ecological Indicators* 38: 54–78.

Madrid, C. and Giampietro M. (2015). The water metabolism of socio-ecological systems: reflections and a conceptual framework. *Journal of Industrial Ecology* (In press). http://onlinelibrary.wiley.com/doi/10.1111/jiec.12340/abstract.

Raymond, C.M., Singh, G.G., Benessaiah, K., Bernhardt, J.R. and Levine, J. (2013). Ecosystem Services and Beyond. *Bio Science*, 63(7): 536–546. doi:10.1525/bio.2013.63.7.7.

Rajagopal S., Gupta H.V., Troch P.A., Dominguez F. and Castro C.L. (2014). Physical mechanisms related to climate-induced drying of two semi-arid watersheds in the southwest US. *Journal of Hydrometeorology*, 15(4): 1404–1418.

Raymond, C.M., Singh, G.G., Benessaiah, K., Bernhardt, J.R. and Levine, J. (2013). Ecosystem Services and Beyond. *Bio Science*, 63(7): 536–546. doi:10.1525/bio.2013.63.7.7.

Rees W.E. (1992) Ecological footprints and appropriated carrying capacity: what urban economics leaves out. *Environment and Urbanization* 4: 121–130.

Serrat-Capdevila, A., Valdés, J.B., González Pérez, J., Baird, K., Mata, L.J. and Maddock III, T. (2007) Modeling climate change impacts—and uncertainty - on the hydrology of a riparian system: the San Pedro Basin (Arizona/Sonora). *Journal of Hydrology*, 347, 48–66. doi: 10.1016/j.jhydrol.2007.08.028.

Serrat-Capdevila A., Dominguez F., Valdes J.B., Rajagopal S. and Gupta H. (2013). Characterizing the Droughts of the new century in the US Southwest: A comprehensive assessment from state-of-the-art climate model projections. *International Journal of Water Resources Development*, 29(2): 152–171. http://dx.doi.org/10.1080/07900627.2012.721717.

Wiedmann T. and Minx, J. (2007) *A Definition of Carbon Footprint*, ISA Research & Consulting, Durham, United Kingdom.

Chapter 22

Next steps: Collaborative research and training towards transdisciplinarity

The SWAN Consortium

1 TOWARDS A NEW WATER CULTURE

The prevailing water management paradigm: One of the main interests of the case study, and of the transdisciplinary perspective presented in this book, is to provide new insights on more general debates related to water policy. First, the complexity of local water issues questions the prevailing *"state-engineering paradigm"* which has contributed to the implementation of specific technical systems (in Europe and United States) characterized by a uniform and universal model of distribution and treatment of urban potable water (Cf. *Chapter 1*). This *"public-private"* distribution model is typically presented as the best (or even the only) way to implement the objective of *"water for everyone"*. The model implements a *"supply network"* over a territory and managed by an operator, and requires very specific skills in terms of administration and engineering, such that water management now constitutes an area of study that blurs the distinctions between science, action and expertise. Meanwhile, the legitimacy of this model is reinforced by public teaching institutions, research and development services, private enterprise, and consultancy companies (*Bakker 2005*).

Challenges to the prevailing paradigm: The relevance of this model (that favours a constant extension of universal services) is now being questioned. In several places, people have shown the negative economic, social and/or environmental impacts of such water supply projects. Such questioning is based on the failures of the accepted model to provide a universal service and, particularly, on its inability to i) effectively serve peri-urban and rural areas; ii) preserve local ecosystems. This has led to proposals for alternative or hybrid models, based essentially on a decentralisation of technical and administrative services, enhanced symbiosis between technical systems (such as treatment and energy production), and the replacement of monopolistic organisations by a multiplicity of smaller institutions. As one example, the *"Integrated Water Resources Management (IWRM)"* paradigm has recently emerged (in the 1990's) to address some of the failings of the state-engineering paradigm. By promoting '*the coordinated development and management of water, land and related resources*' (*GWP 2000: 22*), the IWRM approach seeks to maximize economic and social welfare in an equitable manner, without compromising the sustainability of vital ecosystems.

The pressing need for transdisciplinarity: Given such challenges to the prevailing model, it seems increasingly clear that a transdisciplinary approach to dealing with the scientific and political challenges of water management is necessary. The essential challenge is not so much that *"people"* (i.e., their needs and perspectives) need to be incorporated into a new water paradigm, but rather that the terms of their inclusion in the socio-natural paradigm be restructured and revitalized. In this regard, IWRM has come under the criticism that it does not escape from the gravitational pull of the state-engineering paradigm sufficiently to be able to achieve the desired equity and welfare goals. Further, IWRM (and its spin-offs, such as Ecosystems Services, Adaptive Water Management, etc.) are perceived as not sufficiently engaging people as effective actors in the decisions that affect them.

Democratizing decision-making: So, whereas the state-engineering model treats people as passive beneficiaries of water services, *Del Moral et al. (2014)* refer to the need to bring people into the heart of what they call a *"new water culture"*, in which public and stakeholder voices are brought centrally into the decision process (see also *Pedregal et al. 2015*). This can be achieved in a number of ways including a) the opening of decision-making forums to non-expert water users, b) the recognition that people are not just passive consumers of a commodified water services product, and c) recognition of the importance of the non-monetary values associated with *"water"*. The central point is that the lives of humans are inextricably intertwined with the dynamics of water, from which a key implication is that decision-making needs to be much more completely democratized.

Increasingly, given the pressures of climate change, population growth and concern for environmental degradation, public institutions now view the participation of citizens as a key factor in the development and legitimization of difficult and contentious decisions. However, the institutionalized methods for such participation at the national and sub-national level have been criticized as being multifarious, complex and ambiguous (*Arnstein 1969, Pretty 1995, White 1996, Callon 1999, Cornwall 2008*). At the administrative level, participation is often limited by institutional cost, susceptible to paralysis by endless deliberations amongst stakeholders with conflicting values and interests, and subject to strategic manipulation by powerful institutions (*Bickerstaff & Walker 2005, Staddon & James 2013*).

The possibility of a new mode of governance: It therefore remains to be seen whether a new and truly democratic mode of governance over our essential natural resources such as water will be realized in the near future. In hinting at a radically different mode of governance, *Swyngedouw (2005, 2011)* points out that the *"public-private partnership"* model discussed above is characteristic of the form of governance implemented by Western democracies such as the US and the EU, in which the argumentative confrontation central to truly democratic politics is marginalized, and the resulting political vacuum is filled by *"opinions on anything imaginable (as long as it does not question fundamentally the existing state of the neoliberal political economic configuration)"*. The consequence is that public participation is robbed of its truly deliberative and policy-shaping function, and reduced solely to its legitimizing function. Ultimately, a truly transdisciplinary approach to dealing with the scientific and political challenges of water management will need to find ways to address this foundational issue of the structure by which democratic participation in decision making is realized.

2 TRAINING TOWARDS TRANSDISCIPLINARITY

Merging perspectives: Our experience over four years of the SWAN project has provided a clear direction regarding how the practical implementation of a major international scientific network to fully explore the many interrelated facets of water management can be achieved. As shown by this book, the first steps have been taken towards a transdisciplinary water research strategy and agenda by researchers from Europe and the United States. No question can exist that, given the complexity of the issues involved, a transdisciplinary approach must be applied to water issues. This requires merging the expertise, knowledge and efforts of political and physical scientists, students, and policy makers, together with industry, agricultural and environmental organizations, to identify and implement workable solutions.

Achieving successful integration: To be successful, a transdisciplinary project must fully integrate the expertise and experience of the researchers' and all affected stakeholders, so as to guide the investigation towards a holistic solution for the given problem. This marks a sharp contrast to the conventional approach in which researchers work in parallel – or worse, in isolation – within the strictures of their respective disciplines; such an approach has time and again failed to develop and implement solutions that are able to comprehensively address a given water problem. The important contribution of a scientific network (such as the one generated by the SWAN project) is that it can foster and facilitate the kinds of necessary interaction between stakeholders and researchers from different disciplines that is a necessary condition to be able to resolve complex problems.

The SWAN team greatly benefitted from the project structure, which permitted lengthy periods of daily interactions among the researchers in the framework of the UMI iGLOBES (Interdisciplinary and Global Environmental Research, CNRS/University of Arizona). European students and researchers lived in Tucson, Arizona for periods of three to six months, giving them the opportunity to meet, discuss and work together. In addition, week-long biannual meetings of the entire SWAN team provided forums for concentrated collaboration and development of the project. While the SWAN team also conducted weekly online meetings of the EU and US researchers, it is clear that even the most advanced communication technologies cannot substitute for in-person meetings and workshops, or for the close daily collaboration that is possible when researchers work in close physical proximity to each other. As discussed in the literature, many barriers exist to achieving the recursive integration necessary to a successful transdisciplinary research project, but surprisingly little attention has been paid to one of the major characteristics of the SWAN project, i.e., the importance of having a physical location for real (i.e., non virtual) interaction, on-going research, and periodic meetings. This single component greatly enhances the opportunities for collaboration and eliminates the significant barriers created by distance, multiple time zones, and technological hiccups that can render a meeting or conversation useless.

The need for a trained facilitator: Realistically, a transdisciplinary project should also include a director or facilitator trained or experienced in collaborative research. Just as a symphony requires a conductor, and the construction of a skyscraper requires a project manager, a properly funded and staffed transdisciplinary team requires a coordinator in order to improve efficiency, identify the entire class of relevant stakeholders, and achieve more integrated results. Team participants can focus on the water problems that affect many people and the environment and not be distracted by administrative issues, or hampered by a lack of coordinated efforts, pointless internal conflicts or a failure to completely engage. An effective manager or facilitator can enable the team to accomplish work in a timely manner, productively incorporate the efforts of students (whose participation is necessarily transient), and constructively resolve conflicts. In addition, support is required from a staff able to fully engage stakeholders to obtain more than just the views of the stakeholders; it needs to also obtain their commitment and efforts towards the implementation of well-researched, well-reasoned solutions.

The value of engaging the public and role of new technologies: Engagement can occur through a range of methods, from involvement in projects focused on problems affecting a specific group of stakeholders to broader structures for engagement with broader impacts (*Jasanoff 2007*). Today's rapidly growing fields of open knowledge and citizen science research provide new opportunities for the development of the kinds of collaboration necessary for transdisciplinary investigation (*Giampetro et al., 2012*). Citizen scientist projects are those in which a variety of citizens (from all walks of life) can collaborate with governmental and non governmental organizations, universities and communities, to participate in environmental research projects that may require significant amounts of repeated observational data over large geographic areas or long periods of time, or both.

In this regard, the possibilities created by new technologies and "big data" need to be taken into account. As shown by *Hernandez-Mora et al.* (2015) in Spain, the involvement of citizen and social movements in water management can be enhanced by using such technologies to improve coordination, to aid in evaluation of administrative data, and to facilitate production of new data. While such projects can generate large amounts of extremely valuable information (*Edwards 2013*), they can also motivate a wide variety of stakeholders to go into the field and the laboratories to understand the science underlying societal and environmental issues.

Towards truly transdisciplinary solutions: Reflecting upon our experiences over the four years of the SWAN project, we are now convinced that the most effective way forward would be the establishment of a new "*Institute for Open Knowledge*". Building upon the experiences and insights described above, such an organization would provide an effective and efficient mechanism by which to produce collaborative research, and to generate truly transdisciplinary solutions to pressing human problems.

One of the main issues that must necessarily inform the design and structure of such an Institute is that a major obstacle to the realization of open collaborative research on water (the new water culture referred to above) is the fact that most researchers are poorly prepared to operate in a transdisciplinary manner. Not only do scientists, engineers and managers struggle when required to work across multiple disciplinary contexts, they are also not well versed in strategies for engaging with citizens from non academic backgrounds.

So the most important challenge may actually be the implementation of a mechanism for *"training in transdisciplinarity"*. This will involve, of course, an education that exposes *"students"* to concepts from multiple disciplines. But, more importantly, it pre-supposes a real change in the dispositions of the people involved—a rupture with the kind of *"scholastic bias"* (*Bourdieu 2000*) that causes every scholar to tend to engage in the *"mind projection fallacy"* (*Jaynes 2003*) that social life operates with the same logic that is used in the development of theory. Only then, perhaps, can we be able to break with the viewpoint of the *"new engineers"* of the early 20th century, who believed that any water problem in Southwestern USA could be solved through the application of technology and science.

Finally, beyond an educational experience that opens the mind to multiple alternative perspectives and points of view, and beyond an understanding that an effective search for solutions (based on a comprehensive understanding of the problem, and its greater context) must necessarily involve the participation of a variety of stakeholder interests (policy makers, water managers, scientists, the public, etc.), the search for truly transdisciplinary solutions requires at least one more mental shift. And that is the realization that the important role of science (albeit disciplinary or multi-disciplinary) in helping to solve complex human problems is not found in the traditional "*search for concrete solutions*". Rather it is instead in the "*co-production of knowledge and new ways of thinking*" about the challenges posed by evolving human needs. The fact is that water managers typically produce enough data and technical solutions that they usually do not need the academic to do this for them. Instead, an invigorated collaboration leads to a revaluing of the coproduction of knowledge, and a rupture with traditional, institutionalized, ways of thinking.

REFERENCES

Arnstein, S.R. (1969). A ladder of citizen participation. *Journal of the American Institute of Planners*, 35(4): 216–224.

Bakker, K. (2005). Neoliberalizing nature? market environmentalism in water supply in England and Wales. *Annals of the Association of American Geographers*, 95: 542–565.

Bickerstaff, K. and Walker, G. (2005). Shared visions, unholy alliances: power, governance and deliberative processes in local transport planning. *Urban Studies*, 42(12): 2123–2144.

Bourdieu, P. (2000). *Pascalian Meditations*, Stanford. Stanford University Press.

Callon M. (1999). The lay people in the production and dissemination of scientific knowledge. *Science Technology Society*, 4(1): 81–94.

Cornwall, A. (2008). Unpacking 'Participation': models, meanings and practices, *Community Development Journal*, 43(3): 269–283.

Del Moral, L. and Do O.A. (2014). Water governance and scalar politics across multiple-boundary river basins: States, catchments and territorial powers in the Iberian Peninsula. *Water International*, 39(3): 333–347.

Edwards, P. (2013). *A Vast Machine. Computer Models, Climate Data and the Politics of Global Warming*, MIT Press.

Giampietro, M., Mayumi, K. and Sorman, A.H. (2012). *The Metabolic Pattern of Societies. Where Economists Fall Short*. London and New York: Routledge.

Global Water Partnership (2000). Integrated Water Resources Management, *Global Water Partnership Technical Advisory Committee, Background Paper* no.4.

Hernández-Mora, N., Cabello, V., De Stefano, L. and Del Moral, L. (2015). Networked water citizen organizations in Spain: Potential for transformation of existing power structures in water management, *Water Alternatives*, 8(2): 99–124.

Jasanoff S. (2007). *Designs on Nature. Science and Democracy in Europe and the United States*, Princeton & Oxford, Princeton University Press.

Jaynes, E.T. (2003). *Probability Theory: The Logic of Science*. New York, NY: Cambridge University Press

Pedregal, B., Cabello, V., Hernández-Mora, N., Limones, N. and Del Moral, L. (2015). Information and knowledge for water governance in the networked society. *Water Alternatives* 8(2): 1–19.

Pretty, J.N. (1995). Participatory learning for sustainable agriculture. *World Development*, 23(8): 1247–1263.

Staddon, C. and James, N. (2013). Water Security: genealogy of a new paradigm for water management? in, G Schneier-Madanes (Ed.) *Globalised Water,* Dortrecht, Springer: 261–276.

Swyngedouw, E. (2005). Governance innovation and the citizen: The Janus face of governance-beyond-the-state. *Urban Studies,* 42(11): 1992–2006.

Swyngedouw, E. (2011). Interrogating post-democratization: Reclaiming egalitarian political spaces. *Political Geography,* 30(7): 370–380.

White, S.C. (1996). Depoliticising development: the uses and abuses of participation, *Development in Practice*, 6(1): 6–15.

Contents (full titles and authorship)

Preface — EDITORS — ix

Introduction — 1

1 The idea of a transatlantic dialogue — 3
THE SWAN CONSORTIUM

2 Organization of the book and mind map — 7
EDITORS

Maps — 17

Socio-historic perspectives on water in the American southwest — 25

3 The Tucson basin: Natural and human history — 27
ALEIX SERRAT-CAPDEVILA

4 Laws of the river: Conflict and cooperation on the Colorado River — 45
BRIAN O'NEILL, FRANCK POUPEAU, MURIELLE COEURDRAY & JOAN CORTINAS

5 Water for a new America: The policy coalitions of the Central Arizona Project (Part 1) — 65
JOAN CORTINAS, MURIELLE COEURDRAY, FRANCK POUPEAU & BRIAN O'NEILL

6 Sharing the Colorado River: The policy coalitions of the Central Arizona Project (Part 2) — 79
MURIELLE COEURDRAY, JOAN CORTINAS, BRIAN O'NEILL & FRANCK POUPEAU

7 The making of water policy in the American southwest:
   Environmental sociology and its tools   101
   FRANCK POUPEAU, MURIELLE COEURDRAY, JOAN CORTINAS & BRIAN O'NEILL

Narratives of urban growth   119

8 The social logic of urban sprawl: Arizona cities under
   environmental pressure   121
   ELIZA BENITES-GAMBIRAZIO

9 Water and urban development challenges in the Tucson
   metropolitan area: An interdisciplinary perspective   141
   GRACIELA SCHNEIER-MADANES, JUAN B. VALDES, EDWARD F. CURLEY,
   THOMAS MADDOCK III, STUART E. MARSH & KYLE A. HARTFIELD

10 Comprehensive urban planning: Implications
   for water management in Pima County (Arizona)   159
   SERGIO SEGURA CALERO

11 Potential impacts of the continuing urbanization on regional climate:
   The developing Phoenix-Tucson "Sun Corridor"   179
   ZHAO YANG, FRANCINA DOMINGUEZ, HOSHIN GUPTA, XUBIN ZENG & LAURA NORMAN

Ecosystem services and biodiversity   195

12 Quantification of water-related ecosystem services in the Upper
   Santa Cruz watershed   197
   KREMENA BOYANOVA, REWATI NIRAULA, FRANCINA DOMINGUEZ,
   HOSHIN GUPTA & STOYAN NEDKOV

13 Qualitative assessment of the supply and demand of ecosystem
   services in the Pantano Wash watershed   223
   ROSITSA YANEVA

14 The role of biodiversity in the hydrological cycle
   The case of the American Southwest   249
   MARIA A. SANS-FUENTES & THOMAS MEIXNER

Water use and groundwater management — 289

15 Implications of spatially neutral groundwater management: Water use and sustainability in the Tucson basin — 291
VIOLETA CABELLO, NURIA HERNÁNDEZ-MORA, ALEIX SERRAT-CAPDEVILA, LEANDRO DEL MORAL & EDWARD F. CURLEY

16 Groundwater dynamics: How is Tucson affected by meteorological drought? — 321
NATALIA LIMONES

17 Alternative water sources towards increased resilience in the Tucson region: Could we do more? — 337
KRISTIN KUHN, ALEIX SERRAT-CAPDEVILA, EDWARD F. CURLEY & LÁSZLÓ G. HAYDE

18 Differentiated approaches of groundwater management: A comparison between the TAMA and the San Pedro basin — 363
SUSAN HARRIS

Stakeholders' perspectives — 379

19 Presentation — 381
ALEIX SERRAT-CAPDEVILA, EDWARD F. CURLEY & ALBA BALLESTER CIURÓ

20 Texts — 383

    20a Water Sustainability Program & Pima Association of Governments — 383
    CLAIRE L. ZUCKER

    20b Tucson Water — 385
    WALLY R. WILSON

    20c Central Arizona Project — 387
    DAVID MODEER

    20d Southern Arizona Home Builders Association — 389
    DAVID GODLEWSKI

    20e Pima County — 391
    JULIA FONSECA

20f Southwest Conservation Advocate, Center for Biological Diversity   393
    RANDY SERRAGLIO

20g Bureau of Reclamation   395
    EVE HALPER

20h Arizona Department of Water Resources   397
    TOM BUSCHATZKE

Conclusion   399

21 Bringing all the stories together: Beyond the Tucson case study   401
    ALEIX SERRAT-CAPDEVILA, VIOLETA CABELLO, KREMENA BOYANOVA,
    FRANCK POUPEAU, NURIA HERNÁNDEZ-MORA, HOSHIN GUPTA,
    ZHAO YANG, NATALIA LIMONES, BRIAN O'NEILL, SERGIO SEGURA CALERO,
    ELIZA BENITES-GAMBIRAZIO, ROSITSA YANEVA, MURIELLE COEURDRAY,
    JOAN CORTINAS, KRISTIN KUHN, SUSAN HARRIS, MARIA SANS-FUENTES,
    EDWARD F. CURLEY, LEANDRO DEL MORAL, JUAN VALDES, GRACIELA SCHNEIER-MADANES,
    OWEN KING, DULCE B.B. RODRIGUES & GLORIA SALMORAL

22 Next Steps: Collaborative research and training towards
    transdisciplinarity   417
    THE SWAN CONSORTIUM

Acknowledgments

The SWAN Consortium is grateful to the professors and researchers who have accepted to review different chapters of this book: *Bernard Barraqué (CNRS), Robert Glennon (University of Arizona), Stephanie Pincetl (UCLA), Tom Maddock University of Arizona), Sharon Medgal (University of Arizona), Tom Meixner (University of Arizona), Mitchel Pavao-Zucherman (University of Maryland), Stéphanie Pincetl (University of California, Los Angeles), Chris Scott (University of Arizona), Robert Varady (University of Arizona), Xubin Zeng (University of Arizona).*

The Tucson case study has also beneficiated of the contributions from *Gita Bodner (The Nature Conservancy), David Brookshire (University of New Mexico), Tom Buschatzke (ADWR), Evan Canfield (Pima County), Kathy Chavez (Pima County), Edward F. Curley (Pima County), Pierre Deymier (University of Arizona), Charlie Ester (Salt River Project), Tom Evans (University of Bloomington), Julia Fonseca (Pima County), David Godlewsky (Southern Arizona Home Builders Association), David Goodrich (United States Department of Agriculture Agricultural Research Service), Eve Halper (Bureau of Reclamation), Greg Harris (Lewis Rocca Rothgerber, LLP), Brad Lancaster (Desert Harversters), Michael McNulty (Office of the Pima County Attorney), Ralph Marra (Consultant), Mohammed Mammou (Central Arizona Project), Mead Mier (Pima Association of Governments), Michelle Moreno (ADWR), Brian Powell (Pima County), Linda Stitzer (Western Resources), Jeff Tannler (ADWR), David White (Arizona State University), Wally Wilson (Tucson Water), Claire Zucker (Pima Association of Goverments), Claire Zugmeier (Sonoran Institute).*

The development of the SWAN project would not have been possible without the active support of several persons, from:

- the CNRS: *Patrice Bourdelais, Colette Doressoundiram, Hélène Leniston, Pascal Marty, Hélène Naftalski, Marie-Louise Perez, Marie Weiss;*
- the University of Arizona: *Thomas Alvarez, Andrew Comrie, Candice Crossey, Rhonda Dwyer, Olivia Hanson, Pierre Meystre, Joaquin Ruiz, Erma Santander, Julie Ann Stringer, Larry Winter, Judith-Ann Yandow.*

Subject Index

Active Management Areas (AMAs) 56–57, 93–94, 152–153, 198, 294, 296, 365, 383, 391
  Critical Management Area 56
Agua Caliente Area Plan 133
Akin, Wayne 81
All American Canal 50, 52
alternative water sources 345–349
  characterization of 340–342
  cost estimations 349–350
  graywater 345–346
  overview 337–338
  rainwater 346
  research methodology 339–340
    data collection 339–340
    sustainability assessment 339
  stormwater 346–349
  sustainability analysis 342–350
  and traditional water sources 343–345
    CAP 344
    groundwater 343–344
    reclaimed water 344–345
  for urban greenery 350–353
    cultural services 351–353
    provisioning services 350
    regulating services 350–351
  and water community 353–357
    economic dimension 355–356
    environmental dimension 356–357
    social dimension 353–355
Ambassador Program 97
American Civil War 39
Anamax Mining Company 56
Andrus, Cecil 54, 56
APS, *see* Arizona Public Service (APS)
aquifers 27–28, 33, 40–42, 56–57, 62, 81, 90, 146–147, 153, 171–172, 176, 198, 209, 228, 251, 291–295, 298–300, 311–313, 315, 322–326, 328, 330, 332, 334–335, 337, 339, 341–343, 345, 350, 353, 356, 358, 363, 365, 367–369, 371–372, 374, 384, 391–392, 408, 410, 412–414

Archaic Period 34–35
Areawide Wastewater Treatment Management Program (208) 161–162
Arizona, *see also* Tucson City
  comprehensive urban planning in 163–166
  groundwater struggle of 54–55
  institutional context for water management in 294–296
  urban sprawl of 127–135
    land and planning 129–130
    Phoenix and pro-growth coalitions 130–131
    suburbanization in 129
    technology and water supply 134–135
    and Tucson 131–134
  water management tools of 55–56
Arizona Department of Environmental Quality (ADEQ) 95
Arizona Department of Water Resources (ADWR) 33, 57–58, 94, 153, 160, 165, 187, 199, 293, 295, 335, 365, 367, 383, 397–398, 414
Arizona Game Protective Association 55
Arizona Highline Reclamation Association 84
Arizona Interstate Commission 81
Arizona Interstate Stream Commission 84
Arizona Meteorological Network (AZMET) 182
Arizona National Guard 80
Arizona Public Service (APS) 182, 188
Arizona Revised Statutes (ARS) 49, 57, 163–168, 346, 356, 365
Arizona Salt River Project 74
Arizona Sun Corridor, *see* Phoenix-Tucson Sun Corridor
Arizona v. California 45–47, 53, 79–80
Arizona Water Banking Authority (AWBA) 57, 96
Arizona Water Commission 57
Arizona Wildlife Federation 55

Subject Index

Aspinall, Wayne 85
Assured Water Supply (AWS) Certificate 153, 296, 367
autogenic engineers 256

Babbitt, Bruce 54–57, 93–94
basin and range landscape, of Tucson City 27–28
Beaudry, Bob 95
big data 4
biodiversity, in hydrological cycle
  overview 249–252
  rodents as ecosystem engineers 255–258
    in deserts of Southwest 258–268
      ground squirrels as 258–261
      interdependence of 271–280
      kangaroo rats as 266–268
      marmot as 261–262
      pocket gophers as 264–266
      prairie dogs as 262–264
    in Southwestern US 252–255
biological soil crusts (BSC) 253
bioturbation 256
Black Canyon Dam, *see* Hoover Dam
Board of the Phoenix Chamber of Commerce 81
Boulder Canyon Project Act (BCPA) 52
Boulder Dam, *see* Hoover Dam
Boulder Dam Bill 75–76
Brown, Michael 96
Bureau of Indian Affairs 41
Bureau of Reclamation (BOR) 50–51, 69, 73, 75, 77, 84, 86–88, 94, 101, 105, 107–111, 113–115, 312, 353, 372, 395
Buschatzke, Thomas 397–398
Bush, James 90

Cadillac Desert 102
California *see* Arizona v. California
Carpenter, Delphus Emory 51–53, 55–56
Central Arizona Groundwater Replenishment Districts (CAGRD) 57, 93, 296
Central Arizona Project (CAP) 383
  and Colorado River 80–84
    Arizona v. California 79–80
    and federal government 88–93
    groundwater regulations 88–93
    legislation and environmental pressures 84–88
    and urban development 93–97
  policy coalitions of
    changing 76–77
    and Depression 74–76
    and drought 65–66
    irrigation projects and economic crisis 66–69

New Engineers 71–74
Reclamation Act of 1902 69–71
as traditional water sources 344
Central Arizona Water Conservation District (CAWCD) 295
Central Arizona Water Control Study (CAWCS) 113–114
Centre National de la Recherche Scientifique (CNRS) 3
Chandler, Thomas 90
Citizen's Growth Management Initiative (CGMI) 132
Citizens Water Protection Initiative 95
City Beautiful Movement 161
The City in History (Mumford) 162
City of Tucson General Plan 2001 170
Click, Jim 95–97
climate, of Tucson City 28–31
climate impact
  urbanization, and Tuscon
    methodology 181–182
    observational temperature data 182–183
    overview 179–181
    on precipitation 186–187
    results 183–190
    on temperature 183–186
    on water and energy demand 187–190
    WRF model and configuration 182
Coastal Zone Management Act (1972) 162
Colorado River Compact/Laws of the River/water laws
  Arizona
    groundwater struggle of 54–55
    water management tools of 55–56
  Arizona v. California 53, 79–80
  and CAP association 80–84
  doctrines 48–50
  and federal government 88–93
  frontier 46–47
  groundwater regulations 88–93
  interstate conflict 52–61
  legislation and environmental pressures 84–88
  origin of 50–52
  overview 45–46
  shortage sharing agreement of 2007 58–61
  and urban development 93–97
Colorado River Basin Project Act of 1968 48, 54, 59, 86
Common International Classification of Ecosystem Services (CICES) 201
comprehensive urban planning
  in Arizona 163–166
    characterization of 165–166

Subject Index 431

and smart growth regulations 164–165
and City of Tucson 170–174
overview 159
and Pima County comprehensive plan 167–168
and Pima Prospers 168–170
spatial and water planning 159–160
in United States 161–163
Conservation Lands System (CLS) 167
Continental Ranch development 153–155
cost estimation, of alternative water sources 349–350
Critical Groundwater Code 56, 89

data analysis, for ES 233–236
data collection
　alternative water sources 339–340
　drought, in Tuscon 326
Davis, Arthur Powell 51, 71
DeBolske, Jack 90
Department of the Interior (DOI) 57, 59
Depression, and CAP 74–76
desert landscapes, and water policies 103–105
Digital Elevation Model (DEM) 205
doctrines, of Laws of the River 48–50
drought, in Tuscon
　and CAP 65–66
　data sources 326
　hydrogeological studies in 323–324
　methodology 325–326
　overview 321–322
　precipitation and water tables
　　monthly series of 330–332
　　regimes and time evolution 327–330
　Precipitation Drought Index (PDI) 326–327
　　results of application 332–334
　study area and time span 324–325
　synthesis 334–335
　types 322–323

Ecological-Value Submodel (EVM) 207
Ecology Law Quarterly (McBride) 54
economic crisis, and CAP 66–69
economic dimension, and water sources 355–356
The Economics of Ecosystems and Biodiversity (TEEB) 201
ecosystem services (ES) 197
　concept of 223–224
　map-based supply and demand for 244–247
　　assessment approach 228
　　concept 223–224
　　data analysis 233–236
　　and definitions 243

　　identifying opportunities 239
　　and matrix model 229–231
　　overview 224–225
　　Pantano Wash watershed case study 226–228
　　participants and opinions 237–239
　　problem identification 225–226
　　qualitative assessment 231–233
　　relevance of results 236–237
　　results 236
　　uncertainties and limitations 239–240
　quantification and mapping 201–203
　　hydrological modeling for 203–205
　　and SWAT model 205–207
Eller, Karl 97
Ellis, Brock 90
El Niño 32, 39, 40
Ely, Sims 52
Endangered Species Act (ESA) 272
Endangered Species and Antiquities Acts 55
energy demand, urbanization impact on 187–190
engineers 71–77, 102, 106, 113–116, 123–125, 162, 395, 420
　administrator-engineers 107–108
　allogenic engineers 256
　bioengineering 264
　ecosystem engineers 255, 271
　　ground squirrels as 258–261
　　kangaroo rats as 266–268
　　marmot as 261–262
　　pocket gophers as 264–266
　　prairie dogs as 262–264
　Corps of Engineers 87, 107
　New Engineers 71–74
　state-engineering paradigm 5
environmental dimension, and water sources 356–357
Environmental Impact Statement (EIS) 113
environmental sociological tools, and water policies 116
equitable apportionment doctrine 52
ES, *see* ecosystem services (ES)
ESA, *see* Endangered Species Act (ESA)
Espeland, Wendy Nelson 112
European Union Biodiversity Strategy 224
EVM, *see* Ecological-Value Submodel (EVM)

Farmers Investment Company (FICO) 90, 93
Farmer's Investment Company v. Bettwy 54, 56, 365
Federal Real Estate Profile 124
Federal Reserved Rights Doctrine, *see* Winters Doctrine

FICO, *see* Farmers Investment Company (FICO)
Flood Control Act 75–76
Flowing Wells Irrigation District 148
Fonseca, Julia 391–392
Fort Belknap Reservation 47
Freitas, Chuck 95

Gadsden Purchase 37, 39
Garden City urban movement 161
Geographical Information System (GIS) 211
Gila River 37, 92, 198
Godlewski, David 389–390
government power, and water policies 110–112
    internal struggles within 112–116
Grand Canyon National Park 85–87, 103
Grant, Douglas 58
graywater 345–346
Great Basin Lunch Mess 72
green infrastructure, water sources for 350–353
    cultural services 351–353
    provisioning services 350
    regulating services 350–351
Green River 103
Green Street Policy 348
ground squirrels, as ecosystem engineers 258–261
groundwater, *see also* water management
    Arizona's struggle of 54–55
    management
        comparison of results 372–375
        development of approaches 364–366
        overview 363–364
        spatial assessment of 310–313
        and Tucson Active Management Area (TAMA) 366–368
        and Upper San Pedro Basin 368–372
    regulations, and CAP 88–93
    as traditional water sources 343–344
Groundwater, Climate and Stakeholder Engagement (GCASE) 324
Groundwater Code 89, 397–398
Groundwater Management Act of 1980 (GMA) 46, 54–57, 94, 135, 294–297, 302, 307, 309, 316, 340, 409–410
    and water/wastewater 152–153
Groundwater Saving Facilities (GSFs) 295
Groundwater Users Advisory Councils (GUAC) 296
Growing Smarter Act of 1998 164
Growing Smarter Plus Act of 2000 164

Halper, Eve 395–396
Harvard Water Program 107
Hayden, Carl 81–85
Hohokam culture 35–36
homeowners associations (HOAs) 130
Hoover, Herbert 52
Hoover Dam 41, 50, 53, 79, 92, 349
House Interior Committee 85
Howard, Ebenezer 161
Howell Code 48
human history, in Tucson City 34–43
    arrival of Spanish 36–37
    current 41–43
    incorporation into US 37–41
    prehistoric period 34–36
Human Well-Being Submodel (HWB) 207
Hunt, George W. P. 84
Hurley v. Abbott 49
hydraulic society, and water policies 106–110
hydrogeological studies, of drought 323–324
Hydrological Response Units (HRUs) 205
hydrology, of Tucson City 28–31

Imperial Irrigation District (IID) 91–92
Institutional and Policy Advisory Group (IPAG) 296
institutional context for water management, in Arizona 294–296
International Union for Conservation of Nature (IUCN) 272
International 3–4, 162, 201, 272, 324, 353, 401, 419
Irrigation and Reclamation Subcommittee 85
Irrigation Non-expansion Areas (INAs) 383
irrigation projects, and CAP 66–69

Johnson, Hiram 75
Johnson, James 90
Johnson, Lyndon 86–87
Jones, John S. 95
Juliani, Gerald "Jerry" 95–96

kangaroo rats, as ecosystem engineers 266–268
Kansas v. Colorado 52
Keating, Charles 130
Kennedy, John F. 85
Kent, Edward 48
Kent Decree 48–49
Kibbey, H. 48
Klamath Project 74
Kyl, Jon 90

Lafferty, John 130
land surface model (LSM) 182
land use and land cover (LULC) 225

Subject Index

Land Use Intensity Legend 167
land-use planning 129–130
League of 14 81
"leap-frog" development 150–151
legislation and environmental pressures, of Colorado River 84–88
long-term storage credits (LTSC) 305

Maas, Arthur 107
Management District 131
map-based supply and demand, for ES
  assessment approach 228
  concept 223–224
  data analysis 233–236
  and definitions 243
  identifying opportunities 239
  and matrix model 229–231
  overview 224–225
  Pantano Wash watershed case study 226–228
  participants and opinions 237–239
  problem identification 225–226
  qualitative assessment 231–233
  relevance of results 236–237
  results 236
  uncertainties and limitations 239–240
Market Land-Price Submodel (MLP) 207
marmot, as ecosystem engineers 261–262
matrix model, for ES 229–231
Maxwell, George Hebard 67–71, 77, 82
McBride, Lawrence 54
McFarland, Ernest 81
McKasson, Molly 95–97, 96
Metropolitan Water District 51, 80
Metropolitan Water Improvement District 148
Mexican-American War 123
Mexican Revolution 40
Meyers, Charles J. 45
Milk River 46–47
Millennium Ecosystem Assessment (MA) 201
million acre feet (MAF) 52
Mission Gardens Project 38
Modeer, David 387–388
Modeling
  hydrological modeling 203–206, 401, 403
  Modeling Hydrology Unit 335
Moeur, Benjamin Baker 84
Muddy Waters (Maas) 107
Multiscale Integrated Analysis of Societal and Ecosystems Metabolism (MuSIASEM) 297
Mumford, Lewis 162
municipal corporation 366
Murray Springs 34

National Environmental Policy Act (NEPA) 59, 86, 162
National Environmental Protection Agency 87, 95, 162, 202
National Historic Preservation Act (1966) 162
National Irrigation Association 67–68
National Irrigation Congress 68
New Deal 77, 79–81, 105, 111, 130
Newell, Frederick H. 71–72, 77
New Guard 113–115
Newlands, Francis G. 67, 69, 71, 73–74
North American Regional Reanalysis (NARR) 182
Norton, Gale 59
"not in my back yard" (NIMBY) 133

oasis effect 191
Old Guard 113–115
Olmsted, Frederick Law 161
Orland Project 74
Orme Dam 113–114
Osborn, Sidney 84

Pacific Southwest Water Plan (PSWP) 85
Pantano Wash watershed, and ES 226–228
Parker Dam 80
Phoenix Active Management Area (PAMA) 187–188
Phoenix-Tucson Sun Corridor
  and urbanization impact 130–131
    methodology 181–182
    observational temperature data 182–183
    overview 179–181
    on precipitation 186–187
    results 183–190
    on temperature 183–186
    on water and energy demand 187–190
    WRF model and configuration 182
physical setting, of Tucson City 27–34
  climate, hydrology and vegetation 28–31
  semi-arid basin and range landscape 27–28
  streamflow and recharge 32–34
Pima County comprehensive plan 167–168
Pima County Oral History Project 96
Pima Prospers 168–170
Planning Program (701) 161
Plan Tucson (2013) 170
pocket gophers, as ecosystem engineers 264–266
policy coalitions, of CAP, *see also* water policies
  changing 76–77
  and Depression 74–76

Subject Index

and drought 65–66
irrigation projects and economic crisis 66–69
New Engineers 71–74
Reclamation Act of 1902 69–71
Powell, John Wesley 72, 103–104
prairie dogs, as ecosystem engineers 262–264
precipitation
and urbanization impact 186–187
and water tables
monthly series of 330–332
regimes and time evolution 327–330
Precipitation Drought Index (PDI) 326–327
results of application 332–334
Pro-Development Coalition 97
Progressive Era 72
pro-growth coalitions, and Phoenix 130–131
Public Land Review Commission 124
Public Values Assessment (PVA) 113

qualitative assessment, for ES 231–233
quantification and mapping, and ES 201–203
hydrological modeling for 203–205

rainwater harvesting (RWH) 346
reclaimed water, as traditional water sources 344–345
Reclamation Act of 1902 66, 68–72, 102, 104, 108
Reclamation Service 51, 69–77
Record of Decision (ROD) 59
Regional Atmospheric Modeling System (RAMS) 181
Regional Groundwater Issues and Drought Planning 324
rodents, as ecosystem engineers 255–258
ground squirrels as 258–261
interdependence of 271–280
kangaroo rats as 266–268
marmot as 261–262
pocket gophers as 264–266
prairie dogs as 262–264
in WRES 255–258

Safe Drinking Water Act (SDWA) 95
safe yield 42, 172, 367
Safe Yield Task Force 42, 296, 302
Sahuarita-Continental Critical Groundwater Area 56
Salt River Project (SRP) 182, 188
Salt River Users Association 48
Salt River Valley Land 48
Sanitary Reform Movement 161

Santa Cruz River 33, 35–38, 40, 219, 293–294, 337, 341–343, 345, 356
Santa Cruz Watershed Ecosystem Portfolio Model (SCWEPM) 207, 219
Saylor, John 85
Serraglio, Randy 393–394
Shortage Sharing Agreement of 2007 58–61
Smart Growth Legislation of Arizona 170
Smart Growth regulations 164–165
snowbirds 152
social dimension 353–355
societal metabolism
evolution of 305–307
and water use 297–298
Soil and Water Assessment Tool (SWAT) 197–198
and WRES 205–207
Soil Conservation Service 41
Sonoran Desert Conservation Plan (SDCP) 167
Southern Arizona Leadership Council (SALC) 96
Southern Pacific Railway Company (SPRC) 39
spatial assessment, of groundwater management 310–316
spatial planning 159–160
stakeholder's views, on water management 381
Buschatzke, Thomas 397–398
Fonseca, Julia 391–392
Godlewski, David 389–390
Halper, Eve 395–396
Modeer, David 387–388
Serraglio, Randy 393–394
Wally R. Wilson 385–386
Zucker, Claire L. 383–384
Standard City Planning Enabling Act of 1928 161, 163
Standard State Zoning Enabling Act of 1926 161
State of Arizona Hazard Mitigation Plan 323
Stevens, Bill 90
stormwater source 346–349
streamflow and recharge, of Tucson City 32–34
The Struggle for Water (Espeland) 112
suburbanization, in Arizona 129
suburban real estate development industry 123–127
Sun Corridor, see Phoenix-Tucson Sun Corridor
supply classes, WRES
per indicator 211–213
per service 213–215

supply quantities, per indicator 209–210
sustainability
  of alternative water sources 339, 342–350
  in Tucson City
    characteristics of 293–294
    institutional context for 294–296
    methods 297–302
    overview 291–293
    results 302–313
    and spatially neutral groundwater management 314–316
Sustainable Water Action Network (SWAN) Project 15, 224, 228, 232, 236–238, 266, 316, 381, 401–404, 418–420
Swing, Phil 75

technology, and water supply 134–135
temperature, urbanization impact on climate 183–186
Texas Pacific Company 39
Tohono O'odham Tribe 38
Total Dissolved Solids (TDS) 344
traditional water sources 343–345
  CAP 344
  groundwater 343–344
  reclaimed water 344–345
Transatlantic Dialogue on Water (TDW) 3–5
transdisciplinarity 4–5, 7, 13–14, 417–421
  - and water management 418–421
Tubbs, Michael 96
Tucson Active Management Area (TAMA) 33, 42, 152, 187, 198, 236, 294, 322, 340, 363, 366–368
Tucson Chamber of Commerce 131
Tucson City, *see also* Arizona
  and alternative water sources 345–349
    characterization of 340–342
    cost estimations 349–350
    data collection 339–340
    overview 337–338
    research methodology 339–340
    sustainability analysis 342–350
    sustainability assessment 339
    traditional water sources 343–345
    for urban greenery 350–353
    and water community 353–357
  and comprehensive urban planning 170–174
  and drought
    data sources 326
    hydrogeological studies in 323–324
    methodology 325–326
    monthly series of precipitation and water tables 330–332
    overview 321–322
    Precipitation Drought Index (PDI) 326–327
    regimes and time evolution 327–330
    results of PDI application 332–334
    study area and time span 324–325
    synthesis 334–335
    types 322–323
  human history in 34–43
    arrival of Spanish 36–37
    current 41–43
    incorporation into US 37–41
    prehistoric period 34–36
  metropolitan area 144–145
  new water culture 417–418
  physical setting 27–34
    climate, hydrology and vegetation 28–31
    semi-arid basin and range landscape 27–28
    streamflow and recharge 32–34
  and transdisciplinarity 418–421
  and urban growth (*see* urbanization, and Tuscon)
  and urban sprawl in Arizona 131–134
  water use and sustainability
    characteristics of 293–294
    institutional context for 294–296
    methods 297–302
    overview 291–293
    results 302–313
    and spatially neutral groundwater management 314–316
Tucson Electric Power (TEP) 182, 188

Udall, Morris 85
Udall, Stewart 85–87
UK National Ecosystem Assessment (UK NEA) 224
UMI iGLOBES 3–4
Underground Storage Facilities (USFs) 295
United States (U.S.)
  comprehensive urban planning in 161–163
  hydraulic society, and water policies 106–110
  new water culture 417–418
  rodents as ecosystem engineers in deserts 258–268
    ground squirrels as 258–261
    interdependence of 271–280
    kangaroo rats as 266–268
    marmot as 261–262
    pocket gophers as 264–266
    prairie dogs as 262–264
  soil biodiversity in Southwestern 252–255
  Tuscon basin, incorporation into 37–41

Subject Index

Upper San Pedro basin, and groundwater management 368–372
Upper Santa Cruz watershed, and WRES 198–200
urban canopy model (UCM) 181–182
urban expansion development 150
urban greenery, *see* green infrastructure, water sources for
urbanization, and Tuscon
   climate impact
      methodology 181–182
      observational temperature data 182–183
      overview 179–181
      on precipitation 186–187
      results 183–190
      on temperature 183–186
      on water and energy demand 187–190
      WRF model and configuration 182
   and Colorado River 93–97
   and water/wastewater 145–152
      Continental Ranch development 153–155
      and GMA 152–153
      "leap-frog" development 150–151
      metropolitan area 144–145
      overview 141–144
      patterns of 149–152
      regulations 152–155
      resources and access 146–148
      urban expansion development 150
      "wildcat" development 151–152
urban sprawl
   Arizona 127–135
      land and planning 129–130
      Phoenix and pro-growth coalitions 130–131
      suburbanization in 129
      technology and water supply 134–135
      and Tucson 131–134
   overview 121–122
   since 1950's 122–127
      description 122–123
      suburban real estate development industry 123–127
Urban State Trust Lands 164–165
US Ecosystem Services Research Program 224

vegetation, of Tucson City 28–31

Walkup, Bob 97
Wally R. Wilson 385–386
water accounting areas (WAAs) 313
water community, and sources 353–357
   economic dimension 355–356
   environmental dimension 356–357
   social dimension 353–355
water conservation, and water use 307–310
Water Consumer Protection Act (WCPA) 95–96
Water Consumer Protection Coalition 96
water demand, urbanization impact on 187–190
Water Framework Directive (WFD) 160
water management, *see also* groundwater Arizona
   institutional context in 294–296
   tools of 55–56
   collaborative research 401–403
   new water culture 417–418
   results and insights for 406–415
   scientific lessons 403–406
   stakeholder's views on 381
      Buschatzke, Thomas 397–398
      Fonseca, Julia 391–392
      Godlewski, David 389–390
      Halper, Eve 395–396
      Modeer, David 387–388
      Serraglio, Randy 393–394
      Wally R. Wilson 385–386
      Zucker, Claire L. 383–384
   and transdisciplinarity 418–421
water metabolism 297
water planning 159–160
water policies, *see also* policy coalitions, of CAP
   and collective mobilizations 110–112
   and desert landscapes 103–105
   and environmental sociological tools 116
   and government power 110–112
      internal struggles within 112–116
   hydraulic society and 106–110
   overview 101–103
water regulations 152–155
water-related ecosystem services (WRES)
   bundles supplied by land use types 215–217
   land use change and influence 207–208
      hypothetical influence of 217–219
   materials and methods 198–208
   overview 197–198
   quantification and mapping 201–203
      hydrological modeling for 203–205
   results and discussion 209–219
   rodents as ecosystem engineers 255–258
      in deserts of Southwest 258–268
   supply classes
      per indicator 211–213
      per service 213–215
   supply quantities per indicator 209–210
   and SWAT model 205–207
   Upper Santa Cruz watershed case study 198–200